Essential Statistics

Understanding and Using Data

Sheeny Behmard

Essential Statistics: Understanding and Using Data
© 2025 by Chemeketa Community College

ISBN-13: 978-1-955499-48-4

Chemeketa Press
Chemeketa Community College
4000 Lancaster Dr NE
Salem, Oregon 97305
collegepress@chemeketa.edu
chemeketapress.org

Cover design by Ron Cox
Interior design by Ron Cox

This book is a remix of original content by Sheeny Behmard and Chemeketa Press; OpenStax's *Introductory Business Statistics* by Alexander Holmes, Barbara Illowsky, and Susan Dean (CC BY 4.0); OpenStax's *Introductory Statistics* by Barbara Illowsky and Susan Dean (CC BY 4.0); *Probability and Statistics* developed by the UCLA Statistics Online Computational Resource (CC BY 4.0).

References to website URLs were accurate at the time of writing. Neither the author nor Chemeketa Press is responsible for URLs that have changed or expired since the manuscript was prepared.

Printed in the United States of America.

Land Acknowledgment
Chemeketa Press is located on the land of the Kalapuya, who today are represented by the Confederated Tribes of the Grand Ronde and the Confederated Tribes of the Siletz Indians, whose relationship with this land continues to this day. We offer gratitude for the land itself, for those who have stewarded it for generations, and for the opportunity to study, learn, work, and be in community on this land. We acknowledge that our College's history, like many others, is fundamentally tied to the first colonial developments in the Willamette Valley in Oregon. Finally, we respectfully acknowledge and honor past, present, and future Indigenous students of Chemeketa Community College.

Contents

Chapter 1
Introduction to Statistics

You are probably thinking, "When and where will I use statistics?" Well, you already use statistics in your everyday life. You see statistical information if you read any news, watch television, or use the internet. There are statistics about crime, sports, education, politics, real estate, the stock market, health care, banking, and more. When you read or hear a statistic, you are given sample information about a larger topic. This information lets you decide if the statement, claim, or so-called fact is correct. Statistics helps make complicated issues understandable by connecting them to your everyday life.

Since you already interact with statistics every day, you should be able to analyze the information thoughtfully. Beyond that, economics, business, psychology, education, biology, law, computer science, political science, and early childhood development require at least one course in statistics. Understanding the science of statistics helps deepen your knowledge of your chosen profession. It also helps you manage a monthly budget, make health care decisions, or even decide which smartphone to buy.

This chapter includes the basic ideas and key terms of statistics. You will soon understand that statistics and probability work together. You will also learn how data are gathered and how to determine whether data are reliable, accessible, accurate, consistent, and complete.

After reading this chapter, you will be able to do the following:

1. Describe the difference between parameters and statistics
2. Identify quantitative and qualitative data
3. Describe the four levels of measurement
4. Describe sampling methods and sampling bias
5. Understand the difference between sampling with replacement and sampling without replacement
6. Distinguish a sampling error from a nonsampling error
7. Construct a frequency table
8. Describe the difference between relative frequency and cumulative relative frequency
9. Understand the important components of experimental design
10. Understand the ethical responsibilities of statistics

1.1 Basics of Statistics

Overview

The science of statistics deals with collecting, analyzing, interpreting, and presenting information. **Data** are what we call the information collected for analysis and interpretation. We see and use data every day. You use statistics and data when you bring an umbrella because the forecast calls for rain. You use statistics and data when you make a monthly budget based on potential expenses. Even when considering which career to enter, you use statistics and data.

Like all sciences, statistics has a specialized vocabulary. In this section, we'll learn the basic key terms and concepts. We'll learn about the different types of data so that you can determine the best way to collect and interpret it down the road.

Definitions and Key Terms

Maria needs to buy a new phone. Before she goes to the store, she wants to get a sense of how much a standard smartphone costs so she doesn't end up paying more than she should. There are a lot of phones available, though. How can she find their average value? She knows she can't find *all* the prices of *all* the phones, then add them together and then divide by the number of phones on the market—that's impossible, and she has better things to do with her time. The best she can do is estimate the average price.

One way for her to find the average price is to select a smaller number of phones. Let's say she picks 20 phones. Now, she only has to find the prices of 20 phones, add those prices up, and divide the total by 20 to find the average smartphone price today.

In statistics, a **population** is the entire group of objects you want to study. Maria wants to study the whole population of smartphones available. But because that's too big of a task, the 20 phones she selects from the population are a **sample**, a specific group of objects taken from the larger population.

When Maria writes down each phone's price, she determines its measurement, which is the number given to each element of a population or sample. The average value is a **parameter**, which is a number that describes some feature of the whole population. For Maria, the parameter describes the average price of all smartphones.

Maria's list of 20 phone prices is her sample data, which is the factual information collected about her sample. When she calculates the average smartphone price, she calculates a statistic about smartphones. A **statistic** is a number that describes the sample data. For Maria, the average smartphone price is a statistic that can help her make informed choices when shopping.

Be Careful!

The term "statistics" can mean two things. It can mean "a number calculated from sample data," like Maria's average smartphone price. But it can also mean the broader science of statistics, which includes planning studies and experiments, gathering data, analyzing data, and drawing conclusions.

A **variable** is a characteristic or measurement that can be determined for each member of a population and may be numerical or categorical. Variables are usually notated by capital letters like X and Y. They may be numbers or words. Numerical variables have values with equal units, such as weight in pounds and time in hours. Categorical variables place the person or thing into a category. If X equals the number of smartphones with cameras, then X is a numerical value. If Y equals smartphone manufacturers, then some values of Y include Apple, Samsung, Google, LG, or Motorola, and Y is a categorical variable. We can do math with values of X (calculate the percentage of smartphones with cameras, for example). However, it makes no sense to do math with values of Y (calculating an average smartphone manufacturer makes no sense).

Example 1

Match these key terms to the elements in the scenario: data, parameter, population, sample, statistic, variable.

We want to know the average amount of money first-year students at Wells Community College spend on school supplies, excluding books. We randomly surveyed 100 first-year students at the college. Three of the 100 students spent $150, $200, and $225, respectively.

Solution 1

The **population** is all first-year students attending Wells Community College this term.

The **sample** is the 100 first-year students surveyed at the college.

The **parameter** is the average (mean) amount of money spent on school supplies by first-year college students at Wells Community College this term.

The **statistic** is the average (mean) amount of money spent on school supplies by the 100 first-year students attending Wells Community College.

The **variable** could be the amount of money spent on school supplies by the first-year student. Let X = the amount of money spent on school supplies by one first-year student attending Wells Community College.

The **data** are the dollar amounts spent by the first-year students. Examples of the data are $150, $200, and $225.

Two final important key terms to understand now are mean and proportion. Mean is the average of the data points in a data set. In statistics, "mean" is often used instead of "average." If you took three exams in your math classes and earned scores of 86, 75, and 92, you would calculate your mean score by adding the exam scores together and dividing the answer by 3. Your mean score on the exams is 84.3.

Proportion is a fraction of the entire group with a particular characteristic. In a math class, there are 40 students. Twenty identify as male, 17 as female, and 3 as nonbinary. The proportion of male students $\frac{20}{40}$, the proportion of female students is $\frac{17}{40}$, and the proportion of nonbinary students is $\frac{3}{40}$.

Example 2

A study was conducted at a local college to analyze the average cumulative Grade Point Averages (GPAs) of students who graduated last year. Match the key term to the phrase that describes it.

1. Population
2. Statistic
3. Parameter

4. Sample
5. Variable
6. Data

a. all students who attended the college last year
b. the cumulative GPA of one student who graduated from the college last year
c. 3.65, 2.80, 1.50, 3.90
d. a group of students who graduated from the college last year, randomly selected
e. the average cumulative GPA of students who graduated from the college last year
f. the average cumulative GPA of students in the study who graduated from the college last year

Solution 2

1. a   4. d
2. f   5. b
3. e   6. c

Try It 1.1

Match these key terms to the elements in the scenario: data, parameter, population, sample, statistic, variable.

We want to know the average (mean) amount of money spent on school uniforms annually by families with children at Knoll Academy. We randomly survey 100 families with children in the school. Three families spent $65, $75, and $95, respectively.

Population Parameters and Sample Statistics

The difference between a parameter and a statistic is a key concept in statistics, so let's explore it more. Both parameters and statistics are numerical measurements, but they describe different things. A parameter describes a population, and a statistic describes a sample. A sample statistic is often used to predict or infer a population parameter. In other words, we might use a sample mean to guess an unknown population mean.

To show this, look at Figure 1.1. The icons in the large box represent parts of the smartphone population available to Maria. There is room for a small number of icons in the figure, but hundreds of smartphones are on the market. The blue icons represent the smartphones that Maria selected to be her sample. Each phone in her sample has a price, such as x_1, x_2, x_3, and so on. All the phones' prices make up the sample data set when collected. From there, we can calculate statistics that describe the sample—sample statistics.

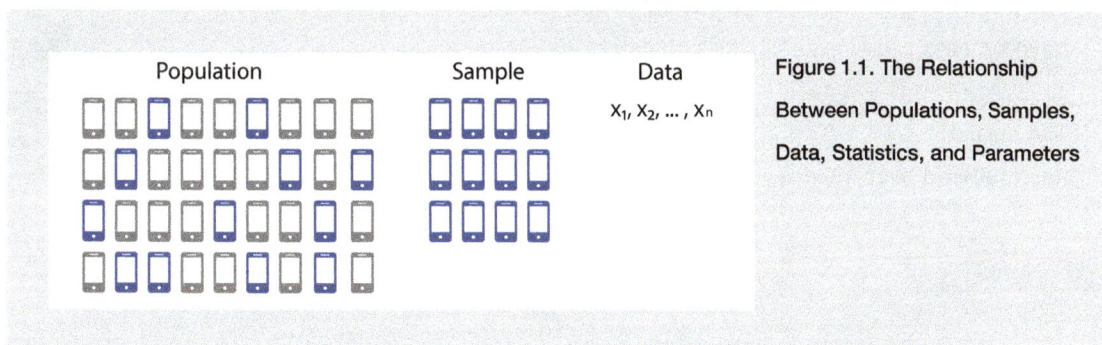

Figure 1.1. The Relationship Between Populations, Samples, Data, Statistics, and Parameters

Population parameters are estimated by converting sample statistics into intelligent guesses about the population. One of the main concerns in statistics is how accurately a statistic estimates a parameter. The accuracy depends on how well the sample represents the population. To be a representative sample, the sample must contain the characteristics of the population.

Example 3

Match these key terms to the elements in the scenario: data, parameter, population, sample, statistic, variable.

As part of a study designed to test the safety of automobiles, the National Transportation Safety Board collected and reviewed data about the effects of an automobile crash on test dummies. Here is the criterion they used:

1. Speed at which cars crashed: 35 miles/hour
2. Location of "driver" (i.e., dummies): front seat

Cars with dummies in the front seats crashed into a wall at a speed of 35 miles per hour. We want to know the proportion of dummies in the driver's seat that would have had head injuries if they had been actual drivers. We start with a simple random sample of 75 cars.

Solution 3

The **population** is all cars containing dummies in the front seat.

The **sample** is the 75 cars, selected by a simple random sample.

The **parameter** is the proportion of driver dummies (if they had been real people) who would have suffered head injuries in the population.

The **statistic** is the proportion of driver dummies (if they had been real people) who would have suffered head injuries in the sample.

The **variable** X is the number of driver dummies (if they had been real people) who would have suffered head injuries.

The **data** are either: yes, had a head injury, or no, did not.

Example 4

Match these key terms to the elements in the scenario: data, parameter, population, sample, statistic, variable.

An insurance company would like to determine the proportion of all medical doctors involved in one or more malpractice lawsuits. The company selects 500 doctors at random from a professional directory and determines the number in the sample who have been involved in a malpractice lawsuit.

Solution 4

The **population** is all medical doctors listed in the professional directory.

The **parameter** is the proportion of medical doctors in the population who have been involved in one or more malpractice suits.

The **sample** is the 500 doctors selected at random from the professional directory.

The **statistic** is the proportion of medical doctors involved in one or more malpractice suits in the sample of 500 doctors.

The **variable** X is the number of medical doctors involved in one or more malpractice suits.

The **data** are either: yes, was involved in one or more malpractice lawsuits, or no, was not.

Types of Data

Qualitative data, often called categorical data, result from categorizing or describing attributes of a population. Hair color, blood type, ethnic group, the car a person drives, and the street a person lives on are examples of qualitative data. Qualitative data are generally described by words or letters. For instance, Omar needs to buy a new car, but he's not sure what type he wants, so he starts to research the types of vehicles: compact, subcompact, sedan, SUV, or truck. He's working with qualitative data, which are names and labels rather than numbers.

Quantitative data are numbers that represent measurements. Quantitative data are the result of counting or measuring attributes of a population. Amount of money, pulse rate, weight, number of people living in a town, and number of students who take statistics are examples of quantitative data. Researchers often prefer to use quantitative data because it's easier to analyze mathematically. For instance, it doesn't make sense to find an average car type, but it does make sense to find an average price.

Note

We may collect data as numbers and report it categorically. For example, an instructor records quiz scores for each student throughout the term. At the end of the term, the quiz scores are reported as A, B, C, D, or F.

Example 5

What are the values of the data below? Are they quantitative or qualitative data?

The data are the colors of backpacks. We sample 5 students. One student has a red backpack, 2 have black backpacks, 1 has a green backpack, and 1 has a gray backpack.

Solution 5

The values are red, black, black, green, and gray. They are qualitative data.

Quantitative data may be either discrete or continuous. **Discrete data** are the result of counting. These data take on only specific numerical values, and the number of values is finite. For instance, we can count the number of phone calls received each day of the week, and the discrete data might have values such as 1, 0, 1, 2, and 3.

Continuous data consist of counting numbers and fractions, decimals, or irrational numbers. Continuous data are often the results of measurements such as lengths, weights, or times, and in theory, there are endless possibilities of values. For example, a list of the length of minutes for all the phone calls made in a week would be continuous data, with numbers like 2.4 minutes or 11 minutes. Another set of continuous data would be the weight in pounds of backpacks with books: 2.3, 7, 3.8, 4.9, 2.9. They are continuous data because there are endless combinations of possible values of backpacks and books.

Example 6
What are the values of the data below? Are the data discrete or continuous?

The data are the number of books students carry in their backpacks. We sample 5 students. Two students carry 3 books, 1 student carries 4 books, 1 student carries 2 books, and 1 student carries 1 book.

Solution 6
The data values are 3, 4, 2, and 1. They are discrete data.

Example 7
What are the values of the data below? Are the data discrete or continuous?

The data are the weights (in pounds) of backpacks with books. We sample the same 5 students. The weights (in pounds) of their backpacks are 6.2, 7, 6.8, 9.1, and 4.3. Notice that backpacks carrying 3 books can have different weights.

Solution 7
The values are 6.2, 7, 6.8, 9.1, 4.3. They are continuous data.

Example 8
We go to the supermarket and purchase 3 cans of soup (19-ounce tomato bisque, 14.1-ounce lentil, and 19-ounce Italian wedding), 2 packages of nuts (walnuts and peanuts),

4 kinds of vegetables (broccoli, cauliflower, spinach, and carrots), and 2 containers of ice cream (16-ounce pistachio ice cream and 32-ounce chocolate chip cookie dough).

Name data sets that are discrete, continuous, and qualitative.

Solution 8

The 3 cans of soup, 2 packages of nuts, 4 kinds of vegetables, and 2 ice creams are discrete data because we count them.

The weights of the soups (19 ounces, 14.1 ounces, 19 ounces) are continuous data because we measure weights as precisely as possible.

Types of soups, nuts, vegetables, and ice creams are qualitative data because they are categorical.

Bonus: Try to identify additional data sets in this example.

Example 9

First, determine whether the data is qualitative or quantitative. Then, indicate whether quantitative data are continuous or discrete.

1. the number of pairs of shoes you own
2. the type of car you drive
3. the distance it is from your home to the nearest grocery store
4. the number of classes you take per school year.
5. the type of calculator you use
6. weights of sumo wrestlers
7. number of correct answers on a quiz
8. IQ scores

Solution 9

Qualitative: 2 and 5

Quantitative discrete: 1, 4, and 7

Quantitative continuous: 3, 6, and 8

Hint

Data that are discrete often start with the words "the number of."

Levels of Measurement

Another way to categorize data is by its level of measurement. The level of measurement describes the type of data in a similar way that data can be continuous or discrete. The level of measurement is important because only some statistical operations can be used with some kinds of data sets. To use the correct statistical operation on a data set, we must know its level of measurement.

Data can be classified into four levels of measurement, ordered from least to most specific, as seen in Figure 1.2. We can also think about the measurement level in increasing detail levels. The higher we are in the levels of the measurement pyramid, the more detail we know about the data.

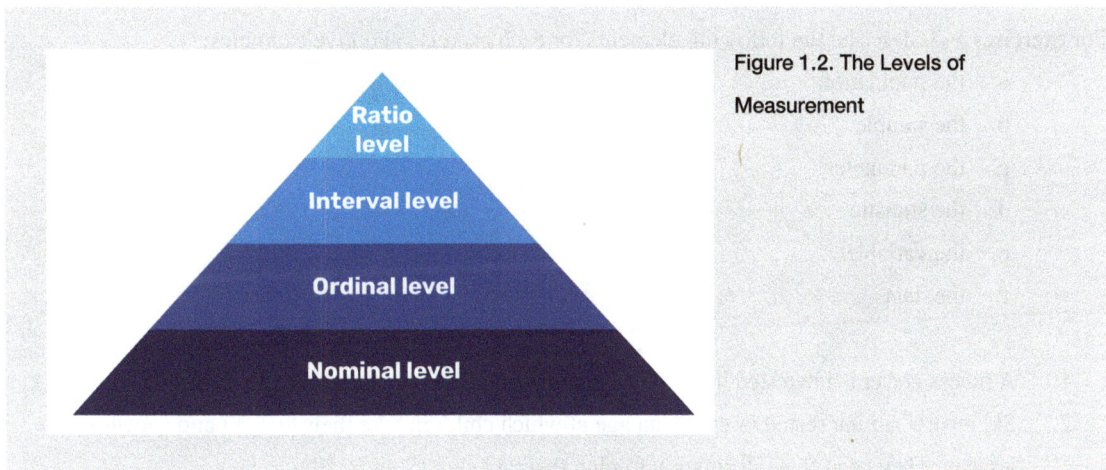

Figure 1.2. The Levels of Measurement

The nominal level of measurement refers to data that is named by categories, names, labels, and yes-or-no responses. Nominal-level data can't be put in any order. For example, classifying people according to their favorite food does not make sense. Putting pizza first and sushi second is not meaningful. Nominal-level data can't be used in calculations.

The ordinal level of measurement refers to data that can be ordered, but the data values are impossible to determine or are meaningless. An example of ordinal-level data is a list of the top 5 national parks in the United States. The parks can be ranked from 1 to 5, but we can't measure differences between the data. Like the nominal-level data, ordinal-level data can't be used in calculations.

The interval level of measurement describes data that can be ordered and has a measurable value difference, but the data values don't have a starting point. An example of interval-level data is body temperatures. If we take the temperatures of two people and get readings of 98.6° F and 99.2° F, we can measure the difference of 0.6° F. But the data doesn't have a real starting point to measure against—we could say 0° F, but we can't compare body temperatures to 0° F because then the people would be icicles. Any starting point would be meaningless. For this reason, interval-level data can't be used to calculate ratios.

The ratio level of measurement describes data that can be ordered and measured and has a zero point from which ratios can be calculated. The ratio level provides the most information about data and can lead to a more detailed data analysis. Test scores are a great example of ratio-level data because they can be ordered from lowest to highest. The differences between the data have meaning, and ratios can be calculated. The score of 92 is more than 68 by 24 points. The smallest possible score is 0, and 80 is 4 times more than 20.

Exercises 1.1

For exercises 1–7, describe the following elements for each exercise and give examples:

    a.   the population
    b.   the sample
    c.   the parameter
    d.   the statistic
    e.   the variable
    f.   the data

1.   A fitness center is interested in the mean amount of time a client exercises in the center each week.
2.   Ski resorts are interested in the mean age at which children take their first ski and snowboard lessons. They need this information to plan their ski classes optimally.
3.   A cardiologist is interested in the mean recovery period of her patients who have had heart attacks.

4.   Insurance companies are interested in the mean health costs of their clients each year so that they can determine the costs of health insurance.
5.   A marriage counselor is interested in the proportion of clients she counsels who stay married.
6.   Political pollsters may be interested in the proportion of people who will vote for a particular cause.
7.   A marketing company is interested in the proportion of people who will buy a particular product.

For exercises 8–10, use the following information to answer: A Chemeketa Community College instructor is interested in the mean number of days Chemeketa Community College math students are absent from class during a quarter.

8.   What is the population she is interested in?
    a.   all Chemeketa Community College students
    b.   all Chemeketa Community College English students
    c.   all Chemeketa Community College students in her classes
    d.   all Chemeketa Community College math students
9.   Consider the following: X = number of days a Chemeketa Community College math student is absent. In this case, what is X an example of?
    a.   variable
    b.   population
    c.   statistic
    d.   data
10.   The instructor's sample produces a mean number of days absent of 3.5 days. This value is an example of a:
    a.   parameter
    b.   data
    c.   statistic
    d.   variable
11.   For the following items, identify the type of data that would be used to describe a response (quantitative discrete, quantitative continuous, or qualitative), and give an example of the data.
    a.   number of tickets sold to a concert
    b.   percent of body fat
    c.   favorite baseball team
    d.   time in line to buy groceries

e. number of students enrolled at Evergreen Valley College

f. most-watched television show

g. brand of toothpaste

h. distance to the closest movie theater

i. age of executives in Fortune 500 companies

j. number of competing computer spreadsheet software packages

For exercises 12 and 13, use the following information to answer: A study was done to measure different variables of resident use of a local park in San Jose, California. Researchers wanted to determine the age of park users, the number of times the park is used per week, and the duration (amount of time) of use. The first house in the neighborhood around the park was selected randomly, and then every 8th house in the neighborhood around the park was interviewed.

12. The phrase "number of times per week" is what type of data?
    a. qualitative (categorical)
    b. quantitative discrete
    c. quantitative continuous

13. The phrase "duration (amount of time)" is what type of data?
    a. qualitative (categorical)
    b. quantitative discrete
    c. quantitative continuous

1.2 Data Basics

Overview

As we've learned, data may come from a population or a sample. The type of data impacts how it's gathered. An important part of the science of statistics is accurately sampling a population. When planning an experiment, sampling methods will determine how well the sample data describes the population. In this section, we'll explore the different sampling methods and when to use them.

Understanding data is about more than just collecting and looking at numbers. A big part of statistics involves displaying data in a way an audience will understand. This section will introduce charts, graphs, and tables that can make data clear and easy to interpret.

Sampling Data

Gathering information about an entire population often costs too much or is virtually impossible. Instead, we use a population sample to find out information about it. Most statisticians use various methods of random sampling to achieve this goal. This section describes some of the most common methods for sampling data.

Random sampling is a technique in which each member of the population has an equal chance of being selected. There are several methods of random sampling, each with pros and cons. In a **simple random sample**, any group of n individuals is equally likely to be chosen as any other group of n individuals. In other words, each sample of the same size has an equal chance of being selected. Simple random sampling is often used when the population is relatively small and homogeneous, and the variable of interest is unrelated to any subgroups. Since each member of the population has an equal chance of being selected for the sample, simple random sampling can provide a representative sample that can be generalized to the entire population.

There are more specific methods to select a random sample. Table 1.1 shows the methods defined here, their main properties, and situations where each could be used. A **stratified sample** is a random sample taken from subgroups of the population called strata. Divide the population into strata, then select a proportionate number from each stratum with the simple random sample technique. A **cluster sample** is formed when the population is divided into clusters (groups), then clusters are randomly selected. All the members from the selected clusters are in the cluster sample. A **systematic sample** is taken from a listing of the population. A starting point is randomly selected and then every n^{th} piece of data is taken from that list.

Table 1.1. Random Sampling Methods

Sampling Method	Description	Situations for Use
Simple random sampling	A sample where each sample of the same size has an equal chance of being selected from the population.	*Conducting a survey of public opinion*: Select a random sample of households from a list of all the households in a city. *Studying the effects of a new teaching method*: Select a random sample of students from a list of all the students in a school.
Stratified sample	A random sample taken from subgroups of the population, often used to compare subgroups with each other.	*Studying the voting behavior of a large city*: Divide the city into strata based on the demographic characteristics of the residents, such as age, gender, race, and income. Select a random sample of voters from each stratum. *Evaluating the effectiveness of a new drug*: Divide patients into strata based on the severity of their illness or the type of treatment they are receiving. Select a random sample of patients from each stratum.
Cluster sample	A random sample taken by dividing the population into clusters (groups) and then randomly select some of the clusters. All the members from the selected clusters are in the cluster sample.	*Studying the quality of education in schools*: Divide the schools in a district into clusters. Select a random sample of schools. Then include all the students in the selected schools in the study. *Evaluating the health status of communities*: Divide communities into clusters. Select a random sample of communities. Then include all the individuals in the selected communities in the study.

Sampling Method	Description	Situations for Use
Systematic sample	A random sample is selected by taking every nth element from a population.	*Evaluating the quality of a production line*: Arrange all items produced on the production line in a sequence. Select every nth item for inspection. *Analyzing the performance of an athlete*: Arrange the performances of an athlete in a series of events in a sequence. Select every nth performance for analysis.

A type of sampling that is non-random is **convenience sampling**. Convenience sampling involves using readily available results. For example, a grocery store conducts a marketing study by interviewing customers who are shopping there. The marketing survey results could be totally different depending on the day of the week, time of day, weather, or any other uncontrollable factor. Convenience sampling can be accurate in some cases and not in others. It should be used with caution. In practice, it is always best for the person conducting the survey or study to select the sample rather than relying on chance.

Example 11

A study is done to determine the average tuition that San Jose State undergraduate students pay per semester. Students in the following samples are asked how much tuition they paid for the Fall semester.

Match the scenario to the type of sampling used.

1. A sample of 100 undergraduate San Jose State students is taken by organizing the students' names by classification (freshman, sophomore, junior, or senior) and then selecting 25 students from each.
2. A random number generator is used to select a student from the alphabetical listing of all undergraduate students in the Fall semester. Starting with that student, every 50th student is chosen until 75 students are included in the sample.
3. A completely random method is used to select 75 students. Each undergraduate student in the fall semester has the same probability of being chosen at any stage of the sampling process.

4. The freshman, sophomore, junior, and senior years are numbered one, two, three, and four, respectively. A random number generator is used to pick two of those years. All students in those two years are in the sample.

5. An administrative assistant is asked to stand in front of the library on Wednesday and to ask the first 100 undergraduate students he encounters what they paid for tuition in the Fall semester. Those 100 students are the sample.

a. systematic
b. cluster
c. simple random

d. stratified
e. convenience

Solution 11

1. d
2. a
3. c

4. b
5. e

Example 12

Determine which type of sampling is used in each scenario. (simple random, stratified, cluster, systematic, convenience).

1. A soccer coach selects six players from a group of boys aged eight to ten, seven from a group of boys aged 11 to 12, and three from a group of boys aged 13 to 14 to form a recreational soccer team.
2. A pollster interviews all human resource personnel in five different high-tech companies.
3. An educational researcher interviewed 50 female and 50 male high school teachers.
4. A medical researcher interviews every third cancer patient from a list of cancer patients at a local hospital.
5. A high school counselor uses a computer to generate 50 random numbers and then picks students whose names correspond to the numbers.
6. A student interviews classmates in his algebra class to determine how many pairs of jeans a student owns on average.

Solution 12

1. Stratified
2. Cluster
3. Stratified
4. Systematic
5. Simple random
6. Convenience

Try It 1.2

Determine the sampling used (simple random, stratified, systematic, cluster, or convenience).

A high school principal polls 50 first-year students, 50 sophomores, 50 juniors, and 50 seniors regarding policy changes for after-school activities.

Sampling with or without Replacement

Actual random sampling is done with replacement. Replacement means that once a member is picked, that member goes back into the population and may be chosen again. However, for practical reasons, simple random sampling is done without replacement in most populations. That is, a member of the population may be chosen only once. Most samples are taken from large populations and tend to be small compared to the population. Since this is the case, sampling without replacement is approximately the same as sampling with replacement because the chance of picking the same individual more than once is very low.

Let's look at a college population of 10,000 students to see the difference between sampling with or without replacement. A random sample of 1,000 students is selected. Table 1.2 shows the impact of sampling with or without replacement on the chances of affecting the survey outcome for any sample of 1,000 students.

Table 1.2. Impact of Replacement on Survey Results

	Sampling with replacement	Sampling without replacement
Chance of picking the first person	1,000 out of 10,000 (0.1000)	1,000 out of 10,000 (0.1000)
Chance of picking a different second person	999 out of 10,000 (0.0999)	999 out of 9,999 (0.0999)

Compare the chance of picking a different second person: $\frac{999}{10,000} = 0.0999$ and $\frac{999}{9,999} = 0.0999$. Even if the decimal is carried four places for accuracy, these numbers are equivalent.

Sampling without replacement becomes a mathematical issue only when the population is small. For example, if the population is 25 students and we want a sample of 10 students, the chances of picking a different second student change. Table 1.3 shows the impact of sampling with or without replacement on the chances of affecting the survey outcome for any sample of 10 students.

Table 1.3. Impact of Replacement on Survey Results

	Sampling with replacement	Sampling without replacement
Chance of picking the first person	10 out of 25 (0.4000)	10 out of 25 (0.4000)
Chance of picking a different second person	9 out of 25 (0.3600)	9 out of 24 (0.3750)

Compare the chances of picking a different second person: $\frac{9}{25} = 0.3600$ and $\frac{9}{24} = 0.3750$. To four decimal places, these numbers are not equivalent and show that whether we sample with or without replacement when the sample size is a larger proportion of the population can influence the survey results.

Example 13

A large community college has 10,000 part-time students (the population). We are interested in the average amount of money a part-time student spends on books in the fall term. Asking all 10,000 students is an almost impossible task. So, we take two different samples.

First, we use convenience sampling and survey 10 students from a first-term organic chemistry class. Many of these students are taking first-term calculus in addition to the organic chemistry class. The amount of money (in dollars) they spend on books is:

128 87 173 116 130 204 147 189 93 153

The second sample is taken using a list of adults over 65 who take physical education classes. We take every fifth person on the list, for a total of 10 adults over 65. They spend the following amounts:

50 40 36 15 50 100 40 53 22 22

It is unlikely that any student is in both samples.

1. Do you think either of these samples is representative of the entire 10,000 part-time student population?
2. If these samples do not represent the entire population, is it wise to use the results to describe the entire population?

Solution 13

1. No. The first sample probably consists of science-oriented students. Besides the chemistry course, some of them are also taking first-term calculus. Books for these classes are expensive. Most of these students are likely paying more than the average part-time student for their books. The second sample is a group of people over 65 who are likely taking courses for health and interest. The amount of money they spend on books is likely much less than the average part-time student. Both samples are biased. Also, not all students can be in either sample in both cases.
2. No. For these samples, each member of the population did not have an equally likely chance of being chosen.

Sampling Errors

A **sampling error** occurs when a sample does not represent the population. Every study will have sampling errors because a sample can never precisely represent a population. However, by choosing the best sampling method and a large sample size, we can reduce the impact of the sampling error on the study. As a rule, the larger the sample, the smaller the sampling error.

There are two types of sampling errors. Random sampling errors occur when there is a difference between the sample and the actual population results. These errors come from sample changes driven by chance. For example, if the sample size is not large enough, even small changes in the population can mean that the sample is no longer representative. **Nonsampling errors** result from human error and aren't related to the sampling process. For example, a defective counting device can cause a nonsampling error. Other nonsampling errors could include simple data entry errors, respondents giving false information, or biased questions.

In statistics, a sampling bias is created when a sample is collected from a population, and some members are less likely to be chosen than others. Each member of the population should have an equally likely chance of being chosen. When a sampling bias happens, incorrect conclusions can be drawn about the population, and the results are biased.

Example 14

Using the information from Example 13, suppose we take a third sample.

We choose 10 part-time students from chemistry, math, English, psychology, sociology, history, nursing, physical education, art, and early childhood development. We assume that these are the only disciplines in which part-time students at this college are enrolled and that an equal number of part-time students are enrolled in each.

Each student is given a unique number and is chosen using simple random sampling. Using a random number generator, the number 102 is selected. A student from history has that corresponding number. The selected students spend the following amounts:

> 180 50 150 85 260 75 180 200 200 150

Is the sample biased?

Solution 14

The sample is unbiased, but a larger sample would be recommended to increase the likelihood that the sample will be close to representative of the population. However, for a biased sampling technique, even a large sample risks not being representative of the population.

Variation in Data and Samples

In statistics, variation describes how much the values in a data set are different from each other. It helps us understand how much the data is spread out, and it helps us learn more about the population by looking at just a part of it. If the numbers are very different from each other, there is a lot of variation. If the numbers are similar to each other, there is little variation. Variation can be present in any set of data. For example, 16-ounce soda cans may contain slightly more or less than 16 ounces of liquid. Manufacturers know there will be some variation and determine a desired range of liquid for a can to pass inspection. They regularly test to see if the amount of liquid falls within the desired range. If the volume of liquid falls within the desired range, the cans pass inspection and end up on the shelves of a grocery store.

If you're gathering data for the same study as another person, the values will vary somewhat from the data the other person collects. This is completely natural. However, if you're both gathering the same data and getting very different results, it is time for you and the others to reevaluate sampling methods and accuracy.

Two or more random samples from the same population with similar population characteristics will differ. Suppose Madeleine and Jung decide to study the average amount of time students sleep each night at their college. They each sample 500 students. Madeleine uses systematic sampling, and Jung uses cluster sampling. Because of their sampling method choice, their samples will be different. Their samples would likely differ even if they used the same sampling method. Neither would be wrong, however.

The sample size (often called the number of observations) is essential. Samples of only a few hundred observations, or even smaller, are sufficient for many purposes. In polling, samples that have 1,200 to 1,500 observations are considered large enough if the survey is random and the sampling method is rigorous. We'll learn why this is large enough when studying confidence intervals in Chapter 5.

Critical Evaluation of Sampling

Critically evaluating statistical studies before accepting their results is important. Data is everywhere, and it is easy to manipulate the results or only tell part of the story.

Table 1.4 shows common statistical problems that can indicate problems with the results of a study.

Table 1.4. Common Statistics Problems That Effect Results

Problem	Description
Problems with samples	A sample must be representative of the population. A sample that is not representative of the population is biased. Biased samples that are not representative of the population give results that are inaccurate and not valid. Self-selected samples, like call-in surveys, are made up of people who choose to respond to the surveys and are often unreliable.
Sample size issues	Samples that are too small may be unreliable. Larger samples are better, if possible. In some situations, having small samples is unavoidable and can still be used to draw conclusions. For examples, crash-testing cars or medical testing for rare conditions often rely on small sample sizes because there are only a small number of available data points.

Problem	Description
Self-funded or self-interest studies	A study performed by a person or organization to support their claim. Ask: Is the study impartial? Or is there evidence of misleading use of data? Read the study carefully and look for discrepancies in graphs, data, or context. Do not automatically assume that the study is good, but don't automatically assume the study is bad either. Evaluate it on its merits and the work done.
Non-response or refusal of subject to participate	The collected responses may no longer be representative of the population. Often, people with strong positive or negative opinions may answer surveys, which can affect the results.
Causality	A relationship between two variables does not mean that one causes the other to occur. They could be related because of their relationship through a different variable.
Confounding	When the effects of multiple factors on a response can't be separated. Confounding makes it difficult or impossible to draw valid conclusions about the effect of each factor.

Try It 1.3

A local radio station has a fan base of 20,000 listeners. The station wants to know if its audience would prefer more music or more talk shows. Asking all 20,000 listeners is an almost impossible task.

The station uses convenience sampling and surveys the first 200 people they meet at one of the station's music concert events. 24 people said they'd prefer more talk shows, and 176 people said they'd prefer more music.

Do you think that this sample is representative of (or is characteristic of) the entire 20,000 listener population?

Displaying Data

We have multiple options for displaying data so that others can understand it. Sometimes, we need to present data in tables with labels, and sometimes, using a graph is more effective. This section will explore the most common ways to make data readable and accessible.

Tables and Graphs

Tables are a good way of organizing and displaying data. They are simple and clean and can display large numbers nicely. Table 1.5 compares the number of part-time and full-time students at Chemeketa Community College and Portland Community College in 2019. The tables show counts and percentages. The percent columns make comparing the same categories in the colleges easier. Displaying percentages along with the numbers is often helpful. However, it is essential when comparing data sets that don't have the same totals, such as the total enrollments for both colleges in this example. Notice how much larger the percentage of part-time students at Portland Community College is compared to Chemeketa Community College.

Table 1.5. 2019 Enrollment, Full-Time vs. Part-Time Students

Chemeketa Community College	Number	Percent	Portland Community College	Number	Percent
Full-time	4,301	46%	Full-time	10,372	37.5%
Part-time	5,046	54%	Part-time	17,278	62.5%
Total	9,347	100%	Total	27,650	100%

While tables are great at organizing data, graphs can be even more helpful in understanding the data. Two graphs that are used to display qualitative data are pie charts and bar graphs. In a pie chart, categories of data are represented by wedges in a circle and are proportional in size to the percentage of individuals in each category. In a bar graph, the length of the bar for each category is proportional to the number or percent of individuals in each category. Bars may be vertical or horizontal.

Looking at the data in different graph styles is a good idea. Depending on the data and the context, one style might make it easier to understand the data. Look at Figures 1.3 and 1.4, which display the same data as Table 1.5. Which type of display shows the comparisons better?

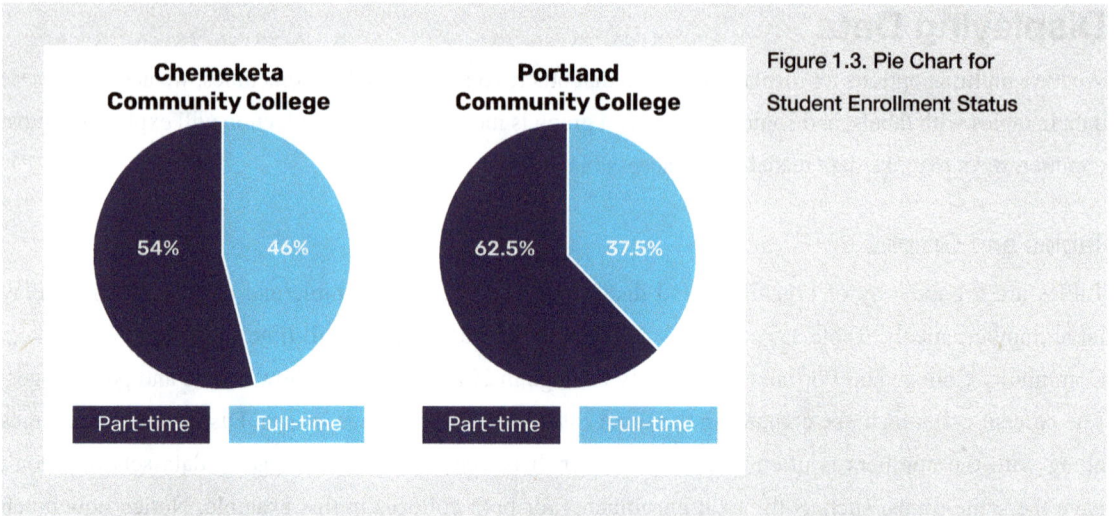

Figure 1.3. Pie Chart for Student Enrollment Status

Chemeketa Community College

54% Part-time 46% Full-time

Portland Community College

62.5% Part-time 37.5% Full-time

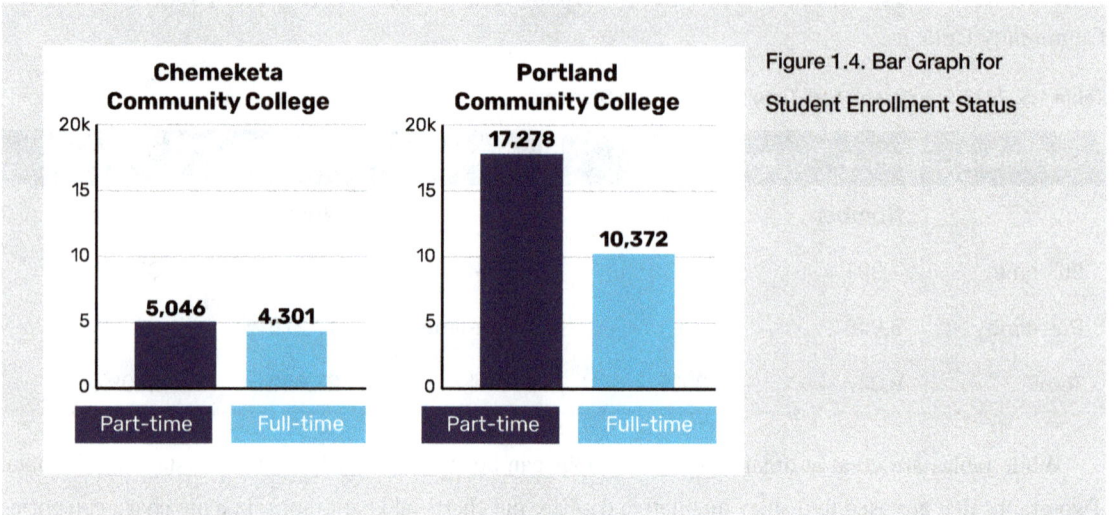

Figure 1.4. Bar Graph for Student Enrollment Status

Chemeketa Community College

Part-time 5,046 Full-time 4,301

Portland Community College

Part-time 17,278 Full-time 10,372

Missing Data

Sometimes, data is left out of a graph or chart. This can make it difficult to draw accurate conclusions about what the data is showing. For example, Table 1.6 displays enrollment by race and ethnicity at Chemeketa Community College but is missing an important category. The Did Not Report/Unknown category contains people who declined to provide their race or ethnicity. Notice that the frequencies don't add up to the total number of students. In this situation, a bar graph is the best way to display the data, as shown in Figure 1.5.

Table 1.6. Enrollment by Race and Ethnicity at Chemeketa Community College in 2019

Ethnicity	Frequency	Percent
American Indian or Alaska Native	97	1.04%
Asian	151	1.62%
Black or African American	83	0.89%
Hispanic or Latino	2,825	30.2%
Native Hawaiian or Other Pacific Islanders	49	0.52%
Nonresident Alien	45	0.01%
Two or More Races	273	2.92%
White	3,625	34.9%
TOTAL	7,148 out of 9,347	72.1% out of 100%

Figure 1.6 displays the same data as Figure 1.5, but the Did Not Report/Unknown percentage (27.9%) has been included. The Did Not Report/Unknown category is larger than others, and excluding it can lead to misinterpretation. It is essential to display all the relevant data when creating charts and graphs.

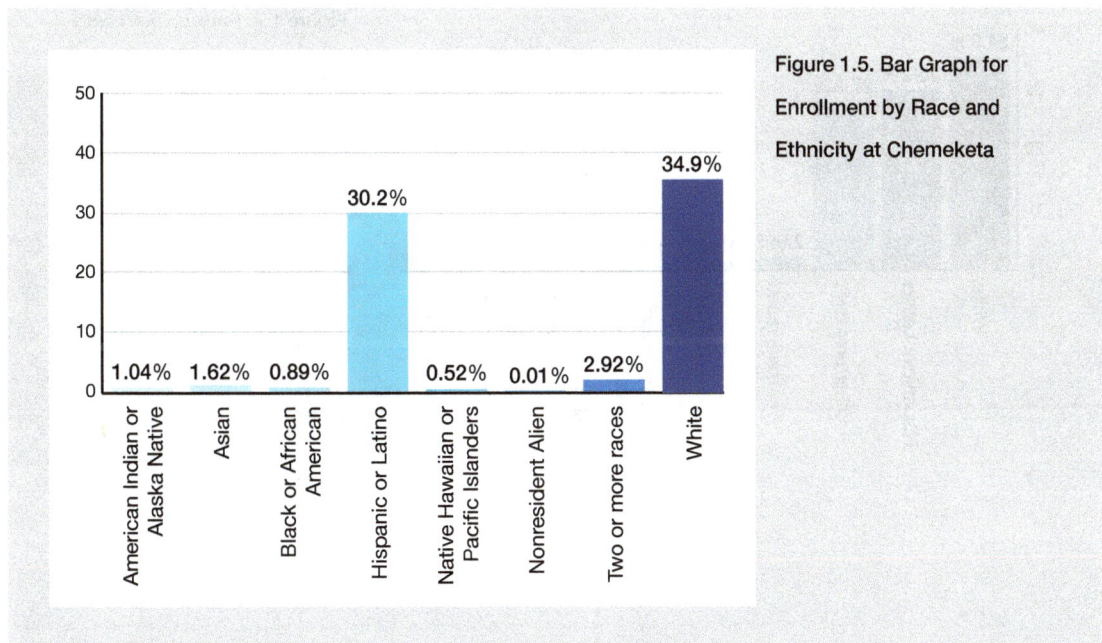

Figure 1.5. Bar Graph for Enrollment by Race and Ethnicity at Chemeketa

Did you notice that the categories in Figure 1.5 are sorted alphabetically? Sometimes, this makes the data difficult to understand. The graph in Figure 1.7 is a **Pareto chart**. A Pareto chart is a bar graph with the bars sorted from largest to smallest, making the data easier to read and interpret.

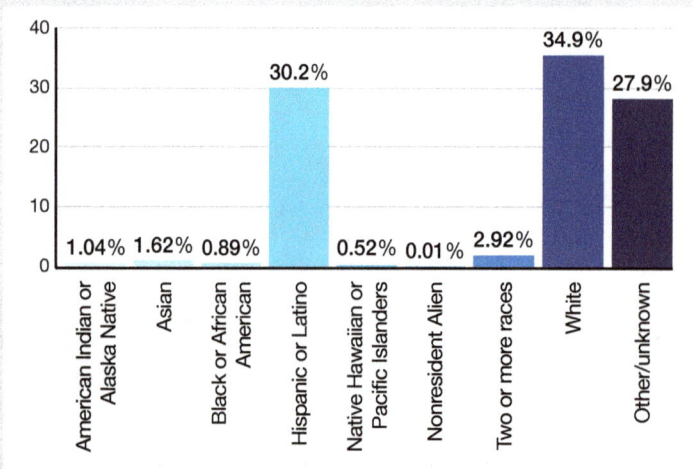

Figure 1.6. Bar Graph with Did Not Report/Unknown Category

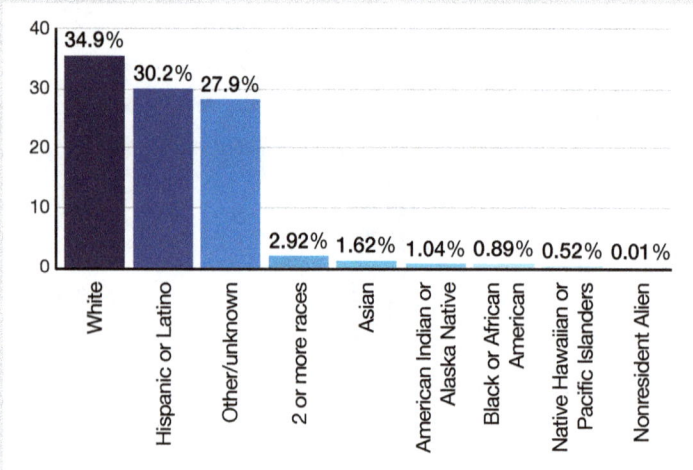

Figure 1.7. Pareto Chart with Bars Sorted by Size

Frequency and Frequency Tables

In statistics, we often want to know how often something happens. When investigating large data sets, we can use frequency and frequency distributions to help organize and summarize how often one thing happens. A frequency is the number of times a specific value in the data occurs. A frequency table organizes values in a data set and tallies how many times each value occurs in the data. Frequency tables help summarize large data sets, see how values are distributed across the data, identify outliers, and provide a basis for designing graphs.

Let's look at an example. Twenty students taking organic chemistry were asked how many hours they studied per day. Their responses, in hours, are as follows, and Table 1.7 lists the different data values in ascending order and their relative and cumulative frequencies:

5 6 3 3 2 4 7 5 2 3 5 6 5 4 4 3 5 2 5 3

Table 1.7. Frequency Table of Student Work Hours

Data Value	Frequency	Relative Frequency	Cumulative Relative Frequency
2	3	$\frac{3}{20}$ or 15%	15%
3	5	$\frac{5}{20}$ or 25%	15% + 25% = 40%
4	3	$\frac{3}{20}$ or 15%	40% + 15% = 55%
5	6	$\frac{6}{20}$ or 30%	55% + 30% = 85%
6	2	$\frac{2}{20}$ or 10%	85% + 10% = 95%
7	1	$\frac{1}{20}$ or 5%	95% + 5% = 100%

According to Table 1.7, there are 3 students who study 2 hours, 5 students who study 3 hours, and so on. The sum of the values in the frequency column, 20, represents the total number of students included in the sample. The frequency table makes a long list of numbers easier to understand and analyze.

A relative frequency compares the number of times a data value occurs in the set of all outcomes to the total number of outcomes. The relative frequency can be expressed as a ratio, fraction, percentage, or decimal. To find the relative frequencies in a data set, divide each frequency by the total number of values in the sample. Table 1.7 shows each data value's frequencies and relative frequencies, expressed as fractions and percentages. The sum of the values in the relative frequency column is $\frac{20}{20}$ or 100%. Relative frequencies are good at showing proportions of the total number of values. For example, we can quickly see that 30% of students study for 5 hours per day, compared to 5% of students who study for 7 hours a day.

The cumulative relative frequency is the total of all the relative frequencies in a data set. It's useful when looking at intervals of data rather than individual values. For example, if we wanted to determine how many students studied more or less than 4 hours a day, finding the cumulative relative frequency will give us that information. In Table 1.7, the cumulative relative frequency column shows that students who studied for 2 or 3 hours a day equal 40% of the students in the class. That means 40% of the class studied for less than 4 hours per day. Since the data set adds up to 100%, the remaining 60% of the class studied for 4 or more hours per day. The last entry of the cumulative relative frequency column is 100%, indicating that 100% of the data has been counted.

When calculating the frequency, we may need to round our answers to be as precise as possible. The most common way to round off answers is to carry the final answer one more decimal place than was present in the original data. For example, if we're looking for the average of the quiz scores 4, 6, and 9, we get the answer 6.333. Round off to the nearest tenth, 6.3, because the data are whole numbers. Refrain from rounding off intermediate results because it will make the final answer less accurate. If it's necessary to round off intermediate results, carry them to at least twice as many decimal places as the final answer.

Note
Because of rounding, the relative frequency column may not always sum to one, and the last entry in the cumulative relative frequency column may not be one. However, they each should be close to one.

Example 15
Table 1.8 shows the amount, in inches, of annual rainfall in a sample of fifty towns. Find the percentage of rainfalls that are less than 9.01 inches.

Table 1.8. Annual Rainfall (inches) From Fifty Towns

Rainfall (Inches)	Frequency	Relative Frequency	Cumulative Relative Frequency
2.95–4.97	6	$\frac{6}{50} = 0.12$	0.12
4.97–6.99	7	$\frac{7}{50} = 0.14$	0.12 + 0.14 = 0.26
6.99–9.01	15	$\frac{15}{50} = 0.30$	0.26 + 0.30 = 0.56

Rainfall (Inches)	Frequency	Relative Frequency	Cumulative Relative Frequency
9.01–11.03	8	$\frac{8}{50} = 0.16$	0.56 + 0.16 = 0.72
11.03–13.05	9	$\frac{9}{50} = 0.18$	0.72 + 0.18 = 0.90
13.05–15.07	5	$\frac{5}{50} = 0.10$	0.90 + 0.10 = 1.00
	Total = 50	Total = 1.00	

Solution 15

Look at the first, second, and third rows. The rainfalls are all less than 9.01 inches. There are 6 + 7 + 15 = 28 towns with rainfalls less than 9.01 inches. The percentage of rainfalls less than 9.01 inches is then $\frac{28}{50}$ or 56%. This percentage is the cumulative relative frequency entry in the third row.

Try It 1.4

From the data in Table 1.8, find the percentage of rainfall between 6.99 and 13.05 inches.

Example 16

Table 1.9 contains the estimated total number of deaths worldwide due to earthquakes from 2009 to 2019.

Table 1.9. Estimated Total Number of Earthquake Deaths Worldwide Between 2009 and 2019

Year	Est. Total Number of Deaths
2009	1,790
2010	226,050
2011	21,942

Year	Est. Total Number of Deaths
2012	689
2013	1,572
2014	756
2015	9,624
2016	1,297
2017	1,012
2018	4,535
2019	224
Total	269,491

1. What is the frequency of deaths measured from 2009 through 2012?
2. What percentage of deaths occurred after 2014?
3. What is the relative frequency of deaths that occurred in 2011 or earlier?
4. What is the percentage of deaths that occurred in 2018?
5. What kind of data are the numbers of deaths?
6. The Richter scale is used to quantify the energy produced by an earthquake. Examples of Richter scale numbers are 2.3, 4.0, 6.1, and 7.0. What kind of data are these numbers?

Solution 16

1. 250,471
2. 6.2%
3. $\frac{249,782}{269,491}$ or 0.927 or 92.7%
4. 1.7%
5. Quantitative discrete
6. Quantitative continuous

Exercises 1.2

1. Airline companies are interested in the consistency of the number of babies on each flight to have adequate safety equipment on board. Suppose an airline conducts a survey. Over Thanksgiving weekend, it surveys 6 flights from Boston to Salt Lake City to determine the number of babies on the flights. It determines the amount of safety equipment needed by the result of that study.

    a. Using complete sentences, list three things wrong with the way the survey was conducted.

    b. Using complete sentences, list three ways that you would improve the survey if it were to be repeated.

2. Suppose you want to determine the mean number of students per statistics class in your state. Describe a possible sampling method in three to five complete sentences. Make the description detailed.

3. Suppose you want to determine the mean number of cans of soda drunk each month by students in their twenties at your school. Describe a possible sampling method in three to five complete sentences. Make the description detailed.

4. List some practical difficulties involved in getting accurate results from a telephone survey.

5. List some practical difficulties involved in getting accurate results from a mailed survey.

6. The instructor takes her sample by gathering data on five randomly selected students from each Lake Tahoe Community College math class. The type of sampling she used is:

    a. cluster sampling

    b. stratified sampling

    c. simple random sampling

    d. convenience sampling

7. A study was done to measure different variables of resident use of a local park in San Jose, California. Researchers wanted to determine the age of park users, the number of times the park is used per week, and the duration (amount of time) of use. The first house in the neighborhood around the park was selected randomly, and then every 8th house in the neighborhood around the park was interviewed. The sampling method was:

    a. simple random sampling

    b. systematic sampling

    c. stratified sampling

    d. cluster sampling

8. Name the sampling method used in each of the following situations:
   a. A woman in the airport is handing out questionnaires to travelers, asking them to evaluate the airport's service. She does not ask travelers who are hurrying through the airport with their hands full of luggage but instead asks all travelers who are sitting near gates and not taking naps while they wait.
   b. A teacher wants to know if her students are doing homework, so she randomly selects rows two and five and then calls on all students in row two and all students in row five to present the solutions to homework problems to the class.
   c. The marketing manager for an electronics chain store wants information about the ages of its customers. Over the next two weeks, at each store location, 100 randomly selected customers are given questionnaires to fill out asking for information about age, as well as about other variables of interest.
   d. The librarian at a public library wants to determine what proportion of the library users are children. The librarian has a tally sheet on which she marks whether books are checked out by an adult or a child. She records this data for every fourth patron who checks out books.
   e. A political party wants to know the reaction of voters to a debate between the candidates. The day after the debate, the party's polling staff calls 1,200 randomly selected phone numbers. If a registered voter answers the phone or is available to come to the phone, that registered voter is asked whom he or she intends to vote for and whether the debate changed his or her opinion of the candidates.

9. The Well-Being Index is a survey that follows trends of United States residents on a regular basis. There are six areas of health and wellness covered in the survey: Life Evaluation, Emotional Health, Physical Health, Healthy Behavior, Work Environment, and Basic Access. Some of the questions used to measure the Index are listed below. Identify the type of data obtained from each question used in this survey: qualitative (categorical), quantitative discrete, or quantitative continuous.
   a. Do you have any health problems that prevent you from doing any of the things people your age can normally do?
   b. During the past 30 days, for about how many days did poor health keep you from doing your usual activities?
   c. In the last seven days, how many days did you exercise for 30 minutes or more?
   d. Do you have health insurance coverage?

10. In advance of the 1936 presidential election, *Literary Digest* released the results of an opinion poll predicting that the Republican candidate Alf Landon would win by a large margin. The magazine sent postcards to approximately 10 million prospective voters. These prospective voters were selected from the subscription list of the magazine, automobile registration lists, phone lists, and club membership lists. Approximately 2.3 million people returned the postcards.

   a. Think about the United States in 1936. Explain why a sample chosen from magazine subscription lists, automobile registration lists, phone books, and club membership lists was not representative of the population of the United States at that time.

   b. What effect does the low response rate have on the reliability of the sample?

   c. Are these problems examples of sampling error or nonsampling error?

   d. During the same year, George Gallup conducted his own poll of 30,000 prospective voters. These researchers used a method they called "quota sampling" to obtain survey answers from specific subsets of the population. Quota sampling is an example of which sampling method described in this section?

1.3 Experimental Design Basics

Overview

The science of statistics meets the real world in studies and experiments. Every career, college discipline, and industry asks questions that require statistical analysis to find an answer. Does aspirin reduce the risk of heart attacks? How much stronger is titanium than steel? Is fatigue as dangerous to a driver as the influence of alcohol? Questions like these are answered using randomized experiments.

Planning and design ensure that data produced from an experiment is accurate and reliable. Experimental design is the process of planning a study or experiment, performing the data collection, and controlling the study parameters for accuracy and consistency. Going through the experimental design process only makes sense for studies where variation, chance, and uncertainty are present and unavoidable. Experimental design helps develop a plan to collect data using a specific process to investigate the relationship between variables. In this section, we'll look at how variables interact, explore the components of experimental design, and discuss the ethical responsibilities of interpreting and sharing data.

Principles of Experimental Design

As we learned earlier, variation in data and samples is expected and acceptable in statistics. There is no way to completely control every variable in a study or experiment. However, if we follow a few fundamental principles of experimental design and use techniques to control variables, we can create a study that produces accurate and reliable data.

The purpose of an experiment is to investigate the relationship between two variables. When one variable affects another, we call the first variable the **explanatory variable**. The affected variable is called the **response variable**. In a randomized experiment, the researcher applies values of the explanatory variable and measures the resulting changes in the response variable. The different values of the explanatory variable are called **treatments**. An experimental unit is a single object or individual to be measured.

Let's look at an example to explain these terms further. A study wanted to examine the effect of sleep deprivation on the ability to drive, and 19 professional drivers participated. The drivers' performance in a driving simulation was measured twice: first after a typical night of sleep and second after 27 hours of sleep deprivation. In this study, the explanatory variable is the amount of sleep. The response variable is the driver's performance in the driving simulation. There are two treatments: normal sleep and 27 hours of sleep deprivation. And the experimental unit being measured is the 19 professional drivers.

Sometimes, additional variables can cloud a study's results. A **lurking variable** is one that is not included in a statistical study, but its existence could influence the relationship between the variables being studied. For example, we want to know if vitamin E effectively prevents disease. We recruit subjects and ask them if they regularly take vitamin E. We notice that the subjects who take vitamin E exhibit better health on average than those who don't. Does this prove that vitamin E is effective in disease prevention? It does not. There are many differences between the two groups beyond vitamin E consumption, including diet, exercise, and family history. As designed, this experiment has too many lurking variables to prove anything about vitamin E consumption. **Confounding** happens when we can see an effect in a study but can't determine what caused it. A randomized study will help avoid confounding.

The best way to counter lurking variables is **randomization**, which is the random assignment of experimental units to treatment groups. When subjects are assigned treatments randomly, all the potential lurking variables are spread equally among the groups. This isolates the explanatory variable by ensuring that the only difference between groups is the one we decide. Then, we know that different outcomes measured in the response variable must be a direct result of the treatments. Then, the experiment can prove a cause-and-effect connection between the explanatory and response variables.

Blocking is the arrangement of participants into groups, or blocks, that are like one another. Blocking is like randomization because it allows treatments to be distributed equally among participants and limits interference from known or irrelevant effects. For example, in an experiment testing a new fertilizer on two varieties of corn, blocking is used to control for soil type, which could influence how the fertilizer affects the corn. The field is divided into blocks of four plots with two plots of each variety. This ensures the soil type is similar within each block.

The power of suggestion is an influential force in experiments. Studies have shown that the expectations of a study participant can be as important as the actual treatment. For example, isolating the explanatory variable's effects can be challenging when studying a participant's physical response, like in the vitamin E health study above. It's possible that the participants who knew they were taking vitamin E were more likely to report good health if they thought that vitamin E would improve their health.

To counter the power of suggestion, researchers make one treatment group the control group. The control group is given a placebo treatment, which is a treatment that can't influence the response variable. For example, the control group in the vitamin E study can be given a pill that looks like the real vitamin E pill but contains no medication. The control group helps researchers balance the effects of participating in an experiment and the effects of the active treatments.

Of course, if you're participating in a study and you know that you're receiving a pill that contains no actual medication, then the power of suggestion isn't a factor. **Blinding** is an experimental design technique where participants don't know who receives an active or placebo treatment. That way, all participants can be free from the power of suggestion. In a double-blind experiment, neither the participant nor the researcher knows which group is receiving the treatment.

Example 17

The Smell & Taste Treatment and Research Foundation conducted a study investigating whether smell affects learning. Subjects completed paper mazes multiple times while wearing masks. They completed mazes three times wearing floral-scented masks and three times with unscented masks. Participants were randomly assigned to wear the floral mask during the first or last three trials. For each trial, researchers recorded the time it took to complete the maze and the subject's impression of the mask's scent: positive, negative, or neutral.

1. Describe the explanatory and response variables in this study.
2. What are the treatments?
3. Identify any lurking variables that could interfere with this study.
4. Is it possible to use blinding in this study?

Solution 17

1. The explanatory variable is scent, and the response variable is the time it takes to complete the maze.
2. There are two treatments: a floral-scented mask and an unscented mask.
3. All subjects experienced both treatments. The order of treatments was randomly assigned, so there were no differences between the treatment groups. Random assignment eliminates the problem of lurking variables.
4. Subjects will know whether they can smell flowers, so subjects can't be blinded in this study. Researchers timing the mazes can be blinded, though. The researcher observing a subject will not know which mask is being worn.

Sometimes, randomized experiments are impractical. **Observational studies** are research studies where the researcher observes and measures the characteristics of a population without intervening. The participants or units in these studies fall naturally within a treatment group. Observational studies can interpret data that has been gathered already or can make predictions based on known factors. Because the researcher can't use randomization, they can't determine how one treatment affects a participant compared to another. Observational studies are often used to explore data over a long period.

Different observational studies are used in many fields, including science, medicine, and education. Retrospective studies look back at past data to find patterns and connections. These studies may use medical records, surveys, or other data sources to investigate previous events. Cross-sectional studies gather data from different groups simultaneously to compare them. These studies may look at things like income, education, or job satisfaction. Prospective studies follow groups of people over time to see how they change or to try to identify what might cause specific outcomes. For example, a study might follow a group of students from kindergarten through high school to see how their academic performance changes over time.

Replication is the assignment of a treatment to many participants, or it can also refer to repeating an experiment more than once. The goal of replication is to ensure that any data variability is managed or measured. The more data collected in a study, the less likely variation will affect the analysis. If the nature of an experiment leads to a small number of participants, repeating the experiment can help distribute any natural variation.

Example 18

Identify the following values for this study: population, sample, experimental units, explanatory variable, response variable, treatments.

Researchers want to investigate whether taking aspirin regularly reduces the risk of heart attack. Four hundred men between the ages of 50 and 84 are recruited as participants. The men are divided randomly into two groups: one will take aspirin, and the other will take a placebo. Each man takes one pill each day for three years, but he does not know whether he is taking aspirin or a placebo. At the end of the study, researchers count the number of men in each group who have had heart attacks.

Solution 18

The **population** is men aged 50 to 84.

The **sample** is the 400 men who participated.

The **experimental units** are the individual men in the study.

The **explanatory variable** is oral medication.

The **treatments** are aspirin and a placebo.

The **response variable** is whether a subject had a heart attack.

Example 19

A researcher wants to study the effects of birth order on personality. Explain why this study could not be conducted as a randomized experiment. What is the main problem in a study that can't be designed as a randomized experiment?

Solution 19

The explanatory variable is birth order. We can't randomly assign a person's birth order. Random assignment eliminates the impact of lurking variables. When we can't randomly assign subjects to treatment groups, there will be differences between groups other than the explanatory variable.

Try It 1.5

We are concerned about the effects of texting on driving performance. Design a study to test drivers' response time while texting and while not texting. How many seconds does it take for a driver to respond when a leading car hits the brakes?

1. Describe the explanatory and response variables in the study.
2. What are the treatments?
3. What should we consider when selecting participants?
4. Our research partner wants to divide participants randomly into two groups: one to drive without distraction and one to text and drive simultaneously. Is this a good idea? Why or why not?
5. Identify any lurking variables that could interfere with this study.
6. How can blinding be used in this study?

Ethics and Statistics

Statistics is the science of variation, randomness, and chance. Statistics is different from other sciences, where experiments lead to concrete, cut-and-dried answers. It provides tools for understanding data while also dealing with uncertainty. Statistical analysis generally doesn't offer strict yes-or-no responses to questions involving random processes. Instead, statistics provides logical, quantitative guesses represented as probability values, confidence or prediction intervals, odds, or chances that may be interpreted in multiple ways.

The possibility of multiple interpretations can be perceived as negative or uncertain. However, these outcomes are beautiful, scientific, and elegant responses to challenging problems. The widespread misuse and misrepresentation of statistical data often gives statistics a bad name. You often hear that "numbers don't lie." But the people who use numbers to support their claims sometimes do.

Researchers have an ethical responsibility to use proper data collection and analysis methods. Many types of statistical fraud are hard to see. Sometimes, researchers stop collecting data once they have enough to prove what they had hoped to show. They don't want to take the chance that more study would complicate their lives by producing data that contradicts what they expected to find. Professional organizations, like the American Statistical Association, clearly define expectations for researchers. There are even federal laws governing the use of research data.

Some recent examples of the misuse or unethical use of statistics might sound familiar. In 2015, Volkswagen was found to have installed "defeat devices" in their diesel cars that allowed them to cheat emissions tests. They used statistical manipulation to make it seem like their vehicles met environmental regulations when they actually emitted up to 40 times the legal limit of nitrogen oxides. In 2019, Facebook was fined $5 billion by the Federal Trade Commission for privacy violations, including the misuse of user data for targeted advertising. Facebook used statistical models to predict user behavior and sell that information to advertisers without users' consent. And in 2020, the *Lancet* medical journal published a study claiming that hydroxychloroquine was linked to increased mortality in COVID-19 patients. The study was retracted shortly after publication when it was revealed that the data was fake and the statistical analysis was deeply flawed. All these examples had real impacts on people just like you. As you learn more about statistics in this course, you will also be more equipped to spot fraud and read statistical studies with a critical eye.

Example 20

Describe the unethical behavior in each example and how it could impact the reliability of the resulting data. Explain how the problem should be corrected.

A researcher is collecting data in a community.

1. She selects a block where she is comfortable walking because she knows many of the people who live there.
2. No one is home at four houses on her route. She doesn't record the addresses and doesn't return later to try to find residents at home.
3. She skips four houses on her route because she's late for an appointment. When she gets home, she fills in the forms by selecting random answers from other residents in the neighborhood.

Solution 20

1. By selecting a convenient sample, the researcher is intentionally selecting a sample that could be biased. Claiming that this sample represents the community is misleading. The researcher needs to select areas in the community at random.
2. Intentionally omitting relevant data will create bias in the sample. Suppose the researcher is gathering information about jobs and childcare. Ignoring people who are not home may be missing data from working families relevant to her study. She needs to try to interview all members of the target sample.
3. It is never acceptable to fake data. Even though the responses she uses are "real" responses provided by other participants, the duplication is fraudulent and can create bias in the data. She needs to work diligently to interview everyone on her route.

Exercises 1.3

1. Design an experiment. Identify the explanatory and response variables. Describe the population being studied and the experimental units. Explain the treatments that will be used and how they will be assigned to the experimental units. Describe how blinding and placebos may be used to counter the power of suggestion.

2. Discuss potential violations of the rule requiring informed consent.
   a. Inmates in a correctional facility are offered good-behavior credit in return for participation in a study.

b. A research study is designed to investigate a new children's allergy medication.

c. Participants in a study are told that the new medication being tested is highly promising, but they are not told that only a small portion of participants will receive the new medication. Others will receive placebo treatments and traditional treatments.

3. How does sleep deprivation affect your ability to drive? A recent study measured the effects on 19 professional drivers. Each driver participated in two experimental sessions: one after normal sleep and one after 27 hours of total sleep deprivation. The treatments were assigned in random order. In each session, performance was measured on a variety of tasks, including a driving simulation. Describe the following elements for this study:

a. Explanatory variable

b. Response variable

c. Treatments

d. Experimental units

e. Lurking variables

f. Random assignment

g. Control/placebo

h. Blinding

4. An advertisement for Acme Investments displays the two graphs to show the value of Acme's product in comparison with Beta Investment's product. As the graphs show, Acme consistently outperforms Beta. Describe the potentially misleading visual effect of these comparison graphs. How can this be corrected?

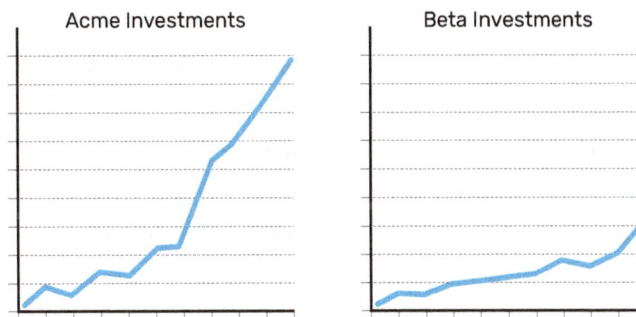

Acme Investments

Beta Investments

5. The graph shows the number of complaints for six different airlines as reported to the US Department of Transportation in February 2013. Alaska, Pinnacle, and Airtrain Airlines have far fewer complaints reported than American, Delta, and United. Can we conclude that American, Delta, and United are the worst airline carriers since they have the most complaints?

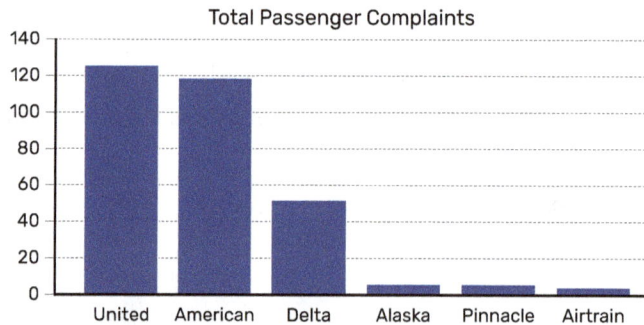

Total Passenger Complaints

6. We want to examine the effectiveness of three programs on the weight loss of men and women ages 40 to 50 years old. There are 150 men and 150 women who participate in the study. Subjects are randomly assigned to the three programs. They spend 3–4 hours in the program per week, and they continue the program for six months. Their weight is recorded before and after the program. Which type of study is this? Choose one answer.

    a. This is an observational study because the subjects may spend less time in the program than they are supposed to.

    b. This is an experiment because the researcher is measuring the weight of the subjects before and after participation in the program.

    c. This is an experimental study because the subjects have been randomly assigned to different treatment groups.

    d. This study is a combination of experimental and observational because we are collecting data on both the experimental and the control group as well as talking to the people.

7. Hospital floors are usually covered by tiles. Carpet would cut down on noise but might be more likely to hold germs. To study this possibility, investigators randomly assigned 8 of 16 available hospital rooms to have carpet installed. The others were left tiled. Later, air from each room was

pumped over a dish of agar. The dish was incubated for a fixed period, and the number of bacteria colonies was counted. Select the appropriate statistical term for the 8 rooms left bare.

a. Treatments

b. Experimental units

c. Control group

d. Response

8. Scientists are interested in the effects of the sun on growth of moss on trees above the Arctic Circle. Twenty-five years of data is collected and then analyzed. The study shows that moss grows the most in the years where there is a moderate amount of sun during the summer and the least in the years where the sun is mostly obscured by clouds during the summer. This is an example of an

a. experimental study from which we can draw causal conclusions cautiously

b. experimental study from which we cannot draw causal conclusions

c. observational study from which we can draw causal conclusions

d. experimental study from which we can draw causal conclusions

e. observational study from which we cannot draw causal conclusions

9. We intend to examine the effect of using a computer lab in statistics classes on the attitudes of students toward statistics. We offer ten classes on statistics in an academic year. Five of these sections are randomly assigned to the experimental group, and the other five are assigned to the control group. The experimental group will go to a lecture, section, and computer lab. The control group will only go to the lecture and section but will not do the computer lab. The attitude of the students toward statistics is measured before and after the course. This study is:

a. A double-blind study

b. A well-designed experiment

c. A blind study

d. Not a randomized experiment

10. What conditions would need to be satisfied to say that a change in variable X causes a change in variable Y?

a. When an experiment reveals that a change in X causes a change in Y.

b. When possible confounding variables have been ruled out.

11. A study indicated that people over the age of 70 who had pets lived longer and became less de-pressed than people over the age of 70 who did not have pets. The data came from the records of 700 people who went to a local clinic for treatment. Based on pre-existing medical records, 400 had pets, and 300 did not.

    a.   This is an observational study because the data are obtained from the pre-existing medical records of the patients who go to the local clinic.

    b.   This is a double-blind study because the patients do not know that they are being studied, and the person in charge of the analysis does not know the names of the patients.

    c.   This is an experiment since treatments were imposed on the patients.

    d.   This is an experiment because the people who have pets represent the experimental group, and those without pets represent the control group.

    e.   This is a blind study because the patients do not know that the researchers are studying the relationship between their pet ownership and whether they feel depressed.

Try It Solutions

Solution 1.1

The **population** is all families with children attending Knoll Academy.

The **sample** is a random selection of 100 families with children attending Knoll Academy.

The **parameter** is the average (mean) amount of money spent on school uniforms by families with children at Knoll Academy.

The **statistic** is the average (mean) amount of money spent on school uniforms by families in the sample.

The **variable** is the amount of money spent by one family. Let X = the amount of money spent on school uniforms by one family with children attending Knoll Academy.

The **data** are the dollar amounts spent by the families. Examples of the data are $65, $75, and $95.

Solution 1.2

Stratified sample

Solution 1.3

No, this sample is not representative of the entire 20,000-listener population. Since the sample of 200 people is from the station's concert events, most of them are more likely to prefer music rather than talk shows. The sample is biased.

Solution 1.4

64%

Solution 1.5

1. The explanatory variable is whether a driver will text while driving. The response variable is the time it takes for the driver to respond when a leading car hits the brakes.
2. The treatments are texting or not texting while driving.
3. When selecting participants, the researchers need to select a random, unbiased sample representing the drivers' general population.
4. It is a good idea to divide participants randomly into two groups: driving without texting and driving while texting simultaneously. By doing so, we eliminate the problem of lurking variables.
5. All treatment subjects were randomly assigned, so there were no differences between the treatment groups. Random assignment eliminates the issue of lurking variables.
6. Subjects will know whether they can text, so subjects can't be blinded in this study. Researchers recording the response time can be blinded, though. The researcher observing a subject will not know whether they are texting.

Chapter 2
Descriptive Statistics

This chapter will study numerical and graphical ways to describe and display data. We'll learn how to calculate and, even more importantly, interpret these measurements and graphs. This area of statistics is called **descriptive statistics**.

Data can be described and presented in many formats. For example, suppose you plan to transfer to a four-year university after earning an associate degree at a community college. You want to study electrical engineering, but you're worried that the cost of tuition will saddle you with a lot of debt. Your adviser helps find a sample data set of all the universities with electrical engineering programs and their tuition. Looking at all the costs in the sample feels overwhelming, though, and isn't helping you decide where to apply. A better way might be to look at these two descriptive statistics: the mean cost of attendance and the variation of costs. Once the data is more descriptive, you can weigh the cost of attendance and other information about the programs to decide where to apply.

Beyond a visual description of data, this chapter will look at some basic ways to describe data by measuring the center of data, the variation of data, the relative position of data, and a few basic tools for understanding what those measurements say about the data.

After reading this chapter, you will be able to do the following:

1. Interpret graphical representations of data
2. Construct stem-and-leaf plots, frequency histograms, and relative frequency histograms
3. Understand the concept of the center of data
4. Describe and compute the mean, the median, and the mode of data
5. Understand the concept of the variability of data
6. Find the range, the variance, and the standard deviation of data
7. Explain when to use the range rule of thumb
8. Recognize the relative position of a data value
9. Find the percentile rank and the z-score of a data value
10. Compute the three quartiles associated with a data set
11. Understand the five-number summary and boxplot

2.1 Popular Data Displays

Overview

Pictures are great at showing a lot of information in one visual message. That's where the phrase "a picture is worth a thousand words" comes from. In the same way, graphical representations make data easier to understand by showing a picture of the data. In this section, we'll look at popular data displays: stem-and-leaf plots, frequency histograms, and frequency polygons.

Stem-and-Leaf Plot

One simple graph, the stem-and-leaf plot, is a good choice when the data set is small. It is a quick way to graph data that gives an exact picture of the data and reveals an overall pattern and any outliers. An **outlier** is an observation of data that does not fit the rest of the data. It is sometimes called an extreme value. Some outliers are due to mistakes (for example, writing down fifty instead of 500), while others may indicate that something unusual is happening. It takes some background information to explain outliers, and we'll cover them in more detail in Chapter 8.

Figure 2.1. Stem-and-Leaf Plot for Test Scores

2	0
3	025
4	11378
5	133467889
6	024559
7	147
8	8
9	
10	00

How to Make a Stem-and-leaf Plot

To create a stem-and-leaf plot, divide each data observation into a stem and a leaf. The leaf consists of a final significant digit. For example, 23 has stem 2 and leaf 3. The number 432 has stem 43 and leaf 2. The decimal 9.3 has stem 9 and leaf 3. Next, write the stems in a vertical line from smallest to largest. Draw a vertical line to the right of the stems. Then, write the leaves in increasing order next to their corresponding stem. Figure 2.1 shows an example of a stem-and-leaf plot for thirty student test scores from a statistics class:

| 20 | 30 | 32 | 35 | 41 | 41 | 43 | 47 | 48 | 51 | 53 | 53 | 54 | 56 | 57 |
| 58 | 58 | 59 | 60 | 62 | 64 | 65 | 65 | 69 | 71 | 74 | 77 | 88 | 100 | 100 |

Example 1

In Vania's pre-calculus class, 30 student test scores for the first exam were as follows (smallest to largest):

33	42	49	49	53	55	55	61	63	67	68	68	69	69	72
73	74	78	80	83	88	88	88	90	92	94	94	94	96	100

Create a stem-and-leaf plot from the data and note anything of interest the plot tells us.

Solution 1

The stem-and-leaf plot in figure 2.2 shows that most scores fell in the 60s, 70s, 80s, and 90s. Seven out of the 30 scores, or approximately 23%, were in the 90s or above, a high number of As.

Figure 2.2. Stem-and-Leaf Plot for Final Exam Scores

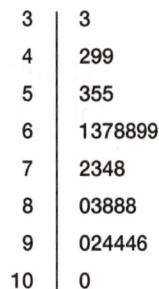

3	3
4	299
5	355
6	1378899
7	2348
8	03888
9	024446
10	0

Try It 2.1

The Park City basketball team scores for the last 30 games were as follows (smallest to largest):

32	32	33	34	38	40	42	42	43	44	46	47	47	48	48
48	49	50	50	51	52	52	52	53	54	56	57	57	60	61

Construct a stem-and-leaf plot for the data.

Example 2

The following data set contains 20 distances (in kilometers) from a home to local supermarkets. Create a stem-and-leaf plot using the data:

$$1.1 \quad 1.5 \quad 2.3 \quad 2.5 \quad 2.7 \quad 3.2 \quad 3.3 \quad 3.3 \quad 3.5 \quad 3.8$$

$$4.0 \quad 4.2 \quad 4.5 \quad 4.5 \quad 4.7 \quad 4.8 \quad 5.5 \quad 5.6 \quad 6.5 \quad 9.8$$

Do the data seem to have any concentration of values?

Solution 2

The stem column in figure 2.3 represents the whole numbers, and the leaf column represents the decimal numbers.

The value 9.8 kilometers may be an outlier. Values appear to concentrate at 3 kilometers and 4 kilometers.

Figure 2.3. Stem-and-Leaf Plot for Distances from Home to Supermarket

Stem	Leaf
1	15
2	357
3	23358
4	025578
5	56
6	5
7	
8	
9	8

Histograms

With big sets of data, making a stem-and-leaf plot isn't practical. A better way to display large data sets is with a histogram. A **histogram** is a graphical representation of the distribution of data. It consists of a series of bars, with the height of each bar representing the frequency of occurrence of values within a certain interval. It has both a horizontal and a vertical axis. The horizontal axis represents what the data is describing (for instance, the distance from your home to school). The vertical axis represents frequencies, either as the frequency or relative frequency of the data. The histogram shows the shape of the data, the center, and the spread of the data. It can also identify any outliers.

In Chapter 1, we learned about frequencies and frequency tables. A histogram is a graphical representation of a frequency table. As a reminder, frequency is the number of times a value occurs in the data. The relative frequency is equal to the frequency for a data value divided by the data set's total number of values.

How to Make a Histogram

To construct histograms, we need to understand the concept of classes. A **class** is a group that represents a range of values. Sometimes, a class is called a bin, and we can think of it as a bin that we use to organize values, just like we might use bins to organize and separate the clothes in the closet. We use classes of data points to construct a histogram. The class midpoint is the value in the middle of the class. It is the average of the upper and lower limit values.

1. **Decide how many classes will represent the data.** Each class will be represented by a bar in the histogram. Many histograms consist of 5 to 15 classes.

2. **Choose a starting point for the first class.** The starting point should be less than the smallest data value. A convenient starting point is a value carried one decimal place lower than the data value with the most decimal places. For example, if the data with the most decimal places has one decimal and the smallest value is 6.1, a convenient starting point is 6.05 ($6.1 - 0.05 = 6.05$). In another example, if the data with the most decimal places has two decimals and the lowest value is 1.5, a convenient starting point is 1.495 ($1.5 - 0.005 = 1.495$). And finally, if the data are all integers and the smallest value is 2, a convenient starting point is 1.5 ($2 - 0.5 = 1.5$).

3. **Choose an ending point for the last class.** The ending point should be more than the largest data value. Use the same method as the one used to choose the starting point but add instead of subtracting. For example, if the data with the most decimal places has one decimal and the largest value is 12.7, a convenient starting point is 12.75 ($12.7 + 0.05 = 12.75$).

4. **Calculate the width of each bar or class interval.** Subtract the starting point from the ending point and divide by the number of bars. For example, the starting point is 6.05, and the ending point is 12.75. With six classes, the class width is:

$$\frac{\text{ending point} - \text{starting point}}{\text{number of classes}} = \frac{12.75 - 6.05}{6} \approx 1.12 \rightarrow \textit{round up to 2}$$

We round the class width up to the next integer for convenience.

Note

When the starting and ending points are carried to one more decimal place, no data value will fall on a boundary.

Example 3

The following data are the heights (in inches to the nearest half inch) of 100 male semiprofessional soccer players. The heights are continuous data because the player's height is measured.

60	60.5	61	61	61.5	63.5	63.5	63.5	64	64	64	64	64	64	64
64.5	64.5	64.5	64.5	64.5	64.5	64.5	64.5	66	66	66	66	66	66	66
66	66	66	66.5	66.5	66.5	66.5	66.5	66.5	66.5	66.5	66.5	66.5	66.5	67
67	67	67	67	67	67	67	67	67	67	67	67.5	67.5	67.5	67.5
67.5	67.5	67.5	68	68	69	69	69	69	69	69	69	69	69	69
69.5	69.5	69.5	69.5	69.5	70	70	70	70	70	70	70.5	70.5	70.5	71
71	71	72	72	72	72.5	72.5	73	73.5	74					

Construct a frequency table and relative frequency histogram from this data using eight classes. See table 2.1 and figure 2.4 for solutions.

Solution 3

Since the data with the most decimal places has one decimal, we want our starting point to have two decimal places. Since the numbers 0.5, 0.05, and 0.005 are convenient, use 0.05 to find the convenient starting and ending points.

The smallest data value is 60.

$$60 - 0.05 = 59.95$$

59.95 is the starting point
The largest value is 74.

$$74 + 0.05 = 74.05$$

74.05 is the ending point.

Next, calculate the width of each class. Subtract the starting point from the ending point and divide by the number of classes.

$$Class\ width = \frac{74.05 - 59.95}{8} = 1.76 \rightarrow round\ up\ to\ 2$$

Table 2.1. Frequency Table for the Heights of 100 Male Semiprofessional Soccer Players

Intervals for Heights (inches)	Frequency	Relative Frequency
59.95–61.95	5	0.05
61.95–63.95	3	0.03
63.95–65.95	15	0.15
65.95–67.95	40	0.4
67.95–69.95	17	0.17
69.95–71.95	12	0.12
71.95–73.95	7	0.07
73.95–75.95	1	0.01

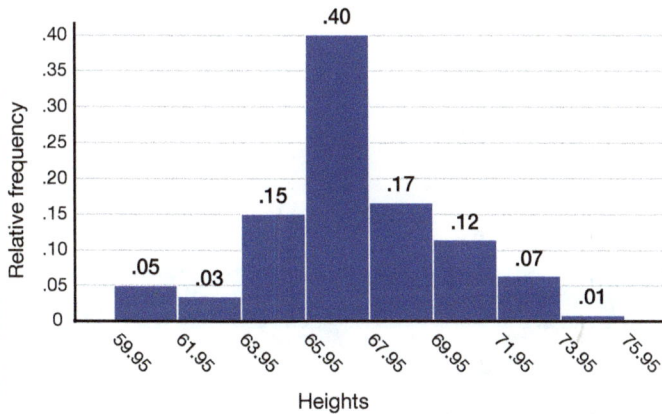

Figure 2.4. Relative Frequency Histogram for the Heights of 100 Male Semiprofessional Soccer Players

Example 4

The following data are the number of times the 24 members of Fremont High's sophomore class check social media each week. Construct a frequency table and histogram for the data using six classes. See table 2.2 and figure 2.5 for solutions.

6	40	88	68	77	83	84	50	90	59	87	15
70	33	75	46	68	34	86	29	56	53	30	92

Solution 4

Since the data are integers, the starting point has one decimal place. Since the numbers 0.5, 0.05, and 0.005 are convenient numbers, we use 0.5 to find the convenient starting point and ending point.

The smallest data value is 6.

$$6 - 0.5 = 5.5$$

5.5 is the starting point.
The largest value is 92.

$$92 + 0.5 = 92.5$$

92.5 is the ending point.
Next, calculate the class width.

$$class\ width = \frac{92.5 - 5.5}{6} = 14.5 \rightarrow 15$$

Table 2.2. Frequency Table of Weekly Social Media Usage

Intervals for Number of Times	Frequency
5.5–20.5	2
20.5–35.5	4
35.5–50.5	3

Intervals for Number of Times	Frequency
50.5–65.5	3
65.5–80.5	5
80.5–95.5	7

Histograms in R

Histograms can be constructed in R programming language with the `hist()` function. This function takes in a vector of values for which the histogram is constructed. In R programming, R chooses the number of classes and creates a frequency histogram. Generally, R calculates the best number of classes. However, we can specify the number of classes, add labels and titles using the arguments `breaks`, `xlab`, `ylab` and `main` to enhance the histogram. We can also add more arguments to histograms like `xlim` and `ylim` to provide a range of the axes, `col` to define color, and more.

There are many ways to enter or import data into R. To plot the histogram for Example 4, we enter the 24 data values as a vector, then plot the histogram with the `hist` function:

```
> Time = c(6, 40, 88, 68, 77, 83, 84, 50, 90, 59, 87, 15, 70, 33,
75, 46, 68, 34, 86, 29, 56, 53, 30, 92)

> hist(Time, xlab = "Number of Times", ylab = "Frequency", main
= "Histogram of Social Media Usage", breaks = c(5.5, 20.5, 35.5,
50.5, 65.5, 80.5, 95.5))
```

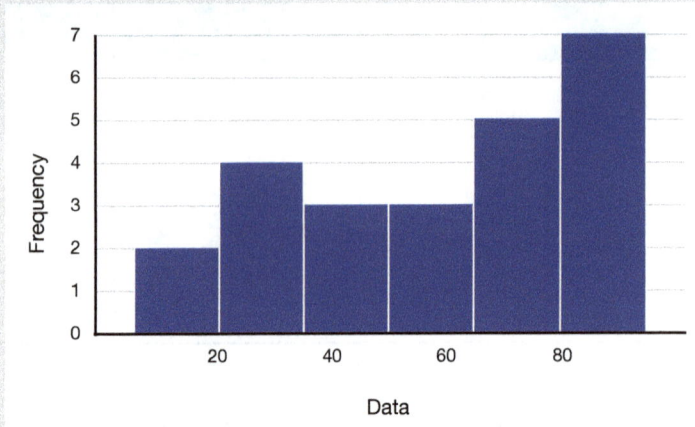

Figure 2.5. Histogram of Weekly Social Media Usage

Frequency Polygons

A frequency polygon is a line graph that plots class midpoints of a data set. It is like a histogram, but rather than bars, it uses points and line segments to make continuous data visually easy to interpret. A relative frequency polygon plots the relative frequency on the y-axis. The following example shows how a frequency polygon is constructed.

Example 5

Table 2.3 is a frequency table that shows the distribution of final test scores in a calculus class. Construct a frequency polygon.

Table 2.3. Frequency Distribution for Calculus Final Test Scores

Intervals for Test Scores	Frequency	Cumulative Frequency
49.5 - 59.5	5	5
59.5 - 69.5	10	15
69.5 - 79.5	30	45
79.5 - 89.5	40	85
89.5 - 99.5	15	100

Solution 5

1. Find the midpoint for each interval. The midpoint is the average of the lower and upper bounds. The midpoint for the first interval is $\frac{49.5 + 59.5}{2} = 54.5$

2. All midpoint values are listed in column 2 of Table 2.4.

3. Record the test scores that fall within each interval. If there are 5 test scores between 49.5 and 59.5, the frequency is 5.

4. Write the midpoint and frequency as an ordered pair so that the first value of the pair is the midpoint, and the second value is the corresponding frequency.

5. Graph each ordered pair on the axes. The horizontal axis is the midpoints, and the vertical axis is the frequencies.

Table 2.4. Values for Example 5

Intervals for Test Scores	Midpoint	Frequency	Ordered Pair
49.5 - 59.5	54.5	5	(54.5, 5)
59.5 - 69.5	64.5	10	(64.5, 10)
69.5 - 79.5	74.5	30	(74.5, 30)
79.5 - 89.5	84.5	40	(84.5, 40)
89.5 - 99.5	94.5	15	(94.5, 15)

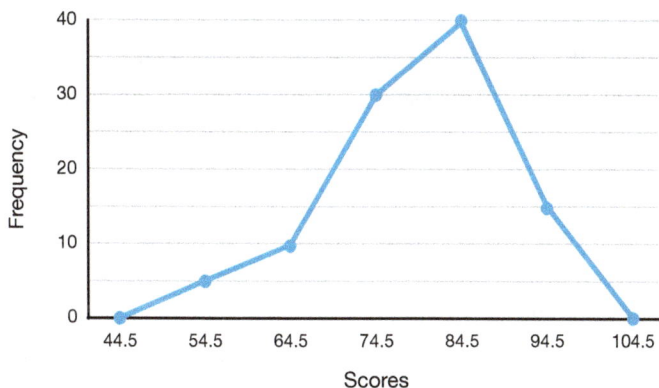

Figure 2.6. Frequency Polygon for Calculus 1 Final Test Scores

The size of a sample will influence the look of its histogram. When working with a small sample and only a few classes, the histogram will look more like a bar graph, like Figure 2.7(a). As the sample size increases, so will the number of classes. It will have narrower columns that start to look more like a curve, as in Figures 2.7(b) and 2.7(c). When the sample size is very large, the histogram's bars become so small that they aren't recognizable as bars anymore and look like a smooth curve, like in Figure 2.7(d).

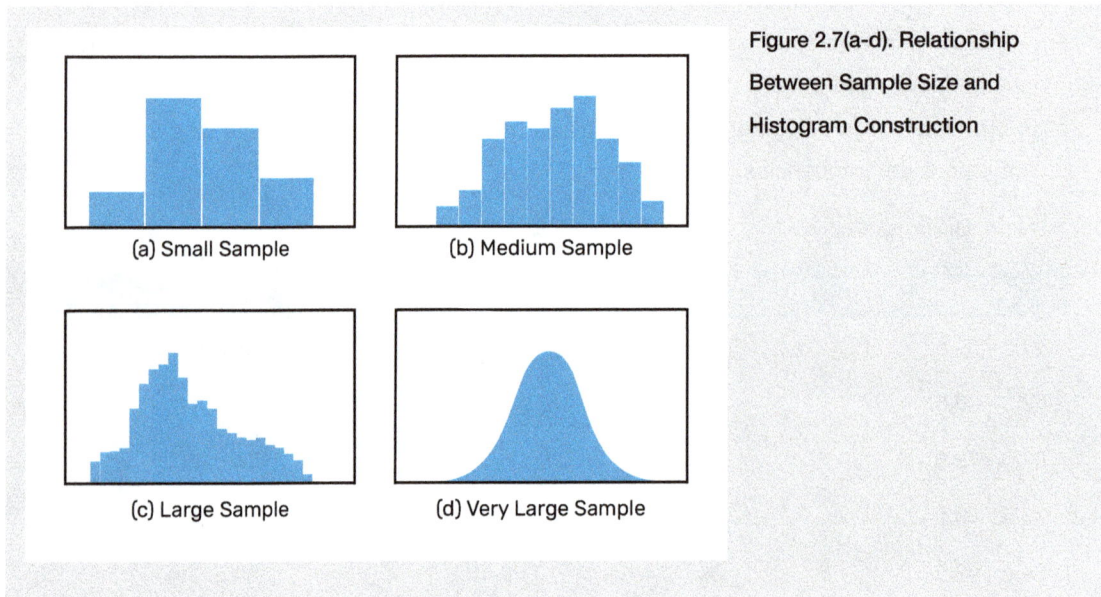

Figure 2.7(a-d). Relationship Between Sample Size and Histogram Construction

(a) Small Sample

(b) Medium Sample

(c) Large Sample

(d) Very Large Sample

Large data sets and populations are often shown as a smooth curve in statistics. What we're seeing is simply a relative frequency histogram with tiny bars. But even if we can't see the edges of the bars, remember that each represents a class. Remember that the total area under the curve of the relative frequency histogram is equal to 1, or 100 percent.

Exercises 2.1

1. The following data are the number of times a group of college sophomores checks social media weekly. Construct a frequency table and histogram for the data. Use 6 classes.

12	45	86	48	79	83	8	18	64	58	80	59	27	15	21	19
40	33	65	46	58	57	34	34	76	39	58	63	80	82	95	20

2. The IQ scores of 20 students randomly selected from an elementary school are given.

108	120	98	84	102	100	99	125	87	105
95	81	107	105	92	88	108	119	118	100

    a. Use groups of 80, 90, 100, 110, and 120 to construct a stem-and-leaf diagram.

    b. Construct a frequency histogram in R using 5 classes.

    c. Construct a relative frequency histogram.

3. During a recent Red Cross blood donation, 500 people came to a donation center. The blood types of these 500 donors are summarized in the following table.

Blood Type	O	A	B	AB
Number of Donors	216	210	62	12

    a. Construct a relative frequency histogram for the data set.

    b. Identify the blood type with the highest relative frequency for these 500 people.

4. The following data are the average annual rainfall (in inches) in 36 towns or cities in Oregon. The rainfalls are continuous data. Construct a frequency table and relative frequency histogram from this data. Use 5 or 6 classes.

10.67	11.10	12.48	13.74	15.95	16.34	17.79	18.91	20.41	22.13
24.56	26.77	29.87	35.45	38.64	41.24	43.67	44.04	51.48	53.84
56.20	58.60	59.50	60.48	61.96	62.50	68.80	70.40	75.90	76.48
77.10	79.90	80.38	81.96	82.59	83.42				

2.2 Measures of Central Location

Overview

The measures of center are some of the most common statistics in this course. Just as a histogram describes data graphically, the center of a data set—the mean and median—describes location. Once a data set's central location is known, we can see how the data is distributed around the center.

Mean

The mean is the most common measure of center. If we say, "It takes me an average of 24 minutes to get to school," what we're talking about is the mean. The **mean** is the average of all the data values. The population mean is calculated if the data comes from a census with a measurement for every element in the population. It is noted by μ, the Greek letter mu. The sample mean is calculated from a population sample and is intended to estimate the population mean. It is denoted by $\bar{x}$, which is said as x-bar. Both values are calculated similarly. In any data set, there is only one mean value.

Population Mean for a Population of Size _N_

$$\mu = \frac{\Sigma x}{N}$$

Sample Mean for a Sample of Size _n_

$$\bar{x} = \frac{\Sigma x}{n}$$

Σ is the Greek uppercase letter sigma, which instructs us to add up whatever values or expressions follow it. In this case, it is followed by x, which stands for all values in the data set n. The mean is a statistic that can apply to a population or a sample.

Hint: Statistic or Parameter?

If a symbol is used over a letter, like $\bar{x}$ or $\hat{p}$, we're dealing with a sample statistic. If it's just a letter, like x or p, we're dealing with a population parameter.

Example 6

For the data set 11, 3, 14, 8, 20, find $\sum x$, $\sum x^2$, $\sum(x-1)^2$, and $\bar{x}$.

Solution 6

$$\sum x = 11 + 3 + 14 + 8 + 20 = 56$$

$$\sum x^2 = 11^2 + 3^2 + 14^2 + 8^2 + 20^2 = 790$$

$$\sum(x-1)^2 = (11-1)^2 + (3-1)^2 + (14-1)^2 + (8-1)^2 + (20-1)^2 = 683$$

$$\bar{x} = \frac{\sum x}{n} = \frac{56}{5} = 11.2$$

Median

The **median** is the middle value when all data values are arranged numerically. The median of a sample can be shown by $\tilde{x}$ (x-tilde), M, or Med. There is no standard notation for a population median. If the total number of values in the sample is odd, the median is the middle value of the ordered data (ordered smallest to largest). If the total number of values in the sample is even, the median equals the two middle values added together and divided by two after the data has been ordered.

The median is a value that splits a histogram in the middle so that half of the data is on the left and half on the right, as in Figure 2.8. In any data set, there is only one median value.

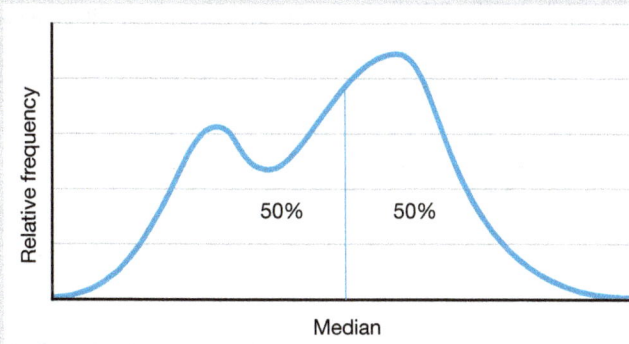

Figure 2.8. Histogram with Median

Example 7

Compute the sample median for the following data.

1.78 1.46 2.35 1.12 2.61 3.12 2.43 4.54 3.71 3.33

Solution 7

Rearrange the data in numerical order.

1.12 1.46 1.78 2.35 2.43 2.61 3.12 3.33 3.71 4.54

Since number of observations is 10, an even number, we will use the two middle measurements, which are 2.53 and 2.71. Therefore, the median of these data is

$$M = \frac{2.43 + 2.61}{2} = 2.52$$

The shape of a relative frequency histogram can tell us a lot about the mean and median. Look at the histograms in Figure 2.9. A symmetrical distribution shows that the mean and the median are equal, shown in Figure 2.9(a-b). A big curve on the left and a long tail on the right shows the distribution is skewed right, and the mean is greater than the median, shown in Figure 2.9(c). A big curve on the right and a long tail to the left show that the distribution is skewed left, and the median is greater than the mean, shown in Figure 2.9(d).

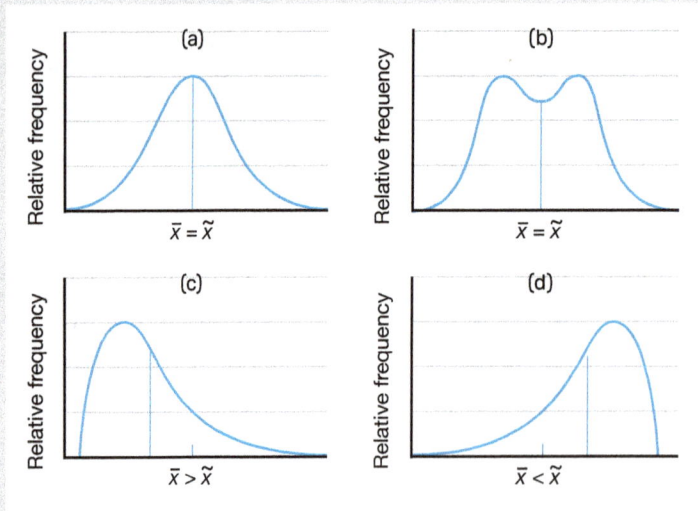

Figure 2.9(a-d). Skewness of Relative Frequency Histograms

Mode

The **mode** is the most frequently occurring value in a data set. The mode is the only measure of center that can be used with qualitative data like names, categories, or labels. To find the mode, determine how many times each value occurs in the data set. A data set can have one mode if only one value has the highest frequency, which is called unimodal. It can have multiple modes if multiple values have the highest frequency, which is called bimodal. And if no value is repeated, we say the data has no mode.

The mode is easy to spot in a relative frequency histogram because it is the highest point on the curve (Figure 2.10).

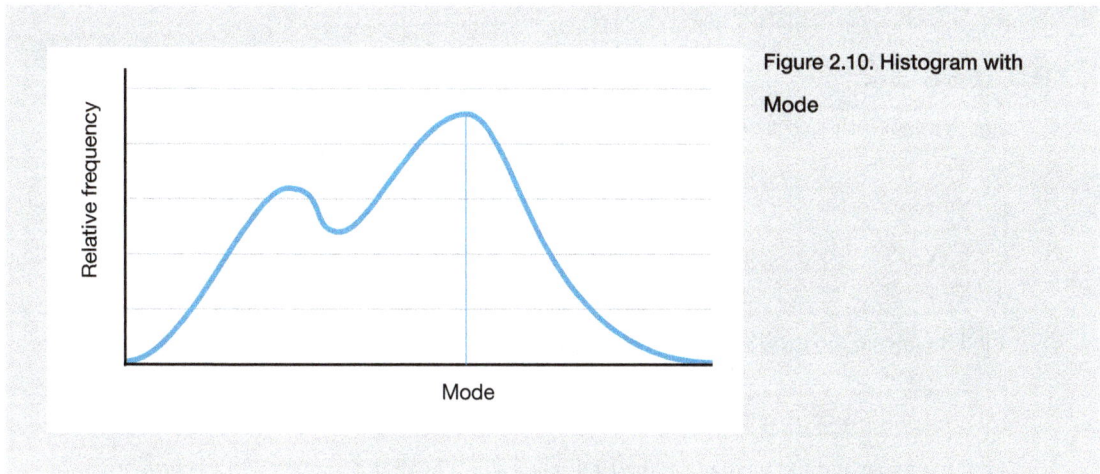

Figure 2.10. Histogram with Mode

Example 8

Find the mode of the following data set.

$$-1 \quad 0 \quad 2 \quad 4 \quad -2 \quad 7 \quad 8 \quad 4$$

Solution 8

The value 4 is most frequently observed, and therefore, the mode is 4.

Note

Not every data set has a mode. There is no value in the data set of Example 8 that is observed more frequently than others. Therefore, there is no mode for this data set.

Midrange

The **midrange** is the average of the lowest and highest data values. Because the midrange only uses the lowest and highest values, it is affected by extreme values in a data set. It's not a reliable measure of center and isn't frequently used in summarizing a data set.

Exercises 2.2

1. For the data set {1, 3, 6, 9} find the following:
   a. $\sum x$
   b. $\sum x^2$
   c. $\sum (x - 3)$
   d. $\sum (x - 3)^2$

2. Find the mean, the median, and the mode for the following data sets:
   a. 2, 1, 2, 7, 12
   b. $-1, 0, 1, 4, 1, 1, 6, 9$

3. Find the mean and the median for the LDL cholesterol level in a sample of 12 heart patients.

132	151	162	133	163	145	148	139	147	160	150	153

4. The number of passengers in each of the 130 randomly observed vehicles during morning rush hour was recorded, with the following results. Find the mean, the median, and the mode of this data set.

Number of passengers in each vehicle	1	2	3	4	5
Number of vehicles	86	32	4	5	3

2.3 Measures of Variability

Overview

An important measure of any data set is the variation in the data, which describes how spread out the data is. In some data sets, the values are concentrated closely near the mean. In others, the values are more widely spread out from the mean. The variation of a data set helps describe the data more accurately than a simple measure of center. The standard deviation is a number that measures how far data values are from their mean and is the most common measure of variation.

Basics of Variation

We learned a little bit about variation in Chapter 1. Variation is one of the most important concepts in statistics. Let's look at a scenario. Professor Shariq's American History class took two quizzes. As a whole, the class did well on the first but had much more mixed results in the second—some students did great, others did terribly, and many students scored in the middle. When Professor Shariq calculated the measure of center for the quizzes, they both had the same mean and mode: 84. But when she made a dot plot for each quiz, shown in Figure 2.11, she could visualize how the quiz scores changed from one to the next. Based on this information, Professor Shariq adjusted the way she covered the information on Quiz 2 in the hopes that students would understand the material better in the future. A dot plot is similar to a histogram. The dots in the dot plot represent the height of the bar in the histogram. Each dot represents a data value.

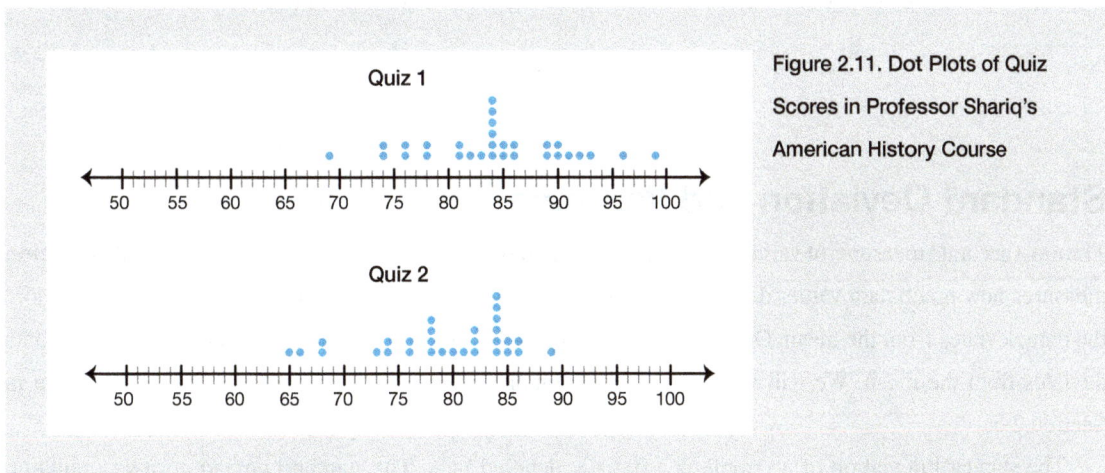

Figure 2.11. Dot Plots of Quiz Scores in Professor Shariq's American History Course

Range

The **range** of a data set is the difference between the highest and lowest values. The range measures variability because it indicates the interval over which the data points are distributed. A smaller range indicates less variability among the data, whereas a larger range indicates the opposite.

Range for a data set

$$R = x_{max} - x_{min}$$

Example 9

Using the data in Table 2.5, calculate the range of the two quiz scores.

Table 2.5. Scores on Quiz 1 and Quiz 2 in American History Course

Quiz 1	81	83	83	84	84	84	84	84	85	87	90
Quiz 2	68	72	75	79	84	84	84	88	91	95	100

Solution 9

For Quiz 1, the maximum is 90, and the minimum is 81. The range is 9.

$$R = 90 - 81 = 9$$

For Quiz 2, the maximum is 100, and the minimum is 68. The range is 32.

$$R = 100 - 68 = 32$$

Standard Deviation and Variance

The most accurate measures of variability are the standard deviation and the variance. The **standard deviation** measures how much data values deviate from the mean. The higher the standard deviation value, the more the data deviates from the mean. On the other hand, the lower the standard deviation value, the less the data deviates from the mean. We will learn more about standard deviation and its relationship to the mean in Section 2.5.

The standard deviation of a sample is a statistic denoted by s. The standard deviation of a population is a parameter denoted by σ, the Greek letter lowercase sigma. The second standard deviation formula of a sample is easier to use. It's called the shortcut formula. Once we calculate the standard deviation, we can determine how far a value is from the mean. In most data sets, most values lie within two standard deviations of the mean according to the empirical rule, which we'll learn later.

Standard Deviation of a Sample

$$S = \sqrt{\frac{\sum(x - \bar{x})^2}{n - 1}}$$

$$S = \sqrt{\frac{\sum x^2 - n(\sum x)^2}{n - 1}}$$

Standard Deviation of a Population

$$\sigma = \sqrt{\frac{\sum(x - \mu)^2}{N}}$$

In the population standard deviation formula, the denominator is N, the total number of observations. Since most data sets are samples, we will work with the sample standard deviation in this book.

The **variance** is a measure of variation that equals the square of the standard deviation. The variance of a sample is a statistic denoted by s^2. The variance of a population is a parameter shown by σ^2. One challenge to understanding the variance is that the variance is expressed in different units than the data. In the American History example, the data is expressed in points. The variation would be in points squared, which is difficult to grasp. Therefore, we will focus on the standard deviation in this textbook.

Variance of a Sample

$$S^2 = \frac{\sum(x - \bar{x})^2}{n - 1}$$

$$S^2 = \frac{\sum x^2 - n(\sum x)^2}{n - 1}$$

Variance of a Population

$$\sigma^2 = \frac{\sum(x - \mu)^2}{N}$$

Example 10

Find the sample variance and the sample standard deviation of Quiz 1 score for American History course in Example 9.

Quiz 1:   81   83   83   84   84   84   84   84   85   87   90

Solution 10

The mean of quiz 1 scores is $\bar{x}$ = 84.46.

$x - \bar{x}$: -3.46, -1.46, -1.46, -0.46, -0.46, -0.46, -0.46, -0.46, 0.54, 2.54, 5.54

$$\Sigma(x - \bar{x})^2 = (-3.46)^2 + (-1.46)^2 + (-1.46)^2 + (-0.46)^2 + (-0.46)^2 + (-0.46)^2$$
$$+ (-0.46)^2 + (-0.46)^2 + 0.54^2 + 2.54^2 + 5.54^2 = 54.73$$

The variance of quiz 1 scores:

$$s^2 = \frac{\Sigma(x - \bar{x})^2}{n - 1} = \frac{54.73}{10} = 5.473$$

The standard deviation of quiz 1 scores:

$$s = \sqrt{5.473} \approx 2.34$$

Graphing Calculator

The sample variance and the sample standard deviation can also be found on a graphing calculator. For Quiz 1 score data from Table 2.5 with a TI83 or TI84 as follows:

Enter the eleven Quiz 1 scores in L1 under $STAT \rightarrow EDIT$

Choose 1-VAR Stats under $STAT \rightarrow CALC$

Enter on "Calculate"

On the calculator display, the symbol for the sample standard deviation is s_x.

The squared value of s_x is the sample variance.

Range Rule of Thumb

We know that the range and the standard deviation are measures of variability. A quick and easy way to estimate the standard deviation is the range rule of thumb. According to the rule, the standard deviation of a sample is approximately equal to one-fourth of the range of the data. This is based on the principle that most values in a data set lie within two standard deviations of the mean. The range rule of thumb is a very straightforward formula to us and should only be used as a rough estimate of the standard deviation.

Range Rule of Thumb

$$s = \frac{Maximum - Minimum}{4}$$

Example 11

Use the range rule of thumb to find the sample standard deviation of Quiz 2 score for American History course in Example 9.

Quiz 2: 68 72 75 79 84 84 84 88 91 95 100

Solution 11

The maximum of the data set is 100.

The minimum of the data set is 68.

The sample standard deviation is

$$s = \frac{100 - 68}{4} = 8$$

According to the range rule of thumb, a **significant value** lies outside of two standard deviations from the mean. There are both significant low and significant high values. In other words, significant low values are values two standard deviations below the mean, and significant high values are values two standard deviations above the mean.

Exercises 2.3

1. Find the range, the variance, and the standard deviation for the following sample.

    $$2 \quad -3 \quad 6 \quad 0 \quad 3 \quad 1$$

2. Find the range, the variance, and the standard deviation for the following sample.

    $$-1 \quad 0 \quad 1 \quad 4 \quad 1 \quad 1$$

3. Find the range, the variance, and the standard deviation for the ten temperature readings randomly selected from an Oregon town in the month of December.

    $$45 \quad 38 \quad 40 \quad 37 \quad 41 \quad 47 \quad 35 \quad 35 \quad 36 \quad 34$$

4. What must be true of a data set if its standard deviation is 0?

2.4 Relative Position of Data

Overview

In some cases, how one data value relates to others is more important than the actual value. In Professor Shariq's quizzes, for example, the mean was 84. If you got a 92, you'd be doing better than the average student. This is called the relative position of data, and it's described by percentiles, quartiles, box plots, and z-scores.

Percentiles and Quartiles

Whenever you take a standardized test, you're given two scores. The first is the actual score, and the second is a percentile ranking of that score. For example, if you took the SAT in 2020 and scored a combined 1200, you would be in the 74th percentile of all test takers. The numerical score shows how well you did on the test, and the percentile rank indicates that 74% of all scores on the SAT were less than or equal to 1200. The percentile rank locates the scores among all the other SAT scores that year.

Percentiles divide ordered data into hundredths. Percentiles are useful for comparing values. For this reason, universities and colleges use percentiles extensively, like when using the SAT to make admissions offers to prospective students. Percentiles are mostly used with very large populations.

How to Find the Percentile

1. **Order the data.** Sort the data into values from lowest to highest.
2. **Determine the position of x.** Count the number of values that are less than x.
3. **Find the percentile of a data value.** Use the Percentile for a Value x formula and round the answer to the nearest whole number. This is the percentile of data value x.

Percentile for a Value x

$$\text{Percentile of } x = \frac{\text{number of values less than } x}{\text{total number of values}} \cdot 100\%$$

Note

If we have a value x in a data set, x is the Pth percentile of the data if the percentage of the data that are less than or equal to x is P. The number P is the percentile rank of x.

Example 12

The following data set has 10 grade point averages (GPAs) for students in a writing class.

2.90 3.00 3.53 3.71 2.42 1.90 2.50 1.49 4.00 3.83

1. What percentile is the value 2.5?
2. Find the GPA value that is 35th percentile of the data.

Solution 12

The data written in increasing order are:

1.49 1.90 2.42 2.5 2.90 3.00 3.53 3.71 3.83 4.00

1. There are 3 data values less than 2.5: 1.49, 1.90, 2.42.

$$\frac{3}{10} = 0.3 \ or \ 30\%$$

The GPA value 2.50 is the 30th percentile.

2. Use the percentile formula to find the value of x such that x is the 35th percentile:

$$0.35 = \frac{\text{number of values less than } x}{10}$$

Solve for the numerator:

$$\text{number of values less than } x = 3.5$$

This means there are 3.5 GPAs below the actual GPA which is the 35th percentile. To find such a GPA, we take the average of the 3rd and the 4th GPA values, $x = \frac{1}{2}(2.42 + 2.5) = 2.46$. So the GPA of 2.46 is the 35th percentile of the data.

Quartiles are numbers that separate the data into quarters and are useful for describing and summarizing data. Quartiles may or may not be part of the data. Figure 2.12 shows how the data in a relative frequency histogram is divided into quartiles. Notice that the median and the second quartile, shown by Q_2, are the same.

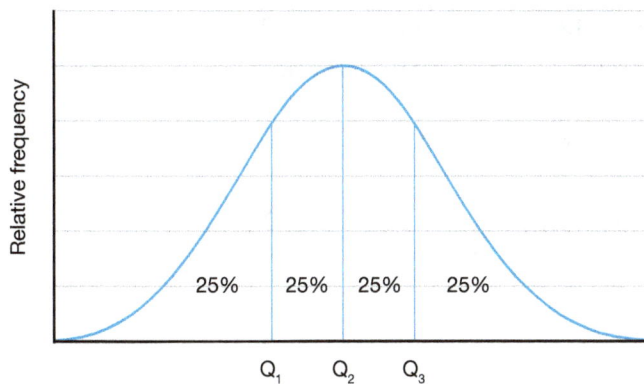

Figure 2.12. Relative Frequency Histogram Divided by Quartile

How to Find the Quartiles

1. **Order the data.** Sort the data into values from lowest to highest.
2. **Find the median.** The median is the same as the second quartile (Q_2).
3. **Divide data into two subsets.** The data below Q_2 is the lower subset. The data above Q_2 is the upper subset.
4. **Find the median of the lower subset.** This median is Q_1.
5. **Find the median of the upper subset.** This median is Q_3.

Example 13

Find the quartiles of the data set of GPAs of Example 12.

Solution 13

First, list the data in numerical order:

1.49   1.90   2.42   2.5   2.90   3.00   3.53   3.71   3.83   4.00

This data set has $n = 10$ observations. Since 10 is an even number, the median is the mean of the two middle observations:

$$\tilde{x} = \frac{2.90 + 3.0}{2} = 2.95$$

The second quartile is $Q_2 = 2.95$. The lower and upper subsets are as follows:

$$\text{Lower: } L = \{\, 1.49 \quad 1.90 \quad 2.42 \quad 2.5 \quad 2.90 \,\}$$
$$\text{Upper: } U = \{\, 3.00 \quad 3.53 \quad 3.71 \quad 3.83 \quad 4.00 \,\}$$

Each has an odd number of elements, so the median of each is its middle observation.

The first quartile and the median of L: $Q_1 = 2.42$

The 2nd quantile: $Q_2 = 2.95$

The third quartile and the median of U: $Q_3 = 3.71$

Box Plots (Box and Whisker Plots)

Box plots are graphs that show a data set on a horizontal line. Figure 2.13 shows five values or the five-number summary: the minimum value (x_{min}), the first quartile (Q_1), the median (Q_2) , the third quartile (Q_3), and the maximum value (x_{max}). We use these values to compare how close other data values are to them. Box plots give a good graphical image of the concentration of the data. They also show how far the extreme values are from most of the data.

To construct a box plot, use a horizontal or vertical number line and a rectangular box. The smallest and largest data values label the endpoints of the line. The first quartile marks one end of the box, and the third quartile marks the other. Approximately the middle 50 percent of the data falls inside the box. The line extends from the ends of the box to the smallest and largest data values. The median or second quartile can be between the first and third quartiles, or it can be one, or the other, or both.

The **interquartile range (IQR)** is a number that indicates the spread of the middle 50 percent of the data. It is the difference between the third quartile (Q_3) and the first quartile (Q_1).

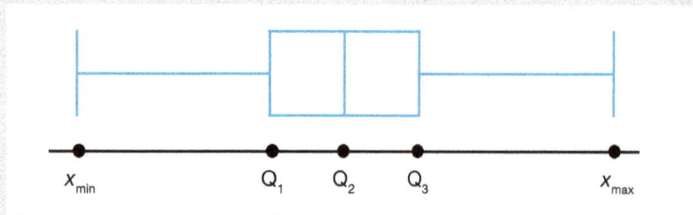

Figure 2.13. Example of a Box Plot

Interquartile Range

$$IQR = Q_3 - Q_1$$

Example 14

Construct a box plot and find the IQR for the data in Example 13.

Solution 14

From Example 13, we know that the five-number summary is as follows:

$$x_{min} = 1.49, \quad Q_1 = 2.42, \quad Q_2 = 2.95, \quad Q_3 = 3.71, \quad x_{max} = 4.00$$

The box plot is shown in Figure 2.14.

The interquartile range is $IQR = 3.71 - 2.42 = 1.29$

Figure 2.14. Box Plot

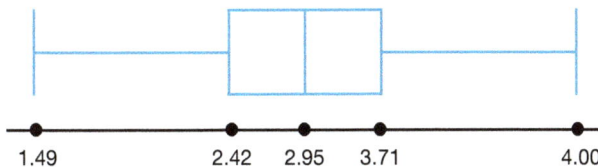

1.49   2.42   2.95   3.71   4.00

Box Plot with R

A box plot can be constructed using R programming. R programming is a free statistical software with many graphing capabilities such as creating boxplots, histograms, and many other graphs. Here, we will learn to create a boxplot for the Grade Point Average (GPA) data set from Example 12. The function accepts data by importing from a spreadsheet or by directly entering as a vector form:

```
> GPA = c(2.9, 3.00, 3.53, 3.71, 2.42, 1.90, 2.50, 1.49, 4.00, 3.38)
```

With the argument, `horizontal` = `true`, we can plot the boxplot horizontally. Additionally, we can add title and label using the `main` and `xlab` argument.

```
> boxplot (GPA, horizontal = TRUE, main = "GPA for Students in a
Writing Class", xlab = "GPA")
```

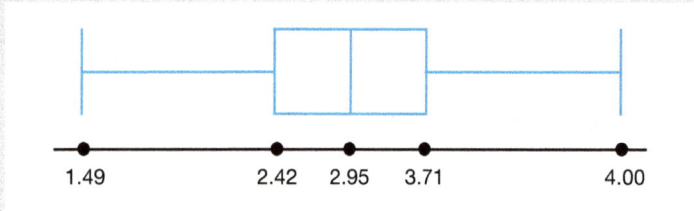

Figure 2.15. Box Plot of Students' GPAs in a Writing Class

1.49   2.42   2.95   3.71   4.00

If we want to find the statistical summary from this boxplot, then `boxplot()$stats` can be used. The argument provides the five-number summary of the data set.

```
> boxplot(GPA)$stats
[1,] 1.49 -- the minimum
[2,] 2.42 -- the first quantile
[3,] 2.95 -- the median
[4,] 3.53 -- the third quantile
[5,] 4.00 -- the maximum
```

z-Scores

The **z-score** indicates how many standard deviations a data point is from the mean. It shows how far the value x is above (to the right of) or below (to the left of) the mean, μ. Values of x larger than the mean have positive z-scores, and values of x smaller than the mean have negative z-scores. If x equals the mean, then x has a z-score of 0. A z-score helps compare and analyze data from different populations or distributions. For example, z-scores can be used to compare the performance of students who take different versions of a standardized test. Figure 2.16 shows how z-scores appear when mapped to a relative frequency histogram.

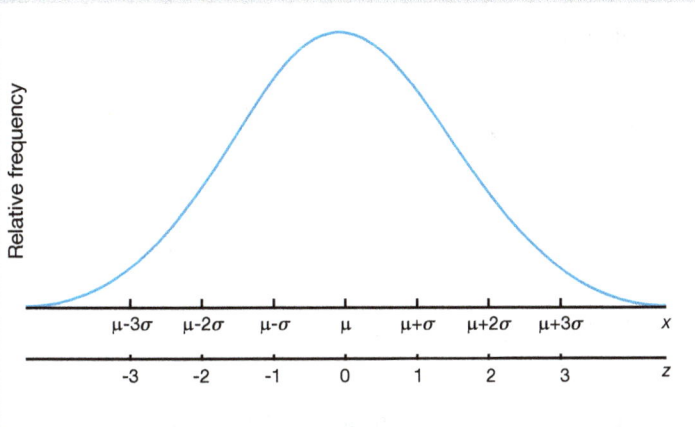

Figure 2.16. z-scores in Relation to Relative Frequency

---

z-Score for Sample

$$z = \frac{x - \bar{x}}{s}$$

---

z-Score for Population

$$z = \frac{x - \mu}{\sigma}$$

Note

These formulas calculate the z-score when x is known. Knowing the z-score, we can find x using these inverse formulas.

Sample value x

$$x = \bar{x} + sz$$

Population value x

$$x = \mu + \sigma z$$

Example 15

Find the z-scores for all ten observations in the GPA sample data in Table 2.1.

$$1.90 \quad 3.00 \quad 2.53 \quad 3.71 \quad 2.12 \quad 1.76 \quad 2.71 \quad 1.39 \quad 4.00 \quad 3.33$$

Solution 15

For these data, $\bar{x} = 2.645$ and $s = 0.8674$. The first observation $x = 1.9$ in the data set has z-score

$$z = \frac{x - \bar{x}}{s} = \frac{1.9 - 2.645}{0.8674} = -0.8589$$

$x = 1.90$ is 0.8589 standard deviations below the sample mean. The second observation $x = 3.00$ has z-score

$$z = \frac{x - \bar{x}}{s} = \frac{3.0 - 2.645}{0.8674} = 0.4093$$

$x = 3.00$ is 0.4093 standard deviations above the sample mean.

Repeating the process for the remaining observations gives the full set of z-scores:

$$-0.86 \quad 0.41 \quad -0.13 \quad 1.23 \quad -0.61 \quad -1.02 \quad 0.07 \quad -1.45 \quad 1.56 \quad 0.79$$

Example 16

Suppose the mean and standard deviation of the GPAs of all currently registered students at a community college are $\mu = 2.68$ and $\sigma = 0.45$. The z-scores of the GPAs of two students, Jose and Jacob, are $z = -0.51$ and $z = 1.38$, respectively. What are their GPAs?

Solution 16

Using the z-score for population formula (when z-score is known), we find:

$$\text{Jose: } x = \mu + z\sigma = 2.68 + (-0.51)(0.45) = 2.4505$$
$$\text{Jacob: } x = \mu + z\sigma = 2.68 + (1.38)(0.45) = 3.301$$

Exercises 2.4

1. Consider the data set:

69	92	68	77	80	93	75	76	82	100
70	85	88	85	96	53	70	70	82	85

    a. Find the percentile rank of 80.

    b. Find the percentile rank of 92.

2. Find the z-score of each measurement in the following sample data set.

$$-5 \quad 4 \quad 2 \quad -1$$

3. A sample with the following data frequency table has a mean $\bar{x} = 1$ and standard deviation $s \approx$ 1.67. Find the z-score for every value in the sample.

x	-1	0	1	4
frequency	1	1	3	1

4. A measurement x in a sample with mean $\bar{x} = 10$ and standard deviation $s = 2$ has z-score $z = 1$. Find x.

5. A measurement x in a population with mean $\mu = 5.3$ and standard deviation $\sigma = 1.3$ has z-score $z = 2$. Find x.

6. The mean score on a standardized exam is 49.6; the standard deviation is 1.85. Mario is told that the z-score of his exam score is 1.19.

    a. Is Mario's score above average or below average?

    b. What was Mario's actual score on the exam?

7. Determine whether the following statement is true. "In any data set, if an observation x_1 is less than another observation x_2, then the z-score of x_1 is less than the z-score of x_2."

2.5 Interpreting the Standard Deviation

Overview

Up to this point, we've learned what different measures tell us about a data set: the shape, the center, the variation, and the position. In this section, we will learn to interpret the standard deviation through three concepts: the empirical rule, Chebyshev's theorem, and the coefficient of variation.

The Empirical Rule

The empirical rule says that in approximately 95% of samples, the sample mean $\bar{x}$ will be within two standard deviations of the population mean μ. The empirical rule applies to approximately bell-shaped distributions. Figure 2.17 shows how the empirical rule maps to a bell-shaped relative frequency histogram. Specifically, the empirical rule states the following:

» About 68% of the x values lie within one standard deviation of the mean ($\bar{x} - s, \bar{x} + s$).

» About 95% of the x values lie within two standard deviations of the mean ($\bar{x} - 2s, \bar{x} + 2s$).

» About 99.7% of the x values lie within three standard deviations of the mean ($\bar{x} - 3s, \bar{x} + 3s$).

Notice that almost all the x values lie within three standard deviations of the mean.

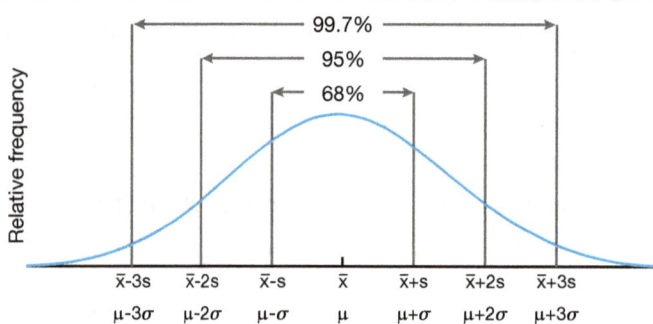

Figure 2.17. The Empirical Rule

Example 17

Daily sleep hours of twenty-year-old college students have a bell-shaped distribution with a mean of 7.5 hours and a standard deviation of 0.15 hours.

1. About what percentage of all 20-year-old college students sleep between 7.35 and 7.65 hours?
2. What interval centered on the mean should contain about 95% of all such students?

Solution 17

A sketch of the distribution of sleep hours is given in Figure 2.18.

1. Since the interval from 7.35 to 7.65 has endpoints $\bar{x} - s$ and $\bar{x} + s$, according to the empirical rule, about 68% of all college students should have daily sleep hours in this range.
2. By the empirical rule, about 95% of the data lie within two standard deviations of the mean.

Since

$$\bar{x} - 2s = 7.5 - 2(0.15) = 7.2$$

and

$$\bar{x} + 2s = 7.5 + 2(0.15) = 7.8$$

the interval (7.2, 7.8) centered on the mean should contain about 95% of all such college students.

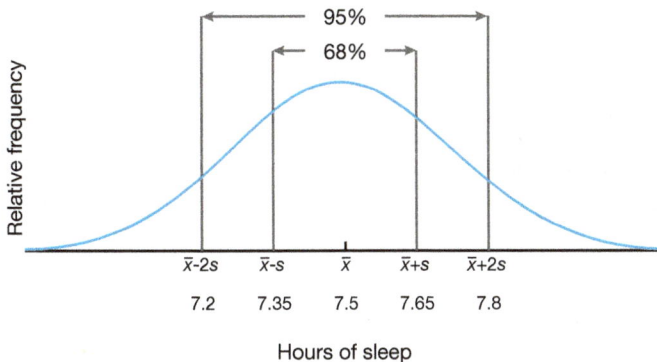

Figure 2.18. Distribution of Sleep Hours Per Day for 20-year-old College Students

Chebyshev's Theorem

Since the empirical rule only applies to bell-shaped data distributions, how can we interpret the standard deviation with data sets that aren't symmetrical or are unevenly distributed? Chebyshev's theorem provides a way to interpret any data set, regardless of shape. Chebyshev's theorem states the following:

» At least 75% of all the data lie within two standard deviations of the mean ($\bar{x} - 2s$, $\bar{x} + 2s$)

» At least 89% of the data lie within three standard deviations of the mean ($\bar{x} - 3s$, $\bar{x} + 3s$)

Like the empirical rule, it provides only approximate results, and we should notice the phrase "at least" means that results are lower limits. Figure 2.19 shows how Chebyshev's theorem can be mapped to a relative frequency histogram.

Chebyshev's Theorem

At least $\left(1 - \frac{1}{k^2}\right) \times 100\%$ of the data lie within k standard deviations of the mean. In the interval with endpoints $\bar{x} \pm ks$ for samples and with endpoints $\mu \pm k\sigma$ for populations, where k is any positive whole number that is greater than 1.

$$k = \frac{distance\ above\ or\ below\ the\ mean}{standard\ deviation}$$

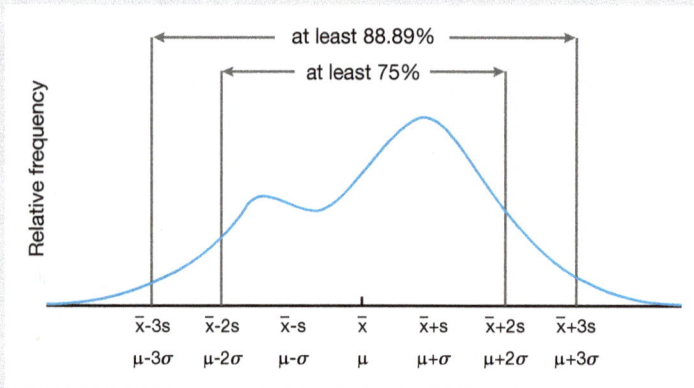

Figure 2.19. Chebyshev's Theorem

Example 18

A sample of size $n = 50$ has mean $\bar{x} = 28$ and standard deviation $s = 3$. Without knowing anything else about the sample, what can be said about the number of observations in the interval (22, 34)? What can be said about the number of observations outside that interval?

Solution 18

The interval (22, 34) is the one that is formed by adding and subtracting two standard deviations from the mean. By Chebyshev's theorem, at least 75% of the data are within this interval. Since 75% of 50 is 37.5, at least 37.5 observations are in the interval. But we can't take a fractional observation, so we conclude that at least 38 observations must lie inside the interval (22, 34), as seen in Figure 2.20(a).

If at least 75% of the observations are in the interval, then at most 25% of them are outside it. Since 25% of 50 is 12.5, at most 12.5 observations are outside the interval. Since, again, a fraction of an observation is impossible, at most 12 observations are outside the interval, as seen in Figure 2.20(b).

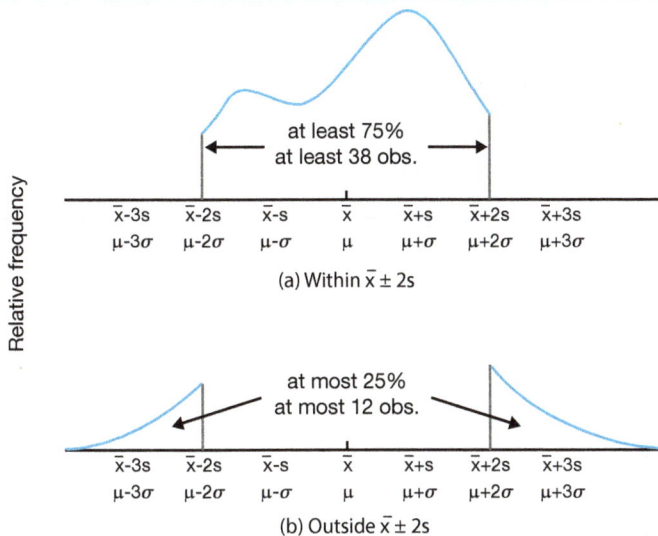

Figure 2.20(a-b). Chebyshev's Theorem

(a) Within $\bar{x} \pm 2s$

(b) Outside $\bar{x} \pm 2s$

Example 19

A data set has a mean of 15 and a standard deviation of 2. Use Chebyshev's theorem to find the percentage of values that fall between 10 and 20.

Solution 19

The distances of the two numbers, 10 and 20, from the mean are $15 - 5 = 10$ and $15 + 5 = 20$.

These numbers show that the values between 10 and 20 are within 5 units of the mean. Therefore,

$$k = \frac{\textit{distance above or below the mean}}{\textit{standard deviation}} = \frac{5}{2} = 2.5$$

$$1 - \frac{1}{k^2} = 1 - \frac{1}{2.5^2} = 0.84 = 84\%$$

Thus 84% of the data values fall between 10 and 20.

The Coefficient of Variation

The **coefficient of variation** (CV) measures relative variability that describes the standard deviation relative to the mean. It is expressed in a percent. The CV is handy when comparing results from two groups with very different means or scales. For example, the CV can be used to compare the variability of salaries across job categories, where the mean and units of measurement may differ. A lower CV indicates less variability relative to the mean, while a higher CV indicates more variability relative to the mean.

Coefficient of Variation for Sample

$$CV = \frac{s}{\bar{x}} \times 100\%$$

Coefficient of Variation for Population

$$CV = \frac{\sigma}{\mu} \times 100\%$$

Example 20

Table 2.6 shows information about two common stocks (A and B) listed on the New York Stock Exchange. The mean return and standard deviations are listed. Which stock is more volatile?

Table 2.6. Mean and Standard Deviation of Stocks A and B

	Stock A	Stock B
Mean Return	$45	$24
Standard Deviation	$9	$7

Solution 20

$$\text{Stock A: CV} = \frac{s}{\bar{x}} \times 100\% = \frac{9}{45} \times 100\% = 20\%$$

$$\text{Stock B: CV} = \frac{s}{\bar{x}} \times 100\% = \frac{7}{24} \times 100\% = 29.2\%$$

Since stock B has a CV of 29.2% and stock A has a CV of 20%, we would say that stock B shows more volatility or variation relative to its mean return.

Exercises 2.5

1. A sample data set with a bell-shaped distribution has a mean $\bar{x} = 6$ and a standard deviation $s = 2$. Find the approximate proportion of observations in the data set that lie between 4 and 8.

2. A population data set with a bell-shaped distribution has mean $\mu = 2$ and standard deviation $\sigma = 1.1$. Find the approximate proportion of observations in the data set that lie:
    a. below 0.9
    b. above 3.1
    c. between 0.9 and 3.1

3. A population data set of size $N = 400$ has mean $\mu = 5.2$ and standard deviation $\sigma = 1.1$. Use Chebyshev's theorem to find the minimum number of observations in the data set that must lie between 3 and 7.4.

4. Scores on a final exam taken by 1,200 students have a bell-shaped distribution with a mean of 72 and a standard deviation of 9.
   a. What is the median score on the exam?
   b. About how many students scored between 63 and 81?
   c. About how many students scored between 72 and 90?
   d. About how many students scored below 54?

5. Lengths of fish caught by a commercial fishing boat off Depoe Bay, Oregon, have a bell-shaped distribution with a mean of 33 inches and a standard deviation of 3 inches.
   a. About what proportion of all fish caught are between 30 inches and 36 inches long?
   b. About what proportion of all fish caught are between 30 inches and 33 inches long?
   c. About how long is the longest fish caught?

6. Recent statistics midterm scores had a bell-shaped distribution with a mean of 75 and a standard deviation 5.
   a. What is the median score?
   b. Approximately what proportion of students in the class scored between 70 and 80?
   c. Approximately what proportion of students in the class scored above 85?
   d. What is the percentile rank of the score 85?

7. The GPAs of all currently registered students at a large university have a bell-shaped distribution with a mean of 2.7 and a standard deviation of 0.6. Students with a GPA below 1.5 are placed on academic probation. Approximately what percentage of currently registered students at the university are on academic probation?

8. Twenty-six math students took the final exam on which the average was 78, and the standard deviation was 5. A rumor says that six students had scored 61 or below. Can this be true? Why or why not?

Try It Solution

Solution 2.1

```
3 |  22348
4 |  022346778889
5 |  00122234677
6 |  01
```

Figure 2.21. Stem-and-leaf plot for basketball game scores

Chapter 3
Probability Topics

It is often necessary to guess about the outcome of an event to make a decision. Politicians study polls to guess their likelihood of winning an election. Teachers choose a textbook based on what they think students will understand and enjoy. Doctors choose the treatments needed for diseases based on their assessment of likely results. Maybe you've gone to a casino to play blackjack because you think the chance of winning is good. You probably chose your college degree based on the likely availability of jobs.

Just like statistics, we have all used probability in our lives. For example, we know there's a 50% chance of getting heads when flipping a coin. We weigh the odds of doing homework or studying for an exam. These are everyday examples of probability, and they relate to decision-making.

Probability describes the likelihood that an event will occur. For instance, meteorologists use weather patterns to predict the probability of rain. Epidemiologists and doctors use probability to understand the relationship between the risk of health effects and virus exposure. An understanding of probability is important because the concept is used in many areas of your everyday life and in the larger study of statistics. In this chapter, we will learn basic terms, the concept of probability, and how to solve probability problems using various rules.

After reading this chapter, you will be able to do the following:

1. Understand the concept of probability
2. Interpret the probability value of an event as the likelihood that it will occur
3. Define the union, intersection, and complement of an event
4. Classify events as independent or dependent on one another and understand mutually exclusive events
5. Understand the Addition Rule and the Multiplication Rule and use them to calculate probabilities of events
6. Understand conditional probability and find conditional probability from the contingency tables
7. Use counting techniques to find probabilities
8. Use permutations and combinations formulas in problem-solving

3.1 Probability Basics

Overview

Probability gives numerical descriptions of how likely it is for an event to occur. Understanding probability's basic concepts and terminology is the first step toward solving probability problems. We're familiar with some of these concepts because we hear about the odds of winning the lottery, or placing a bet on a sporting event, or the chances of rain for today. There are more formal ways to understand these concepts, and we'll learn about them here.

Key Concepts

Probability is a measure of how certain we are of outcomes. An **experiment** is a planned operation carried out under controlled conditions. If we don't know the result in advance, the experiment is called a **chance** experiment. Flipping a coin is an example of an experiment. A result of an experiment is called an **outcome**.

The experiment's **sample space** is the set of all possible outcomes. The uppercase letter S is used to denote the sample space. For example, in a coin-flip experiment, the outcomes are either heads (H) or tails (T). The sample space is written as $S = \{H, T\}$.

An **event** is any combination of outcomes. Uppercase letters like A and B represent events. For example, if a fair coin is flipped, event A could be flipping heads. The probability of an event A is written $P(A)$. A **simple event** is an event that can't be broken down into smaller parts. For example, if you flip a coin 3 times, the outcome of 2 heads *followed by* 1 tail is a simple event because it can't be broken down further. On the other hand, the outcome of 2 heads and 1 tail *with no specified order* isn't a simple event because we could flip 2 heads, then 1 tail, or 1 tail, then 2 heads, or 1 head, then 1 tail, then 1 head, and so on.

The probability of any outcome is the long-term relative frequency of that outcome. Probabilities are between 0 and 1, inclusive, which means the probability could be 0, 1, or all numbers between 0 and 1. $P(A) = 0$ means that event A can never happen. $P(A) = 1$ means that event A always happens. $P(A) = 0.5$ means that event A is equally likely to occur or not to occur. For example, if we flip a fair coin repeatedly (from 20 to 2,000 to 20,000 times), the relative frequency of getting heads approaches 0.5, which is also the probability of getting tails.

Equally likely means that each outcome of an experiment occurs with equal probability. For example, if we toss a fair, six-sided die, each face number (1, 2, 3, 4, 5, or 6) is as likely to occur as any other face.

If we toss a fair coin, a head (H) and a tail (T) are equally likely to occur. If we randomly guess the answer to a true/false question on an exam, you are equally likely to select a correct answer or an incorrect answer.

To calculate the probability of event A when all outcomes in the sample space are equally likely, count the number of outcomes for event A and divide by the total number of outcomes in the sample space. For example, if we toss a fair coin twice, the sample space is $\{HH, TH, HT, TT\}$ where $T =$ tails and $H =$ heads. The sample space has four outcomes. $A =$ getting 1 head. Two outcomes meet this condition: $\{HT, TH\}$, so $P(A) = \frac{2}{4} = 0.5$.

Suppose we roll one six-sided die with the numbers 1, 2, 3, 4, 5, and 6 on its faces. Let event E equal rolling a number that is at least 5. Two outcomes would meet these criteria (rolling a 5 or a 6), so $P(E) = \frac{2}{6} = \frac{1}{3}$.

If we rolled the die only a few times, we wouldn't be surprised if the results didn't match the probability. If we rolled the die many times, we'd expect that about one-third of the rolls would result in an outcome of at least 5. This means overall, the long-term relative frequency of rolling at least 5 would get closer to $\frac{1}{3}$ as the number of rolls gets larger and larger.

This characteristic of probability experiments is called the **law of large numbers**, which says that as the number of repetitions of an experiment increases, the relative frequency of the experiment's outcome tends to get closer and closer to the theoretical probability even though the outcomes don't happen according to any set pattern or order.

In many situations, outcomes are not equally likely, though. A coin or die may be unfair or biased. Two math professors in Europe had their statistics students test the Belgian one Euro coin and discovered that in 250 trials, a head was flipped 56% of the time, and a tail was flipped 44% of the time. The data seem to show that the coin is not a fair coin. However, more repetitions would be helpful to draw a more accurate conclusion about such bias.

Some dice may be biased. Look at the dice in a game. The spots on each face are usually small holes carved out and then painted to make the spots visible. It's possible that the outcomes would be affected by the slight weight differences due to the different numbers of holes in the faces. Later, we will learn techniques to use with probabilities for events that are not equally likely.

The Union, Intersection, and Complement of Events

Sometimes, it's useful to look at events in terms of how they relate to other events. Say we want to find the probability that a student at a college works part-time *and* takes a full credit load. That probability problem would require finding the union and intersection event.

A **union** ($\cup$) of events happens if the outcome is either in A or is in B, or is in both A and B. Consider the union as the word "or" when we describe it. The union of events A and B is written as $A \cup B$, which we say

"A union B." For example, let $A = \{1, 2, 3, 4, 5\}$ and $B = \{4, 5, 6, 7, 8\}$, then $A \cup B = \{1, 2, 3, 4, 5, 6, 7, 8\}$. Notice that 4 and 5 are *not* listed twice, even though they appear in both A and B.

An **intersection** ($\cap$) event happens if the outcome is in both A and B at the same time. To describe an intersection event, we write $A \cap B$ and say as "A intersect B." Consider the intersection as the word "and" when describing the event. For example, let $A = \{1, 2, 3, 4, 5\}$ and $B = \{4, 5, 6, 7, 8\}$, then $A \cap B = \{4, 5\}$.

The **complement** of A consists of all outcomes that are *not* in A. We write the complement of event A as A', which we say as "A prime." The probability of an event plus the complement of the event equals 1 or $P(A) + P(A') = 1$. For example, if $S = \{1, 2, 3, 4, 5, 6\}$ and $A = \{1, 2, 3, 4\}$, then $A' = \{5, 6\}$, $P(A) = \frac{4}{6}$, $P(A') = \frac{2}{6}$ and $P(A) + P(A') = \frac{4}{6} + \frac{2}{6} = 1$.

Complement

$$P(A) + P(A') = 1$$

The **conditional probability** measures the probability of an event occurring when another event is known to have already occurred. We denote the conditional probability as $P(A \mid B)$, which is said as "the probability of A given B."

Conditional Probability

$$P(A \mid B) = \frac{P(A \cap B)}{P(B)}$$

where

$$P(B) > 0$$

For example, suppose we toss a fair, six-sided die. The sample space $S = \{1, 2, 3, 4, 5, 6\}$. Let $A =$ rolling a 2 or a 3, and $B =$ rolling an even number. Applying the formula, we have the following conditional probability:

$$P(A \mid B) = \frac{P(A \cap B)}{P(B)} = \frac{\dfrac{\textit{the number of outcomes that are 2 or 3 and even in S}}{6}}{\dfrac{\textit{the number of outcomes that are even in S}}{6}} = \frac{1/6}{3/6} = \frac{1}{3}$$

We'd interpret the conditional probability like this: The probability of getting a 2 or a 3, given an even number has already occurred, is $\frac{1}{3}$.

Note

It's important to read each problem carefully. Make sure you understand what the events are before trying to solve the problem. Reread the problem and pay close attention to the wording. Clearly identify the event of interest. Determine whether there is a condition stated in the wording which indicates that the probability is conditional. A key word, "given," means we're dealing with a conditional probability. The event that comes after "given" is always the first event—the one that has already occurred. The event whose probability is of interest comes next. The probability of the second event happening is conditional on the occurrence of the first event.

Example 1

For this example, we define the sample space S to be a set of whole numbers from 1 to 20 as follows:

$$S = \{1, 2, 3, 4, 5, 6, 7, 8, 9, 10, 11, 12, 13, 14, 15, 16, 17, 18, 19, 20\}$$

Let event A = the even numbers, and event B = numbers greater than 13.

$$A = \{2, 4, 6, 8, 10, 12, 14, 16, 18, 20\}, \text{ and } B = \{14, 15, 16, 17, 18, 19, 20\}$$

Find the following:

1. Probability of A, $P(A)$
2. Probability of B, $P(B)$
3. The set for $A \cap B$
4. The set for $A \cup B$
5. Probability of $A \cap B$
6. Probability of B', $P(B')$
7. Conditional probability $P(A \mid B)$
8. Conditional probability of $P(B \mid A)$

Solution 1

1. There are 10 numbers in set A, so $P(A) = \frac{10}{20} = \frac{1}{2}$

2. There are 7 numbers in set B, so $P(B) = \frac{7}{20}$

3. The set for $A \cap B$ is the set containing all numbers of $A \cap B$, so $A \cap B = \{14, 16, 18, 20\}$

4. The set for $A \cup B$ is the set containing all numbers of $A \cup B$, so $A \cup B = \{2, 4, 6, 8, 10, 12, 14, 15, 16, 17, 18, 19, 20\}$

5. $P(A \cap B) = P(\{14, 16, 18, 20\}) = \frac{4}{20} = \frac{1}{5}$

6. $P(B') = 1 - P(B) = 1 - \frac{7}{20} = \frac{13}{20}$

7. $P(A \mid B) = \frac{P(A \cap B)}{P(B)} = \frac{^1/_5}{^7/_{20}} = \frac{4}{7}$

8. $P(A \mid B) = \frac{P(A \cap B)}{P(A)} = \frac{^1/_5}{^1/_2} = \frac{2}{5}$

Example 2

A fair, six-sided die is rolled. Let A = the event of rolling an even number, and B = the event of rolling a number that is less than 3. Find the following:

1. $P(A)$
2. $P(B)$
3. $P(A \cap B)$
4. Probability of rolling an even number given a number less than 4 has been rolled
5. Probability of rolling a number less than 4 given an even number has been rolled

Solution 2

1. $A = \{2, 4, 6\}$, $P(A) = \frac{1}{5} = 0.5$
2. $B = \{1, 2\}$, $P(B) = \frac{2}{6} = \frac{1}{3} = 0.333$
3. $P(A \cap B) = P(\{2\}) = \frac{1}{6} = 0.1667$
4. This is the conditional probability:

    P(Roll an even number | Less than a 4 already occurred)

$$= P(A \mid B) = \frac{P(A \cap B)}{P(B)} = \frac{^1/_6}{^1/_3} = \frac{1}{2} = 0.5$$

5. This is the conditional probability:

P(Less than a 4 already occurred | Roll an even number)

$$= P(B \mid A) = \frac{P(A \cap B)}{P(A)} = \frac{1/6}{1/2} = \frac{1}{3} = 0.333$$

Note: Conditional probabilities $P(A \mid B)$ and $P(B \mid A)$ are not equal because the "given" information is different, which means we're working with different conditional probabilities.

Odds

The **odds** for an event indicate the probability that the event will occur. You've probably heard that the odds of winning the Powerball if you play one set of numbers are 1 in 292,201,338. Did you know that you have a better chance of being killed by a vending machine than you do of winning the Powerball?

We use the term odds to describe the likelihood of an event occurring. The term odds against describes the likelihood of an event not occurring. For example, suppose there are 20 marbles in a bag. Eight are blue, six are red, eight are yellow, and the rest of the marbles are green. If one marble is selected randomly, the odds of getting a red marble is 6/20 or 3:10. Unlike probability, which is expressed as a decimal, odds are expressed as a ratio. Calculating odds can be useful in gambling or sports betting, but it doesn't help understand probability or statistical theory.

Odds in Favor of Event A

$$\frac{P(A)}{1 - P(A)}$$

Odds Against Event A

$$\frac{1 - P(A)}{P(A)}$$

Example 3

Suppose there are 20 marbles in a bag. Seven are red, 6 are blue, 4 are yellow, and 3 are green. If 1 marble is picked at random, find the following:

1. The odds in favor of a green marble.
2. The odds against a green marble.

Solution 3

1. Let A = the event of picking a green marble. Then

$$P(A) = \frac{3}{20}$$

$$\frac{P(A)}{1 - P(A)} = \frac{{}^3/_{20}}{(1-3)/_{20}} = \frac{{}^3/_{20}}{{}^{17}/_{20}} = \frac{3}{17} \text{ or } 3:17$$

The odds in favor of a green marble are 3 to 17 or $\frac{3}{17}$.

2. Use the formula of odds against an event, we have

$$\frac{1 - P(A)}{P(A)} = \frac{{}^{17}/_{20}}{{}^3/_{20}} = \frac{17}{3} \text{ or } 17:3$$

The odds against a green marble are 17 to 3 or $\frac{17}{3}$.

Exercises 3.1

1. A box is filled with several party favors. It contains 12 hats, 15 noisemakers, 10 finger traps, and 5 bags of confetti. One party favor is chosen from the box at random.

    Let H = the event of getting a hat.

    Let N = the event of getting a noisemaker.

    Let F = the event of getting a finger trap.

    Let C = the event of getting a bag of confetti.

    a. Find $P(H)$

    b. Find $P(N)$

    c. Find $P(F)$

    d. Find $P(C)$.

2. You draw a card from a standard deck of 52 cards. Find the probability of

    a. drawing a red card

    b. drawing a club card

3. You see a game at the state fair. You have to throw a dart at a color wheel like Figure 3.1. Each section on the color wheel is equal in area.

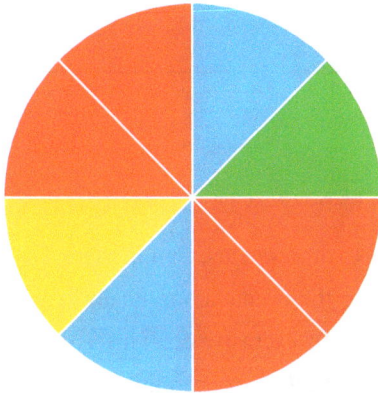

Figure 3.1. Color Wheel Dart Board

Let B = the event of landing on blue.

Let R = the event of landing on red.

Let G = the event of landing on green.

Let Y = the event of landing on yellow.

  a.   If you land on Y, you get the biggest prize. Find $P(Y)$.

  b.   If you land on red, you don't get a prize. Find $P(R)$.

  c.   What is the odds of landing on red?

4.   A shelf holds 12 books. Eight are fiction, and the rest are nonfiction. Each is a different book with a unique title. The fiction books are numbered one to eight. The nonfiction books are numbered one to four. Randomly select one book.

Let F = event that book is fiction

Let N = event that book is nonfiction

  a.   What is the sample space?

  b.   What is the sum of the probabilities of an event and its complement?

  c.   What is the odds of selecting a nonfiction book?

5.   You are rolling a fair, six-sided number cube. Let E = the event that it lands on an even number. Let M = the event that it lands on a multiple of three.

  a.   What does $P(E \mid M)$ mean in words?

  b.   What does $P(E \cup M)$ mean in words?

3.2 Independent and Mutually Exclusive Events

Overview

When looking at the probability of events in relation to other events, an important determination is whether the events are independent or dependent and whether the events are mutually or not mutually exclusive. Even though they sound similar, independent and mutually exclusive don't mean the same thing, and verifying each requires different probability rules. We introduce these probability rules here.

Addition Rule and Multiplication Rule

Two basic rules in probability are the addition rule and the multiplication rule. These rules help explain the interaction between compound events. A **compound event** is made up of more than one simple event.

The **addition rule** allows you to find $P(A \text{ or } B)$, which is the probability that either event A or event B occurs, or that both events A and B occur. If A and B are defined on a sample space, then $P(A \cup B) = P(A) + P(B) - P(A \cap B)$. Remember, you can think of the union symbol substituting for the word "or." Subtracting the intersection of A and B is to keep from double counting elements that are in both sets A and B.

Addition Rule

$$P(A \cup B) = P(A) + P(B) - P(A \cap B)$$

If A and B are mutually exclusive, then

$$P(A \cap B) = 0$$
$$P(A \cup B) = P(A) + P(B)$$

Example 4

Klaus is trying to choose where to go on vacation. His two choices are A = New Zealand and B = Alaska. Klaus can only afford one vacation. The probability that he chooses A is $P(A) = 0.6$ and the probability that he chooses B is $P(B) = 0.35$. Find the probability $P(A \cup B)$, the probability that he chooses either New Zealand or Alaska.

Solution 4

This problem wants you to find the probability of the union of A and B. So, we use the Addition Rule:

$$P(A \cup B) = P(A) + P(B) - P(A \cap B)$$

First, we find $P(A \cap B)$ because Klaus can only afford to take one vacation. Therefore,

$$P(A \cap B) = 0$$

This means that event A and B are mutually exclusive. The probability that he chooses either New Zealand or Alaska is

$$P(A \cup B) = P(A) + P(B) = 0.6 + 0.35 = 0.95$$

Notice that the probability of going either New Zealand or Alaska but not both is 0.95. That means there is a 0.05 probability that Klaus will choose to go nowhere on vacation.

Try It 3.1

Zadie is deciding which route to take to work. Her choices are I = the interstate and F = Fifth Street.

$P(I) = 0.44$ and $P(F) = 0.56$

$P(I \cap F) = 0$ because Zadie will take only one route to work. What is the probability of $P(I \cup F)$?

The **multiplication rule** allows you to find $P(A$ and $B)$, which is the probability that both events A and B occur. If A and B are two events defined on a sample space, then $P(A \cap B) = P(B) \times P(A \mid B)$. Remember, you can think of the intersection symbol as substituting for the word "and." One easy way to remember the multiplication rule is that the word "and" means that the event must satisfy two conditions (A and B). Suppose a teacher is drawing a student's name to win prize, but the student must be both a sophomore and have brown hair. It's harder to satisfy two conditions than one. When you multiply fractions the result is always smaller, which makes it harder to satisfy the two conditions.

Multiplication Rule

$$P(A \cap B) = P(A) \cdot P(B \mid A)$$
$$P(A \cap B) = P(B) \cdot P(A \mid B)$$

This rule may also be written as

$$P(A \mid B) = \frac{P(A \cap B)}{P(B)}$$
$$P(B \mid A) = \frac{P(A \cap B)}{P(A)}$$

If A and B are independent, then

$$P(A \mid B) = P(A)$$

$$P(B \mid A) = P(B)$$

$$P(A \cap B) = P(A) \cdot P(B)$$

Example 5

Carlos plays college soccer. He scores a goal 65% of the time he shoots. Suppose Carlos attempts 2 goals in a row in his next game. The probability that he succeeds on the second attempt given that he already succeeded on the first attempt is 0.9.

1. Are the two events, succeeds on the first and second attempts, independent?
2. What is the probability that he scores both goals?
3. What is the probability that Carlos scores either the first goal or the second goal?
4. Are the two events (succeeds on the first and second attempts) mutually exclusive?

Solution 5

Since Carlos scores a goal 65% of the time that he shoots a ball, the probability that he succeeds on his first attempt is 0.65, the probability that he succeeds on his second attempt is also 0.65. Define the following events:

$$A = \text{Carlos succeeds on first attempt, and } P(A) = 0.65$$
$$B = \text{Carlos succeeds on second attempt, and } P(B) = 0.65$$

$B \mid A$ = Carlos succeeds on the second attempt given that he already succeeded on the first attempt, and $P(B \mid A) = 0.90$

1. To check for independence, we use one of the three formulas. Since

$$P(B \mid A) = 0.9 \neq P(B) = 0.65$$

the two events (succeeds on the first and second attempts) are not independent. The probability that Carlos scores on the second attempt depends on whether he scores on the first attempt.

2. We denote the probability that he scores both goals as $P(A \cap B)$. Using the Multiplication Rule, we find that

$$P(A \cap B) = P(A) \cdot P(B \mid A) = 0.65 \cdot 0.9 = 0.585$$

The probability that Carlos scores on both the first and second attempts is 0.585.

3. Here we're supposed to find the probability of the union, $P(A \cup B)$. Using the Addition Rule, we find that

$$P(A \cup B) = P(A) + P(B) - P(A \cap B) = 0.65 + 0.65 - 0.585 = 0.715$$

The probability that Carlos scores either on the first or on the second attempt is 0.715.

4. To see if the first and second attempts are mutually exclusive, we look at $P(A \cap B)$. To be mutually exclusive, $P(A \cap B)$ must equal 0. Since $P(A \cap B) = 0.585$, then two events are not mutually exclusive.

Try It 3.2

Helen plays basketball. She makes her free throw shot 75% of the time. Helen must now attempt 2 free throws. C = the event that Helen makes the first shot. $P(C) = 0.75$. D = the event Helen makes the second shot. $P(D) = 0.75$. The probability that Helen makes the second free throw given that she made the first is 0.85. What is the probability that Helen makes both free throws?

Example 6

Antonia attends Blue Mountain Community College. The probability that she enrolls in a math class is 0.2 and the probability that she enrolls in a speech class is 0.65. The probability that she enrolls in a math class given that they enroll in speech class is 0.25.

Let: M = math class, S = speech class, $M \mid S$ = math given speech

1. What is the probability that Antonia enrolls in math and speech class?
2. What is the probability that Antonia enrolls in math or speech classes?
3. Are M and S independent?
4. Are M and S mutually exclusive?

Solution 6

1. $P(M) = 0.2$, $P(S) = 0.65$, and $P(M \mid S) = 0.25$. Applying the multiplication rule, the probability that Antonia enrolls in math *and* speech is

$$P(M \cap S) = P(M \mid S) \times P(S) = 0.25 \times 0.65 = 0.1625$$

2. Applying the addition rule, the probability that Antonia enrolls in math *or* speech classes is

$$P(M \cup S) = P(M) + P(S) - P(M \cap S) = 0.2 + .065 - 0.1625 = 0.6875$$

3. If M and S are independent then $P(M \mid S)$ must equal $P(M)$. Since $P(M \mid S) = 0.25$ and $P(M) = 0.2$, $P(M \mid S) \neq P(M)$. So, M and S are not independent.
4. M and S are not mutually exclusive since $P(M \cap S) = 0.1625 \neq 0$.

Independent Events

We have seen some examples of dependent events in Examples 5 and 6. When events are **dependent**, the outcome of the first event affects the probability of the second event occurring. When events are **independent** the outcome of one does not affect the probability the other occurs. For example, the outcomes of two rolls of a fair die are independent events because the outcome of the first roll does not change the probability for the outcome of the second roll.

Two events being dependent doesn't mean that the first event is the cause of the second event, though. Here's an example. If you're driving your car and you get in a car accident, the events are dependent. You couldn't get in an accident unless you were driving. But driving the car wasn't necessarily the cause of the accident. In this case, the accident was dependent on driving, but doesn't have to be the cause.

In Chapter 1, we learned about sampling with or without replacement. When it comes to sampling in the context of probability, we can relate it to independent or dependent events. Here's a quick refresher on sampling with or without replacement. Sampling with replacement means that after a member of a population is selected, that member is put back into the population so it can be selected again. With replacement, events are independent, meaning the result of the first selection will not change the probabilities of the second selection. Sampling without replacement means that each member of a population may be chosen only once. In this case, the probabilities for the second selection are affected by the result of the first selection. The events are dependent. If it isn't known whether A and B are independent or dependent, assume they are dependent until we can show otherwise.

Example 7

A box contains 20 batteries and four of them are defective. Suppose 2 batteries are selected without replacement. Find the probability that they both are defective.

Solution 7

This is sampling without replacement and the events of getting a defective battery on the first selection and on the second selection are dependent.

Let A = the first battery is defective and B = the second battery is defective. To find the probability that both batteries are defective is to find $P(A \cap B)$, and by definition,

$$P(A \cap B) = P(A) \cdot P(B \mid A)$$

where $P(B \mid A)$ is the conditional probability that the second battery is defective given the first selection is already defective.

Because we are sampling without replacement, meaning the first defective battery is set aside after selection and then we draw the second battery from the remaining 19 batteries, of which only three can be defective. Thus,

$$P(B \mid A) = \frac{3}{19}$$

Now, we are ready to calculate $P(A \cap B)$:

$$P(A \cap B) = P(A) \cdot P(B \mid A) = \frac{4}{20} \cdot \frac{3}{19} = \frac{3}{95}$$

Mutually Exclusive Events

Mutually exclusive events can't occur at the same time as we saw in Example 4 where Klaus couldn't choose both New Zealand and Alaska for his vacation. These are also called disjoint events. In other words, if A occurred then B can't occur, and vice versa. This means that A and B don't share any outcomes and $P(A \cap B) = 0$.

For example, a sample space, $S = \{1, 2, 3, 4, 5, 6, 7, 8, 9, 10\}$, events $A = \{1, 2, 3, 4, 5\}$, $B = \{4, 5, 6, 7, 8\}$, and $C = \{7, 9\}$. In this case, $A \cap B = \{4, 5\}$, and $P(A \cap B) = \frac{1}{5}$ and is not equal to 0. Therefore, events A and B are not mutually exclusive. But events A and C don't have any numbers in common so $P(A \cap C) = 0$. Therefore, A and C are mutually exclusive.

The Addition Rule is useful when working with mutually exclusive events. Consider the Addition Rule $P(A \cup B) = P(A) + P(B) - P(A \cap B)$. If events A and B are mutually exclusive, then $P(A \cap B) = 0$, and therefore, $P(A \cup B) = P(A) + P(B)$. Otherwise, $P(A \cap B) \neq 0$. If it's not known whether A and B are mutually exclusive, assume they are not until we can show otherwise.

Example 8
Let event C = taking an English class. Let event D = taking a speech class.
Suppose $P(C) = 0.75$, $P(D) = 0.3$, $P(C \mid D) = 0.75$ and $P(C \cap D) = 0.225$.

1. Are C and D independent?

2. Are C and D mutually exclusive?

3. What is $P(D \mid C)$?

Solution 8

1. Yes, because $P(C \mid D) = P(C)$.

2. No, because $P(C \cap D)$ is not equal to 0.

3. $P(D \mid C) = \dfrac{P(C \cap D)}{P(C)} = \dfrac{0.225}{0.75} = 0.3$

Exercises 3.2

1. Suppose that you have eight cards. Five are green and three are yellow. The five green cards are numbered 1, 2, 3, 4, and 5. The three yellow cards are numbered 1, 2, and 3. The cards are well shuffled. You randomly draw one card. Let G be the event that a card drawn is green and let E be the event that a card drawn is even numbered.

    a. List the sample space.

    b. Find $P(G)$

    c. Find $P(G \mid E)$

    d. Find $P(G \cap E)$

    e. Find $P(G \cup E)$

    f. Are G and E mutually exclusive? Justify your answer numerically.

2. Roll two fair dice separately. Each die has six faces.

    a. List the sample space.

    b. Let A be the event that either a three or four is rolled first, followed by an even number. Find $P(A)$.

    c. Let B be the event that the sum of the two rolls is at most seven. Find $P(B)$.

    d. In words, explain what $P(A \mid B)$ represents. Find $P(A \mid B)$.

    e. Are A and B mutually exclusive events? Explain your answer in one to three complete sentences, including numerical justification.

    f. Are A and B independent events? Explain your answer in one to three complete sentences, including numerical justification.

3.   Consider the following scenario. Let $P(C) = 0.4$, $P(D) = 0.5$, and $P(C \mid D) = 0.6$.
   a.   Find $P(C \cap D)$.
   b.   Are C and D mutually exclusive? Why or why not?
   c.   Are C and D independent events? Why or why not?
   d.   Find $P(C \cup D)$.
   e.   Find $P(D \mid C)$.

4.   Approximately 281,000,000 people over the age of five live in the United States. Of these people, 55,000,000 speak a language other than English at home. Of those who speak another language at home, 62.3% speak Spanish.

   Let E = speaks English at home, E' = speaks another language at home, and S = speaks Spanish. Finish each probability statement in Table 3.1 by matching the probability to the correct answer.

Table 3.1. Probability Statement Matching

Probability Statements	Answers
$P(E') =$	0.8043
$P(E) =$	0.623
$P(S \cap E') =$	0.1957
$P(S \mid E') =$	0.1219

5.   United Blood Services is a blood bank that serves more than 500 hospitals in 18 states. According to their website, a person with type O blood and a negative Rh factor (Rh-) can donate blood to any person with any blood type. Their data show that 43% of people have type O blood, 15% have Rh- factor, and 52% have type O *or* Rh-factor.
   a.   Find the probability that a person has both type O blood and the Rh- factor.
   b.   Find the probability that a person does *not* have both type O blood and the Rh- factor.

6.   At a college, 72% of courses have final exams, and 46% of courses require research papers. Suppose 32% of courses have a research paper and a final exam. Let F be the event that a course has a final exam. Let R be the event that a course requires a research paper.

   a.   Find the probability that a course has a final exam or a research project.

   b.   Find the probability that a course has neither of these two requirements.

7.   In a box of assorted cookies, 36% contain chocolate, and 12% contain nuts. Of those, 8% contain both chocolate and nuts. Sean is allergic to both chocolate and nuts.

   a.   Find the probability that a cookie contains chocolate or nuts (Sean can't eat it).

   b.   Find the probability that a cookie does not contain chocolate or nuts (Sean can eat it).

8.   A college finds that 10% of students have taken a distance learning class and that 40% of students are part-time students. Of the part-time students, 20% have taken a distance learning class. Let D = event that a student takes a distance learning class and E = event that a student is a part-time student.

   a.   Find $P(D \cap E)$.

   b.   Find $P(E \mid D)$.

   c.   Find $P(D \cup E)$.

   d.   Using an appropriate test, show whether D and E are independent.

   e.   Using an appropriate test, show whether D and E are mutually exclusive.

3.3 Conditional Probabilities and Contingency Tables

Overview

Earlier, you learned about conditional probabilities, but here, we'll explore them further in relation to contingency tables. Sometimes, the probability of an event is affected by the knowledge of an event that has already occurred. If we're finding the probability that a person will run a mile in less than six minutes, the results would be affected if you knew the person was a former college track star. In this section, you'll learn about conditional probabilities and contingency tables, which help calculate probabilities of categorical variables.

Conditional Probabilities

You already know that the conditional probability of an event is a probability calculated with the information that another event has already occurred. The conditional probability of A given B, $P(A \mid B)$, is the probability that event A will occur given that event B has already occurred. A conditional probability reduces the sample space, and we can calculate the probability of A from the reduced sample space of B.

Contingency Tables

A **contingency table** is a table that displays the frequency distribution of categorical variables arranged in at least two rows and two columns. It provides a picture of how variables are related and interact. A contingency table is a useful tool to describe relationships between categorical variables and makes calculating probabilities easy. A contingency table displays sample values of two different variables that may be dependent or contingent on one another. One of the variables categorizes the columns, and the other categorizes the rows. Example 9 shows how probabilities are found from the information given in the contingency table.

Example 9

A study of speeding violations and drivers who use cell phones produced the data in Table 3.2.

Table 3.2. Speeding Violation and Cell Phone Usage

Events	Event C = Speeding violation in the last year	Event D = No speeding violation in the last year	Total
Event A = Uses cell phone while driving	25	280	305
Event B = Does not use cell phone while driving	45	405	450
Total	70	685	755

Calculate the following probabilities using the table.

1. Find P (driver is a cell phone user), $P(A)$.
2. Find P (driver had no violation in the last year), $P(D)$.
3. Find P (driver had no violation in the last year AND was a cell phone user), $P(A \cap D)$.
4. Find P (driver is a cell phone user OR driver had no violation in the last year), $P(A \cup D)$.
5. Find P (driver is a cell phone user GIVEN driver had a violation in the last year), $P(A \mid C)$.
6. Find P (driver had no violation last year GIVEN driver was not a cell phone user), $P(D \mid B)$.

Solution 9

The total number of people in the sample is 755. The row totals are 305 and 450. The column totals are 70 and 685. Notice that 305 + 450 = 755 and 70 + 685 = 755.

1. $P(A) = \dfrac{\text{number of cell phone users}}{\text{total number in study}} = \dfrac{305}{755} \approx 0.404$

2. $P(D) = \dfrac{\text{number that had no violations}}{\text{total number in study}} = \dfrac{685}{755} \approx 0.907$

3. The number of drivers that are in the set of $A \cap D$ is 280. Thus

$$P(A \cap D) = \frac{280}{755} \approx 0.371$$

4. Use the Addition Rule:

$$P(A \cup C) = P(A) + P(C) - (A \cap D) = 0.404 + 0.907 - 0.371 = 0.94$$

5. The sample space is reduced to the number of drivers who had a violation. Use the Multi-plication Rule, the conditional probability $P(A \mid C)$ can be written as

$$P(A \mid C) = \frac{P(A \cap C)}{P(C)} = \frac{\frac{25}{755}}{\frac{70}{755}} = \frac{25}{70} \approx 0.3576$$

6. The sample space is reduced to the number of drivers who were not cell phone users.

$$P(D \mid B) = \frac{P(B \cap D)}{P(B)} = \frac{\frac{405}{755}}{\frac{450}{755}} = \frac{405}{450} = 0.9$$

Example 10

Table 3.3 shows a random sample of 100 hikers and their preferred hiking areas.

Table 3.3. Hiker and Hiking Preference (Incomplete)

Sex	The Coastline	Near Lakes and Streams	On Mountain Peaks	Total
Female	18	16	___	45
Male	___	___	14	55
Total	___	41	___	___

Complete the table.

1. Let F = being female and let C = preferring the coastline. Are event F and C independent?
2. Let M = being male and L = prefers hiking near lakes and streams. Find $P(M \mid L)$.
3. Let F = being female, and P = prefers mountain peaks. Find $P(F \cup P)$.

Solution 10

1. Table 3.4. Hiker and Hiking Preference (Complete)

Sex	The Coastline	Near Lakes and Streams	On Mountain Peaks	Total
Female	18	16	11	45
Male	16	25	14	55
Total	34	41	25	100

2. Use

$$P(F \cap C) = \frac{18}{100} = 0.18$$

$$P(F)P(C) = \frac{45}{100} \cdot \frac{34}{100} = 0.153$$

Since $P(F \cap C) \neq P(F)P(C)$, the events F and C are not independent by the criteria for independent events from the Multiplication Rule.

3. This is a conditional probability.

$$P(M \mid L) = \frac{25}{41} \approx 0.6098$$

4. We use the Addition Rule to find $P(F \cup P)$:

$$P(F \cup P) = P(F) + P(P) - P(F \cap P)$$

First, we find the following probabilities from the contingency table:

$$P(F) = \frac{45}{100} = 0.45$$

$$P(P) = \frac{25}{100} = 0.25$$

$$P(F \cap P) = \frac{11}{100} = 0.11$$

Now, we have

$$P(F \cup P) = P(F) + P(P) - P(F \cap P) = 0.45 + 0.25 - 0.11 = 0.59$$

Exercises 3.3

1. Roll a fair, six-sided die. Let A = a prime number of dots is rolled. Let B = an odd number of dots is rolled. Then A = $\{2, 3, 5\}$ and B = $\{1, 3, 5\}$. The sample space for rolling a fair die is S = $\{1, 2, 3, 4, 5, 6\}$. Find:

   a. $A \cap B$

   b. $A \cup B$

2. In a bookstore, the probability that the customer will buy a novel is 0.6, and the probability that the customer will buy a nonfiction book is 0.4. Suppose that the probability that the customer buys both is 0.2. Find the probability that the customer buys either a novel or a nonfiction book.

3. Forty percent of the students at a local college belong to a club, and 50% work part-time. Five percent of the students work part-time and belong to a club. Let C = student belongs to a club, and PT = student works part-time. If a student is selected at random, find the following:

   a. the probability that the student belongs to a club *and* works part-time.

   b. the probability that the student belongs to a club, *given* that the student works part-time.

4. Table 3.5 shows a random sample of 200 cyclists and their preferred routes. Using the table, answer the following questions:

   a. Of the males, what is the probability that the cyclist prefers a hilly path?

   b. Are the events "being male" and "preferring the hilly path" independent events?

Table 3.5. Preferred Cycling Routes by Gender

Gender	Lake Path	Hilly Path	Wooded Path	Total
Female	45	38	27	110
Male	26	52	12	90
Total	71	90	39	200

5. Table 3.6 shows the number of athletes who stretch before exercising and how many had injuries within the past year. Answer the following questions:
    a. What is P(athlete stretches before exercising)?
    b. What is P(athlete stretches before exercising | no injury in the last year)?

Table 3.6. Athletes and Stretching

	Injury in last year	No injury in last year	Total
Stretches	55	295	350
Does not stretch	231	219	450
Total	286	514	800

3.4 Counting Techniques

Overview

Finding the probabilities of complicated events can be challenging. If we want to find the probability of getting a full house in cards (a pair and three-of-a-kind) when five cards are selected at random from a fully shuffled standard deck of fifty-two cards. To find this probability, first, we'd need to calculate the number of possible full houses in five dealt cards, then calculate the number of ways that five cards can be selected from 52 cards. Last, we'd find the ratio of these two numbers for probability. That's a lot of cards and full houses. We need more techniques to solve this kind of problem, and we'll learn some of them here.

Counting: Multiplication Rule and Factorial Rule

Two basic rules that can help count outcomes are the multiplication and factorial rules. The **multiplication counting rule** finds the total number of outcomes from some series of events. To use it, first determine how many outcomes each individual event could have and denote each event as n_1, n_2, n_3, and so on. Then multiply n_1, n_2, n_3,.., n_m to find the total number of outcomes from m series of events.

Multiplication Counting Rule

$$\text{Total number of outcomes} = n_1 \cdot n_2 \cdot n_3 \cdot ... \cdot n_m$$

For example, we want to know how many 6-digit combinations are possible for a phone's password. First, we need to select 6 digits. Next, we know that there are 10 possibilities for each digit (0 to 9). Using the multiplication rule, the total number of possibilities is $10 \times 10 \times 10 \times 10 \times 10 \times 10 = 1{,}000{,}000$.

Example 11

The password for a new bank account must be 5 letters followed by a number from 0 to 9. Letters can be repeated. How many possibilities are there for the password?

Solution 11

Since the 5 letters can be repeated and there are 26 letters, each with 26 possibilities. There are 10 numbers from 0 to 9, each with 9 possibilities. Using the multiplication counting rule, the total number of possibilities is

$$26 \times 26 \times 26 \times 26 \times 26 \times 10 = 118{,}813{,}760$$

The **factorial rule** is used to find the number of ways different items can be arranged. The factorial rule uses a special symbol "!". It means multiplying all the decreasing whole numbers from the one written. For example, $3! = 3 \times 2 \times 1 = 6$. We say this as "Three factorial equals 6."

Factorial Rule

The number of arrangements of n unique items is $n!$

$$n! = n \cdot (n - 1) \cdot (n - 2)\ldots \cdot 1$$

For example, if you wanted to know how many ways a deck of 52 cards could be ordered, you'd solve for $52! = 52 \times 51 \times 50\ldots \times 1 = 8.065801752 \times 10^{67}$. In other words, 8 with 67 zeros after it, or you'd never be able to finish putting cards in different orders.

Example 12

A museum curator is getting ready for an art exhibition. She wants to hang 10 paintings in a line on a wall. How many ways can the 10 paintings be arranged?

Solution 12

Using the factorial rule, we know the number of arrangements of 10 paintings is $10!$

$$10! = 10 \times 9 \times 8 \times 7 \times 6 \times 5 \times 4 \times 3 \times 2 \times 1 = 3{,}628{,}800$$

Permutations and Combinations

When using counting techniques, it's important to know if we're counting arrangements of elements once or multiple times. **Permutations** are different arrangements of the same objects that are counted separately. For example, to find out how many ways the numbers in the set $\{1, 2, 3, 4, 5, 6\}$ can be arranged, we need to make sure each possible ordering is counted separately. There are $6! = 720$ permutations of these numbers, one of which is $\{6, 5, 4, 3, 2, 1\}$. With permutations, the order of the objects always matters. There are three variations on the permutations rule. Whether we are selecting items with or without replacement matters in each variation.

The permutations rule *with replacement* is used when there are *n* different items available to select, we select *r* of them with replacement, and the order matters.

Permutations Rule with Replacement

$$\text{Number of arrangements} = n^r$$

For example, if we have four letters (A, B, C, D) and we want to know how many ways there are to arrange them in three-letter patterns, then you would find $4^3 = 64$ arrangements. In this case, order matters (for example, ABC is different from BAC, so both are included as arrangements), and each letter can be chosen more than once in an arrangement (for example, AAA, BBB, CCC, and DDD are possible).

Example 13

The door on the computer center has a lock with five buttons numbered from 1 to 4. The code of numbers that opens the lock is a sequence of five numbers. How many different codes are possible if the five numbers can be repeated, such as 11244?

Solution 13

Since each digit can be selected from 1 to 4 and numbers can be repeated, there are 4 ways to pick the first digit, and 4 ways to pick the second digit.... Therefore, the number of possible codes is $4^5 = 1024$.

The next two variations of the permutations rule don't use replacement. The permutations rule *with different items* is used when there are *n* different items available to select, we want to select *r* of them without replacement, and the order of the items matters.

Permutations Rule with Different Items without Replacement

$$_nP_r = \frac{n!}{(n-r)!}$$

Say you're in a group of 5 people, and you need to choose 3 out of the 5 to sit at a small table at a busy restaurant. To find how many permutations of your group are possible, you'd discover that there are $\frac{5!}{(5-3)!}$ = 60 ways that your group could be seated at that little table.

If you were in a group where $n = r$ (meaning the number of chosen items is equal to the number of items to choose from), then you'd pick all 5 people. In that case, the number of permutations is 5! as stated in the Factorial Rule. The Factorial Rule is a special case where $n = r$, so we have

$$\frac{n!}{(n-n)!} = \frac{n!}{0!} = n!$$

Note

By definition, 0! = 1

The permutations rule *with some alike items* is used when there are n items to select and some of those items are the same, if r items are selected from n items without replacement, and the order of the items matters. In this case, we group identical items together and count how many of them there are.

Permutations Rule with Some Alike Items without Replacement

$$\frac{n!}{n_1! \cdot n_2! \, ... \, n_d!}$$

where n_1 are alike, n_2 are alike, ..., and n_d are alike.

For example, if we wanted to find out how many ways the letters in the word "Tennessee" can be arranged, we count up the number of each letter, and there are 1 T, 4 E's, 2 N's, and 2 S's. Therefore,

$$\frac{9!}{1! \cdot 4! \cdot 2! \cdot 2!} = \frac{362,880}{96} = 3,780$$

There are 3,780 ways to arrange the letters of Tennessee.

Try It 3.3

Find the number of ways we can arrange the letters in the word BANANA.

In a permutation, the order of the selected items matters. However, in a **combination**, different arrangements of the same objects are considered the same. The order of the items in a combination is not important. Also, the items can't be repeated in a combination. A committee's membership is a good example of a combination: it doesn't matter how we arrange the committee members because no matter what, the committee membership stays the same. Also, we can't count an individual committee member twice.

Combinations are defined by the elements contained in them. In a combination, the set {1, 1, 2} is the same as {2, 1, 1}. For example, any five cards from a fifty-two-card deck can form a valid hand, or combination. The order of the cards doesn't matter, and we can't repeat cards once they've been dealt.

The combinations rule is used when there are n different items to select, we want to select r of them without replacement, and the order of the items doesn't matter.

Combinations Rule

$$_nC_r = \binom{n}{r} = \frac{n!}{r! \cdot (n-r)!}$$

where n is the number of items from which r items are chosen.

For example, if we wish to choose 4 members from a committee of 10 to attend a conference, we would have $\dfrac{10!}{4! \cdot (10-4)!} = \dfrac{3{,}628{,}800}{24 \cdot 720}$ ways to choose. The formula is also used to calculate the number of combinations in a lottery, as the order of the lottery numbers doesn't matter and once drawn, a number isn't replaced.

Example 14

Consider a quality control procedure in which an inspector randomly selects 2 of 5 parts to test for defects. In a group of 5 parts, how many ways can the 2 parts be selected?

Solution 14

The number of parts available is $n = 5$, and the number of parts selected is $r = 2$.

Because the order of the parts doesn't matter, the inspector is selecting without replacement, and all the parts are different, we apply the combinations rule.

$$_5C_2 = \frac{5!}{2! \cdot (5-2)!} = \frac{120}{12} = 10$$

There are 10 ways that the 2 parts can be selected from 5 parts to test for defects.

Example 15

In a group of 10 married couples, 2 people will be selected. How many different choices are possible? If each choice is equally likely, what is the probability that the 2 people selected are married to each other?

Solution 15

First, let's determine the sample space, S. We select any 2 people from 20 without replacement, so we use the combinations rule. The number of ways of selecting 2 from 20 people is 190.

$$_{20}C_2 = \frac{20!}{2! \cdot (20-2)!} = 190$$

Now, define event A = 2 people selected are married to each other. The number of ways event A can occur is 10.

$$n(A) = {}_{10}C_1 = \binom{10}{1} = \frac{10!}{1! \cdot (10-1)!} = 10$$

Finally, we find the probability that the 2 people selected are a married couple.

$$P(A) = \frac{n(A)}{n(S)} = \frac{10}{190} = 0.0526$$

The probability that the 2 people selected are married to each other is 0.0526.

Exercises 3.4

1. In a medical study, patients can be classified according to 4 blood types and whether their blood pressure is low, normal or high. In how many ways can a patient be classified?

2. How many possible license plates could be stamped if each license plate were required to have exactly 3 letters and 4 numbers?

3. The access code for a car's security system consists of four digits. Each digit can be 0 through 9. How many access codes are possible if

   a. each digit can be used only once and not repeated?

   b. each digit can be repeated?

4. How many different 5-digit odd numbers are possible if a digit may not be repeated?

5. An urn contains 10 balls numbered 0, 1, 2, . . ., 9. If four balls are selected one at a time and with replacement, what is the number of possible ordered samples?

6. Shuffle the deck of cards and choose five cards in order. How many outcomes are there?

7. How many ways can four people fill four executive positions: President, Vice President, Treasurer, and Secretary?

8. Maria has three tickets for a concert. She'd like to use one of the tickets herself. She could then

offer the other two tickets to any of four friends (Svetlana, Beth, Chris, Wayne). How many ways can 2 people be selected from 4 to go to the concert?

9. A committee of 4 people is to be selected from a group of 5 men and 7 women. If the selection is made randomly, how many different groups of size 4 can be chosen? What is the probability the committee will consist of 2 men and 2 women?

10. Suppose that 20 members of an organization are to be divided into three committees A, B and C in such a way that each of the committees A and B is to have eight members and committee C is to have four members. How many different ways can the members be assigned to these committees? Note that each of the 20 members gets assigned to one and only one committee.

11. How many different letter arrangements can be made from the letters in the word STATISTICS?

12. In how many ways can a person choose three books from a list of ten bestsellers, assuming that the order in which the books are chosen doesn't matter?

13. A candy bar in a vending machine costs 85 cents. How many ways can a customer put in two quarters, three dimes, and one nickel?

14. A multiple-choice test consists of eight questions, each permitting a choice of three alternatives.

15. In how many ways can one choose an answer to each question?

Try It Solutions

Solution 3.1

Because $P(I \cap F) = 0$, therefore,

$$P(I \cup F) = P(I) + P(F) - P(I \cap F) = 0.44 + 0.56 - 0 = 1$$

Solution 3.2

$$P(D \mid C) = 0.85$$
$$P(C \cap D) = P(D \cap C)$$
$$P(D \cap C) = P(D \mid C)P(C) = (0.85)(0.75) = 0.6375$$

Helen makes the first and second free throws with probability 0.6375.

Solution 3.3

There are 7 letters in BANANA. Some are alike without replacement. There are 1 B, 3 As, 2 Ns.

$$\frac{7!}{1! \cdot 3! \cdot 2!} = \frac{5,040}{12} = 420$$

There are 420 ways to rearrange the letters in BANANA.

Chapter 4
Discrete Probability Distributions

Think about this scenario. Javier manages a grocery delivery service. He wants to know how many more delivery calls come in during the peak time of day compared to the average. If the historical average is twenty delivery calls per day, what is the probability that he'll get more than twenty calls during the peak call time?

This example illustrates a type of probability problem involving discrete random variables. We learned about discrete data in Chapter 1 and know that discrete data can be counted, which means that the data can only take on whole number values. The scenario uses discrete random variables because we can count the expected outcomes. This chapter will explain the basic concepts of random variables, probability distributions, and how to use different methods to calculate the probability of discrete random variables.

After reading this chapter, you will be able to do the following:

1. Understand the concept of a random variable
2. Understand the probability distribution of a discrete random variable
3. Calculate the mean and standard deviation of a discrete random variable
4. Understand the concept of a binomial random variable
5. Calculate the probability of a binomial probability distribution
6. Find the mean and standard deviation of a binomial probability distribution
7. Understand the characteristics of the Poisson probability distribution
8. Calculate the probability of a Poisson probability distribution
9. Find the mean and standard deviation of a Poisson probability distribution

4.1 Basic Concepts of Probability Distributions

Overview

We need to grasp several basic concepts and rules to understand probability distributions. Ideas like discrete data and distributions should be familiar from previous chapters. In this chapter, we'll learn how to combine several of these to get to a new concept: probability distributions. We'll also learn how to calculate important parameters from the probability distribution.

Random Variables

There are two types of random variables: discrete random variables and continuous random variables. A **discrete random variable** takes on a finite set of values. For example, the number of seats in a movie theater, the names of passengers on an airplane, or the types of trees in a park. A **continuous random variable** has only continuous values. Examples of continuous random variables include temperature, the amount of sugar in a dessert, and the time it takes to walk home from school. Usually, a random variable is denoted by X or Y. Lowercase letters like x or y denote the value of a random variable. If X is a random variable, X is written in words, and x is given as a number or description. The random variable value x is also called an **outcome**. A set that contains all possible outcomes from an experiment is the **sample space**.

For example, let $X =$ the number of heads we get when we toss three fair coins. The sample space for the toss of three fair coins is TTT; THH; HTH; HHT; HTT; THT; TTH; HHH. Then, $x = 0, 1, 2, 3$. X is in words, and x is a number. Notice that for this example, the x values are countable outcomes. Because we can count the possible values as whole numbers that X can take on and the outcomes are random (the x values 0, 1, 2, 3), X is a discrete random variable.

Probability and Probability Distribution Basics

Probability describes the likelihood that an event will occur. An understanding of probability is important because the concept is used in so many areas of our everyday life, and in the larger study of statistics. For instance, meteorologists use the weather pattern to predict the probability of rain. Epidemiologists and doctors use probability to understand the relationship between risk of health effects and exposures to viruses. We often must guess the outcome of an event to make a decision. For instance, politicians study polls to guess their likelihood of winning an election. Doctors choose the treatments needed for diseases based on their assessment of likely results. These are everyday examples of probability, and they relate to decision making.

A **probability distribution** $P(X)$ describes the likelihood of different outcomes in a random event. It assigns probabilities to all possible values of a random variable. A discrete probability distribution is one where the random variable can only take a finite set of values, like the number of heads in a coin toss or the number of cars passing by on a road during a time interval. A continuous probability distribution is one where the random variable can take any value within a certain range, like the height of individuals in a population or the time it takes to complete a task. Since there are an infinite number of possible values, the probability that a continuous random variable will take on any one specific value is zero.

We'll focus on discrete probability distribution here. In a coin-toss experiment, we want to know the probability of two heads coming up when we toss three fair coins. So, we write $P(2)$ or $P(X = 2)$. Here, the random variable X represents the number of heads coming up when three fair coins are tossed. There are two requirements that a probability distribution must satisfy:

1. For each value of the random variable X, $P(x)$ is between 0 and 1, inclusive: $0 \le P(x) \le 1$.
2. The sum of all probabilities is 1: $\sum P(x) = 1$.

Example 1

A child psychologist is interested in the number of times a newborn baby's crying wakes its mother after midnight. For a random sample of 100 mothers, the information in Table 4.1 was gathered. Let X = the number of times per night a newborn baby's crying wakes its mother after midnight. Table 4.1 lists six values of X and probabilities associated with these six values.

Table 4.1. Newborn Baby's Crying

x	0	1	2	3	4	5
P(x)	0.04	0.22	0.46	0.18	0.08	0.02

Is this a discrete probability distribution? Why or why not?

Solution 1

The random variable X takes on the values 0, 1, 2, 3, 4, 5, and $P(x)$ = probability that X takes on a value *from 0, 1, 2, 3, 4, or 5.*

This is a discrete probability distribution because X satisfies the two requirements:

Each probability $P(x)$ is between 0 and 1.

$$P(X = 0) = 0.04 \quad P(X = 1) = 0.22 \quad P(X = 2) = 0.46$$
$$P(X = 3) = 0.18 \quad P(X = 4) = 0.08 \quad P(X = 5) = 0.02$$

The sum of the probabilities is 1.

$$\sum P(X) = 0.04 + 0.22 + 0.46 + 0.18 + 0.08 + 0.02 = 1$$

Parameters of Discrete Probability Distributions

Parameters of a probability distribution are numbers that characterize a distribution. Probability distributions have a measure of center (the mean or expected value), a variance, and a standard deviation. Because probability distributions describe a population, not a sample, these measures are parameters. Here we'll learn how to find these parameters of discrete probability distributions.

Mean or Expected Value

The **expected value** is the mean calculated over the long term of doing an experiment again and again. It is the same as the population mean (μ) you learned about in Chapter 2 and is denoted by E, so $E = \mu$. Say you toss a coin and record the result. What is the probability that the result is heads? If we flip a coin twice, does probability tell us that these flips will result in one head and one tail? We might toss a fair coin ten times and record nine heads. But probability doesn't describe the short-term results of an experiment. It gives information about what can be expected in the long term and is described by the law of large numbers.

Formula: Mean or Expected Value of Discrete Probability Distribution

$$\mu = \sum [x \cdot P(x)]$$

Standard Deviation

Recall that the standard deviation measures how much values deviate from the mean. Use the formula below to find the standard deviation of a discrete probability distribution. Standard deviation is the square root of variance. Generally, for probability distributions, use a calculator or a computer to find μ and σ to reduce roundoff error.

Variance σ^2 for a Probability Distribution

$$\sigma^2 = \sum (x - \mu)^2 \, P(x) \quad or \quad \sigma^2 = \sum x^2 P(x) - \mu^2$$

Standard Deviation σ for a Probability Distribution

$$\sigma = \sqrt{\sum (x - \mu)^2 P(x)} \quad or \quad \sigma = \sqrt{\sum x^2 P(x) - \mu^2}$$

Example 2

Emory University's soccer team practices 0, 1, or 2 days a week. The probability that they practice 0 days is 0.2, the probability that they practice 1 day is 0.5, and the probability that they practice 2 days is 0.3.

1. Find the expected number of days the team practices soccer each week.
2. Find the standard deviation of the probability distribution.

Solution 2

1. Let the random variable X = the number of days the team practices soccer per week. X takes on the values 0, 1, 2. The probability, $P(X)$, associated with each x is shown in table 4.2.

Table 4.2. Emory's Soccer Team Probability Distribution Table

X (number of days soccer practice per week)	P(x)	x · P(x)	x² · P(x)
0	0.2	$0 \times 0.2 = 0$	$0^2 \times 0.2 = 0$
1	0.5	$1 \times 0.5 = 0.5$	$1^2 \times 0.5 = 0.5$
2	0.3	$2 \times 0.3 = 0.6$	$2^2 \times 0.3 = 1.2$

Using the formula for expected value, we find the expected number of days the team practices soccer each week:

$$\mu = \sum x \cdot P(x)$$
$$= 0 \cdot 0.2 + 1 \cdot 0.5 + 2 \cdot 0.3 = 1.1 \; days$$

On average, Emory's team would expect to practice soccer 1.1 days per week if they practiced week after week.

2. To the standard deviation calculation, we first find the following:

$$\sum x^2 \cdot P(x) = 0 + 0.5 + 1.2 = 1.7$$

Using the formula for standard deviation, we find the standard deviation of the probability distribution:

$$\sigma = \sqrt{\sum x^2 \cdot P(x) - \mu^2} = \sqrt{1.7 - 1.1^2}$$
$$= 0.7 \text{ days}$$

Example 3

We are interested in playing a game of chance in which 5 numbers are chosen from 0, 1, 2, 3, 4, 5, 6, 7, 8, 9. It costs $2 to play. We could win $100,000 if we match all 5 numbers in order. If we win, we also get your $2 back in addition to the winning amount of $100,000. Over the long term, what is the expected profit of playing the game?

Solution 3

Let X = the amount of money you could make. But be careful: the values of the random variable X are not 0, 1, 2, 3, 4, 5, 6, 7, 8, 9. Since we are interested in profit (or loss), the values of X are 100,000 dollars and –2 dollars.

Since there are ten numbers, and we may choose a number more than once, the probability of choosing one correct number is $\frac{1}{10}$. The probability of choosing all five numbers correctly and in order is

$$\frac{1}{10} \times \frac{1}{10} \times \frac{1}{10} \times \frac{1}{10} \times \frac{1}{10} = \frac{1}{10^5} = 0.00001$$

Therefore, the probability of winning is 0.00001, and the probability of losing is

$$1 - 0.00001 = 0.99999$$

Table 4.3. Expected Value of the Game

	x	P(x)	x P(x)
Loss	-2	0.99999	-2 · 0.99999 = -1.99998
Profit	100,000	0.00001	100,000 · 0.00001 = 1

The expected value

$$\mu = -1.99998 + 1 = 0.99998 \approx 1 \text{ } dollar$$

On average, we expect to lose approximately $1 for each game played. However, each time we play, we could either lose $2 or profit $100,000. The $1 is the average loss per game after playing this game repeatedly over time.

Example 4

To the best knowledge of seismologists, the probability that a magnitude 8.0 earthquake would occur along the Oregon coast in the next 50 years is 0.38. If we make a bet that a magnitude 8.0 earthquake would occur and win, we get $50. If we lose the bet, we pay $20. Let X = the amount of profit from a bet. If we bet many times, will we come out ahead? What is the standard deviation of X?

Solution 4

Let X = the amount of profit from a bet, we have

$$P(X = 50) = 0.38$$

and

$$P(X = -20) = 1 - 0.38 = 0.62$$

The probability and the related calculations are shown in the following table.

Table 4.4. Emory's Soccer Team Probability Distribution Table

	x	$P(x)$	$x \cdot P(x)$	$(x - \mu)^2 \cdot P(x)$
Profit	$50	0.38	$19	$(50 - 6.6)^2 \cdot 0.38 = 715.75$
Loss	-$20	0.62	-$12.40	$(-20 - 6.6)^2 \cdot 0.62 = 438.69$

Using the information from the table, the expected value

$$\mu = 19 - 12.4 = 6.6 \, dollars$$

If we make this bet many times under the same conditions, our long-term outcome will be an average profit of $6.60 per bet.

The standard deviation is:

$$\sigma = \sqrt{\sum(x - \mu)^2 \, P(x)}$$
$$= \sqrt{715.75 + 438.69} \approx 33.98 \, dollars$$

Exercises 4.1

1. A company wants to evaluate its attrition rate, in other words, how long new hires stay with the company. Over the years, they have established the following probability distribution.

   Let X = the number of years a new hire will stay with the company.
   Let $P(x)$ = the probability that a new hire will stay with the company x years.

x	$P(x)$
0	0.12
1	0.18
2	0.30
3	0.15
4	
5	0.10
6	0.05

Use the data provided in the table to find the following.

    a. $P(x = 4)$

    b. $P(x \geq 5)$

    c. On average, how long would you expect a new hire to stay with the company?

    d. What does the column "$P(x)$" sum to?

2. A baker is deciding how many batches of muffins to make to sell in his bakery. He wants to make enough to sell every batch. Through observation, the baker has established a probability distribution.

x	P(x)
1	0.15
2	0.35
3	0.40
4	0.10

    a. Define the random variable X.

    b. What is the probability the baker will sell more than one batch?

    c. What is the probability the baker will sell exactly one batch?

    d. On average, how many batches should the baker make?

3. Ellen has music practice 3 days a week. She practices for all the 3 days 85% of the time, 2 days 8% of the time, 1 day 4% of the time, and 0 days 3% of the time. One week is selected at random.

    a. Define the random variable X.

    b. Construct a probability distribution table for the data.

    c. We know that for a probability distribution function to be discrete, it must have two character-istics. One is that the sum of the probabilities is one. What is the other characteristic?

4. Javier volunteers in community events each month. He does not do more than five events in a month. He attends exactly five events 35% of the time, four events 25% of the time, three events 20% of the time, two events 10% of the time, one event 5% of the time, and no events 5% of the time.

    a. Define the random variable X.

    b. What values does x take on?

    c. Construct a probability distribution table.

    d. Find the probability that Javier volunteers for less than three events each month.

    e. Find the probability that Javier volunteers for at least one event each month.

5. Find the standard deviation using the values from the table.

x	P(x)	x · P(x)	(x – μ) · 2P(x)
2	0.1	2(0.1) = 0.2	(2–5.4)2(0.1) = 1.156
4	0.3	4(0.3) = 1.2	(4–5.4)2(0.3) = 0.588
6	0.4	6(0.4) = 2.4	(6–5.4)2(0.4) = 0.144
8	0.2	8(0.2) = 1.6	(8–5.4)2(0.2) = 1.352

6. A music instructor is interested in knowing what percent of each year's class will continue on to the next, so that she can plan what classes to offer. Over the years, she has established the following probability distribution.

Let X = the number of years a student will study music with the teacher.

Let $P(x)$ = the probability that a student will study music x years.

x	P(x)	x · P(x)
1	0.10	
2	0.05	
3	0.10	
4		
5	0.30	
6	0.20	
7	0.10	

a. In words, define the random variable X.

b. $P(x = 4) = $ _____

c. $P(x < 4) = $ _____

d. On average, how many years would you expect a student to study music with this teacher?

e. What does the column "$P(x)$" sum to and why?

f. What does the column "$x · P(x)$" sum to and why?

7. You are playing a game by drawing a card from a standard deck and replacing it. If the card is a face card, you win $30. If it is not a face card, you pay $2. There are 12 face cards in a deck of 52 cards. What is the expected value of playing the game?

4.2 Binomial Probability Distribution

Overview

A binomial probability distribution is a relatively easy way to find the answer to experiments with only two possible outcomes, like a true-false quiz or a coin-toss experiment.

Binomial Distribution Basics

The **binomial distribution** is a widely used discrete distribution in statistics. It has only two possible results: success or failure. A binomial distribution determines the probability of observing a specific number of successes in a fixed number of trials in the data. The underlying assumptions of binomial distribution are:

1. The number of trials n is fixed.
2. Each trial is independent.
3. There is only one outcome for each trial.
4. The probability of success p is the same for each trial.

The binomial distribution is determined by the number of trials n and the probability of success p. Once we know we're working with a binomial random variable, we can use the binomial probability formula to calculate the probability of x successes in a fixed number of trials n.

Binomial Probability Distribution

$$P(x) = \frac{n!}{(n-x)!\,x!}\, p^x q^{n-x}$$

$n!$ is the factorial of n. By definition

$$n! = 1 \cdot 2 \cdot 3 \cdot \dots \cdot n$$

and

$$0! = 1$$

Notation for Binomial Probability Distribution

$X \sim B(n, p)$ is interpreted as "X is a random variable with a binomial distribution." whose parameters are n and p where

x is the number of successes, and $x = 0, 1, 2, ..., n$

n is the number of trials

p is the probability of a success on each trial

q is the probability of a failure on each trial, and $q = 1 - p$

Example 5

Approximately 70% of statistics students turn in their homework on time. Each student does homework independently. A statistics class has 25 students.

3. What is the probability that exactly 23 students will turn in their homework on time?
4. What is the probability that *at least* 23 students will turn in their homework on time? Students are selected randomly.

Solution 5

We define the random variable X = number of students who turn in their homework on time. X takes on the values 0, 1, 2, ..., 25.

There is only a success or a failure, the number of trials is fixed, and the probability of a success is $p = 0.70$. Therefore,

$$X \sim B(25, 0.7)$$

1. The probability that exactly 23 students will turn in their homework on time:

$$P(X = 23) = \frac{25!}{(25 - 23)! \cdot 23!} \cdot 0.7^{23} \cdot 0.3^2$$
$$= \frac{25!}{2! \cdot 23!} \cdot 0.7^{23} \cdot 0.3^2$$
$$= 0.0074$$

2. The phrase "at least 23" means that $X = 23, 24,$ or 25. Thus, the probability that at least 23 will turn in their homework on time is

$$P(X = 23) + P(X = 24) + P(X = 25)$$

$$= \frac{25!}{23! \cdot (25 - 23)!} \cdot 0.7^{23} \cdot 0.3^2 + \frac{25!}{24! \cdot (25 - 24)!} \cdot 0.7^{24} \cdot 0.3^1 + \frac{25!}{25! \cdot (25 - 25)!} \cdot 0.7^{25} \cdot 0.3^0$$

$$= 0.00897$$

We notice the probability that at least 23 students will turn in homework on time is slightly higher than the probability that exactly 23 students will turn in their homework. This is because "at least 23" includes both 24 students and 25 students turning in their homework.

Use Technology to Accomplish the Task

Using the formulas to find the probability of x successes can take a long time. Technology like a TI-84 graphic calculator or R programming can accomplish the task in a much shorter time.

TI-84:

```
binompdf(n, p, x)
binomcdf(n, p, x)
1- binomcdf(n, p, x-1)
```
 # Probability of exact x successes
 # Probability of at most x successes
 # Probability of at least x successes

R function:

```
> dbinom(x, n, p)
> pbinom(x, n, p)
> pbinom(x-1, n, p, lower.tail = FALSE)
```
 # Probability of exact x successes
 # Probability of at most x successes
 # Probability of at least x successes

For practice, we can find the probability that at least 23 students will turn in their homework using the R function:

```
> pbinom(22, 25, 0.7, lower.tail = FALSE)
[1] 0.008960528
```
 # Probability of at least 23 turn in homework

Try It 4.1

Sixty-five percent of people pass the driver's exam on the first try. A group of 50 people who have taken the driver's exam is randomly selected. Give two reasons why this is a binomial problem.

Parameters of Binomial Probability Distribution

The binomial probability distribution has a mean and standard deviation, just like the discrete probability distribution. We can use the formulas from the previous section to find these parameters of discrete probability distributions, but if we already know the values of n and p, there are much simpler formulas to use.

Mean of Binomial Probability Distribution

$$\mu = np$$

Standard Deviation of Binomial Probability Distribution

$$\sigma = \sqrt{npq}$$

Example 6

Accident statistics show that 32% of high school students have texted while driving. Use the binomial probability function formula to construct the probability distribution for the number (X) of students in a random sample of 5 high school students who texted while driving. Find the mean and standard deviation of the random variable.

Solution 6

The random variable X is binomial with parameters $n = 5$ and $p = 0.32$; $q = 1 - p = 0.68$. The possible values of X are 0, 1, 2, 3, 4, and 5.

$$P(X = 0) = \frac{5!}{5! \cdot 0!} \cdot 0.32^0 \cdot 0.68^5 = \frac{1 \cdot 2 \cdot 3 \cdot 4 \cdot 5}{(1 \cdot 2 \cdot 3 \cdot 4 \cdot 5) \cdot (1)} \cdot 1 \cdot (0.3939041) \approx 0.1454$$

$$P(X = 1) = \frac{5!}{4! \cdot 1!} \cdot 0.32^1 \cdot 0.68^4 = \frac{1 \cdot 2 \cdot 3 \cdot 4 \cdot 5}{(1 \cdot 2 \cdot 3 \cdot 4 \cdot 5) \cdot (1)} \cdot 0.32 \cdot (0.2138) \approx 0.3421$$

$$P(X = 2) = \frac{5!}{3! \cdot 2!} \cdot 0.32^2 \cdot 0.68^3 = \frac{1 \cdot 2 \cdot 3 \cdot 4 \cdot 5}{(1 \cdot 2) \cdot (1 \cdot 2 \cdot 3)} \cdot (0.1024) \cdot (0.314432) \approx 0.3220$$

We can also use R to find the above three probabilities:

```
> dbinom(0, 5, 0.32)
[1] 0.1453934

> dbinom(1, 5, 0.32)
[1] 0.342102

> dbinom(2, 5, 0.32)
[1] 0.3219784
```

The remaining three are computed similarly to give the probability distribution in Table 4.6.

Table 4.5. Probability Distribution of Teen Drivers Texting While Driving

x	0	1	2	3	4	5
P(x)	0.1454	0.3421	0.3220	0.1515	0.0356	0.0034

This probability distribution is represented by the histogram in Figure 4.1.

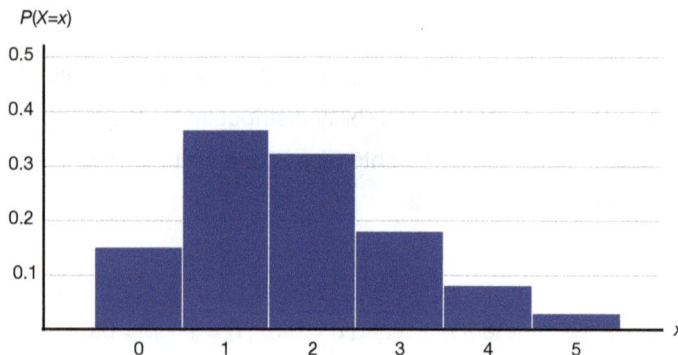

Figure 4.1. Probability Distribution of Teen Drivers

The mean and standard deviation of the random variable

$$\mu = np = 5 \cdot 0.32 = 1.6 \ students$$

$$\sigma = \sqrt{npq} = \sqrt{5 \cdot 0.32 \cdot 0.68} \approx 1.043 \ students$$

On average, 1.6 students texted while driving each day.

Try It 4.2

During the 2013 regular NBA season, DeAndre Jordan of the Los Angeles Clippers had the highest field goal completion rate in the league. Jordan scored on 61.3% of his shots. Suppose we choose a random sample of 80 shots made by Jordan during the 2013 season. Let X = the number of shots that scored points.

1. What is the probability distribution for X?
2. Using the formulas, calculate the mean and standard deviation of X.
3. Find the probability that Jordan scored with 60 of these shots.
4. Find the probability that Jordan scored with more than 50 of these shots.

Cumulative Probability Distribution

Cumulative Probability represents the probabilities that a random variable X takes a value less than or equal to a specific value, x. Cumulative Probability is written as $P(X \leq x)$. In example 6, we found the Probabilities for each of the 5 x values. We can find the cumulative probability by adding up all the individual probabilities for $P(X=0)$, $P(X=1)$, through $P(X=5)$. These values are shown in the table 4.6.

Table 4.6. Cumulative Probability Distribution of Teen Drivers

x	0	1	2	3	4	5
$P(X \leq x)$	0.1454	0.4875	0.8095	0.961	0.9966	1

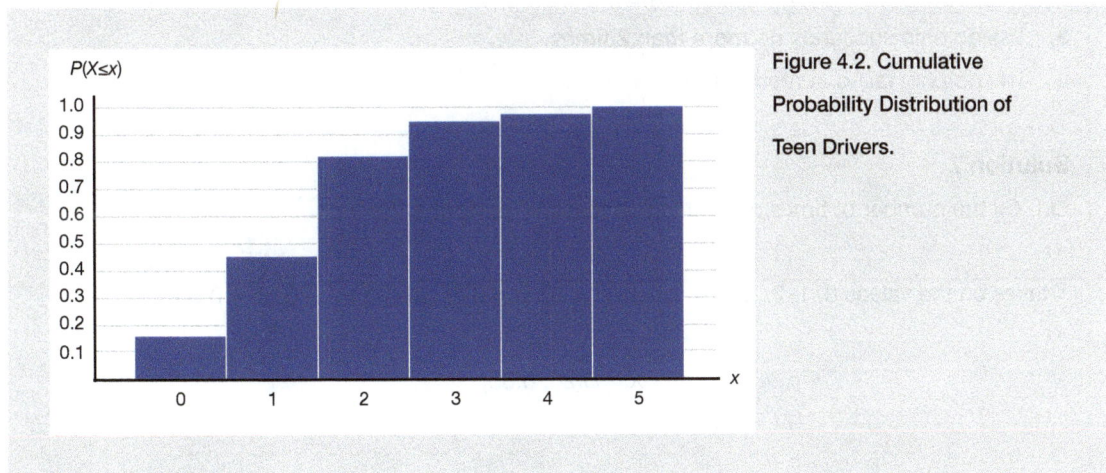

Figure 4.2. Cumulative Probability Distribution of Teen Drivers.

The **cumulative probability distribution** describes the area under the distribution curve or in a histogram. It is used to evaluate probability as area. For example, the histogram in Figure 4.2 illustrates the probability that the random variable X takes on any value no more than x by shading in the bars with values less than or equal to x. In this case, we would write $P(X \leq x)$. Generally, if X is a discrete random variable, then $X \leq x$ is interpreted as "no more than" or "at most," and $X \geq x$ is interpreted as "at least."

Cumulative Probability

$$P(X \leq x) = P(0) + P(1) + \cdots + P(x)$$
$$P(X < x) = P(0) + P(1) + \cdots + P(x - 1)$$

Probability Rule for Complements

$$P(X \geq x) = 1 - P(X < x) = 1 - P(0) - P(1) - \cdots - P(x - 1)$$
$$P(X > x) = 1 - P(X \leq x) = 1 - P(0) - P(1) - \cdots - P(x)$$

Example 7

A trainer is teaching a dolphin to do a trick. The probability that the dolphin successfully performs the trick is 0.35. Out of 20 attempts, we want to find the following probabilities.

1. The dolphin succeeds 12 times.
2. The dolphin succeeds at least 2 times.
3. The dolphin succeeds no more than 2 times.
4. The dolphin failed at most 5 times.

Solution 7

Let X = the number of times a dolphin successfully performs the trick.

X takes on the values 0, 1, 2, ..., 20, where $n = 20$, $p = 0.35$, and $q = 1 - 0.35 = 0.65$.

$$X \sim B(20, 0.35)$$

1. Use the binomial distribution formula:

$$P(X = 12) = \frac{20!}{(20 - 12)! \cdot 12!} \cdot 0.35^{12} \cdot 0.65^{8} = 0.0136$$

Note: The above probability can also be found using TI – 84:

$$P(X = 12) = \text{binompdf}(20, 0.35, 12) = 0.0136$$

2. Use the probability rule for complements:

$$P(X \geq 2) = 1 - P(X \leq 1) = 1 - P(0) - P(1)$$

$$= 1 - \frac{20!}{(20 - 0)! \cdot 0!} \cdot 0.35^{0} \cdot 0.65^{20} - \frac{20!}{(20 - 1)! \cdot 1!} \cdot 0.35^{1} \cdot 0.65^{19}$$

$$= 0.9979$$

Note: Above probability can also be found using TI – 84:

$$P(X \geq 2) = 1 - \text{binomcdf}(20, 0.35, 1) = 0.9979$$

3. The phrase "no more than 2 times" means that the number of times is 0, 1, or 2. Besides the formula, we can use TI-84 to find the probability.

$$P(X \leq 2) = P(X = 0) + P(X = 1) + P(X = 2)$$

$$= \text{binomcdf}(20, 0.35, 2) = 0.01212$$

4. Let Y = the number of times a dolphin failed to perform the trick, we know the random variable Y, just like the random variable X, takes on any values between 0 and 20, where $n = 20$, $p = 0.65$, and $q = 1 - 0.65 = 0.35$. Therefore

$$Y \sim B(20, 0.65)$$

To find the probability that the dolphin failed at most 5 times, we need to find $P(Y \leq 5)$. Without a formula, we can do this using the R function pbinom(5,20,0.65)

$$P(Y \leq 5) = P(Y = 0) + P(Y = 1) + P(Y = 2) + P(Y = 3) + P(Y = 4) + P(Y = 5)$$

$$= \text{pbinom}(5, 20, 0.65)$$

$$= 0.00031$$

Exercises 4.2

1. According to a recent article, the average number of babies born with significant hearing loss is approximately 2 per 1,000 babies in a hospital's healthy baby nursery. The number climbs to an average of 30 per 1,000 babies in a hospital's intensive care nursery. Suppose that 1,000 babies from healthy baby nurseries were randomly surveyed. Find the probability that exactly two babies were born with significant hearing loss.

2. A nurse commented that when a patient calls the medical advice line claiming to have the flu, the chance that they truly have the flu (and not just a nasty cold) is only about 4%. Of the next 25 patients calling in claiming to have the flu, you want to know how many actually have the flu.

    a. Define the random variable and list its possible values.

    b. State the distribution of X.

    c. Find the probability that at least 4 of the 25 patients actually have the flu.

    d. On average, for every 25 patients calling in, how many do you expect to have the flu?

3. According to The World Bank, in 2009, only 9% of the population of Uganda had access to electricity. Suppose we randomly sample 150 people in Uganda. Let $X =$ the number of people who have access to electricity.

    a. What is the probability distribution for X?

    b. Using the formulas, calculate the mean and standard deviation of X.

    c. Use your calculator to find the probability that 15 people in the sample have access to electricity.

    d. Find the probability that at most 10 people in the sample have access to electricity.

    e. Find the probability that more than 25 people in the sample have access to electricity.

4. The probability that a baseball team in Portland, Oregon, will win any given game is 0.3104 based on a 10-year record of 352 wins out of 1,134 games played. An upcoming monthly schedule contains 8 games.

    a. What is the expected number of wins for that upcoming month?

    b. What is the probability that the team wins six games in that upcoming month?

    c. What is the probability that the team wins at least five games in that upcoming month?

5. A student takes a 32-question multiple-choice exam but didn't study and randomly guesses each answer. Each question has 3 possible choices for the answer. Find the probability that the student guesses more than 75% of the questions correctly.

6. More than 96% of colleges offer some online courses. Suppose 13 colleges are randomly selected, we are interested in the number that offer online courses.

    a. In words, define the random variable X.

b. List the values that X may take on.

c. Define the distribution of X.

d. On average, how many colleges would offer online courses?

e. Find the probability that at most 10 colleges offer online courses.

f. Is it more likely that 12 or 13 will offer online courses? Use numbers to justify your answer numerically and answer in a complete sentence.

7. It's estimated that only about 30% of Oregon residents have adequate earthquake supplies. Suppose you randomly survey 11 Oregon residents. You are interested in the number of residents who have adequate earthquake supplies.

a. In words, define the random variable X.

b. List the values that X may take on.

c. Define the distribution of X.

d. What is the probability that at least 8 have adequate earthquake supplies?

e. Is it more likely that none or that all of the residents surveyed will have adequate earthquake supplies? Why?

8. The literacy rate for a nation measures the proportion of people aged 15 and over who can read and write. The literacy rate in Afghanistan is 28.1%. Suppose you choose 15 people in Afghanistan at random. Let X = the number of people who are literate.

a. Sketch a graph of the probability distribution of X.

b. Using the formulas, calculate the (i) mean and (ii) standard deviation of X.

c. Find the probability that more than 5 people in the sample are literate. Is it more likely that 3 people or 4 people are literate.

9. Suppose 1 of every 200 kids prefers broccoli over strawberries. Out of a randomly chosen group of 600 kids, determine the following:

a. In words, define the random variable X.

b. List the values that X may take on.

c. Define the distribution of X.

d. How many are expected to prefer broccoli?

e. Find the probability that no one prefers broccoli.

f. Find the probability that more than 4 prefer broccoli.

4.3 Poisson Probability Distribution

Overview

Another useful probability distribution is the Poisson distribution. It is used for calculating the probability of an event occurring a certain number of times within a given interval of time. Unlike the binomial distribution with two parameters, n and p, the Poisson distribution has only one parameter, μ, which is the mean number of values. The Poisson distribution is used in analyzing traffic flow, predicting randomly occurring accidents, and in other scenarios.

Poisson Distribution Basics

The **Poisson probability distribution** is the discrete probability distribution that gives the probability of probability of an event occurring a given number of times over a specified period of time.

There are three requirements to use the Poisson probability distribution:

1. The random variable X equals the number of occurances in the interval of interest, and the ocurances are evenly distributed over the same interval.
2. The occurances are random.
3. The occurances are independent of each other.

Poisson Probability Distribution

$$P(x) = \frac{\mu^x e^{-\mu}}{x!}$$

Notation for Poisson probability distribution

$P(X)$ is the probability of x occurrences

μ is the mean number of occurrences

x is the number of occurrences per unit

e is a constant and $e \approx 2.71828$

P = Poisson Probability Distribution Function

$$X \sim P(\mu)$$

Read this as "X is a random variable with a Poisson distribution." The parameter μ is the mean for the interval of interest.

To use the Poisson distribution, the probability of success, μ, is unchanged within the interval. The Poisson distribution can be considered a clever way to convert a continuous random variable, usually time, into a discrete random variable by breaking up time into discrete independent intervals. This way of thinking about the Poisson helps us understand why it can be used to estimate the probability for the discrete random variable from the binomial distribution. The Poisson asks for the probability of several successes over time, while the binomial asks for the probability of a certain number of successes for a given number of trials.

Example 8

A bank expects to receive 6 bad checks per day, on average. What is the probability of the bank getting fewer than 2 bad checks on any given day?

Solution 8

Let X = the number of bad checks the bank receives per day. The time interval of interest is 1 day.

Here, we have an average rate of occurrences but no estimate of the probability, so it looks like we have a Poisson distribution with $\mu = 6$.

The probability of getting fewer than 2 bad checks per day is getting 0 or 1 bad check per day. Using the formula for the Poisson probability distribution, we have:

$$P(X < 2) = P(X = 0) + P(X = 1) = \frac{6^0 e^{-6}}{0!} - \frac{6^1 e^{-6}}{1!}$$
$$= 0.0025 + 0.01487 \approx 0.01737$$

Parameters of Poisson Probability Distribution

The Poisson probability distribution has a mean and standard deviation like other probability distributions. The mean (μ) is the average of the distribution, and the standard deviation is easy to calculate from there.

Standard Deviation for Poisson Probability Distribution

$$\sigma = \sqrt{\mu}$$

Example 9

It's 8 a.m. Leah usually receives about 6 emails between 8 a.m. and 10 a.m. What is the probability that Leah receives at least 2 emails in the next 15 minutes? Find the standard deviation of the distribution.

Solution 9

Let X = the number of emails Leah receives in 15 minutes. (The interval of interest is 15 minutes or 0.25 hours).

$$X = 0, 1, 2, 3, ...$$

If Leah receives, on the average, 6 emails in 2 hours, and there are eight 15-minute intervals in 2 hours, then on average, Leah receives $0.125 \times 6 = 0.75$ emails every 15 minutes. So,

$$\mu = 0.75 \quad and \quad X \sim P(0.75)$$

We can use R function `ppois(x,µ)` for the probability that Leah receives at least 2 emails

$$P(X \geq 2) = 1 - P(X \leq 1) = 1 - \texttt{ppois(1,0.75)} = 1 - 0.8266 = 0.1734$$

The standard deviation of the distribution is

$$\sigma = \sqrt{\mu} = \sqrt{0.75} = 0.866$$

The probability that Leah receives more than 2 emails in the next 15 minutes is about 0.1734.

Figure 4.3 shows the graph of $X \sim P(0.75)$. The y-axis contains the probability of x where X = the number of emails in 15 minutes.

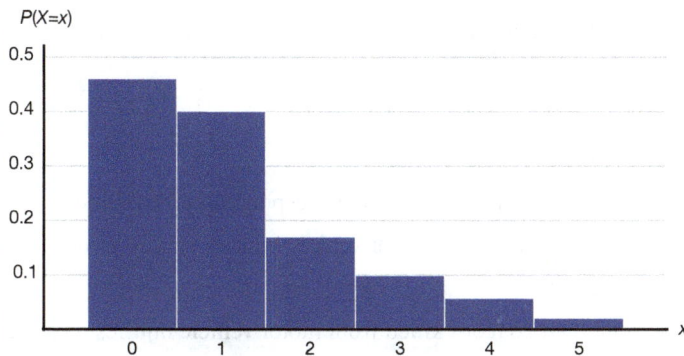

Figure 4.3. Probability Distribution of Emails

P(X=x) axis labels: 0.5, 0.4, 0.3, 0.2, 0.1
x axis labels: 0, 1, 2, 3, 4, 5

Try It 4.3

Atlanta's Hartsfield-Jackson International Airport is the busiest airport in the world. On average, there are 2,500 arrivals and departures each day.

1. How many airplanes arrive and depart the airport per hour?
2. What is the probability that there are exactly 100 arrivals and departures in one hour?
3. What is the probability that there are at most 100 arrivals and departures in one hour?

Exercises 4.3

1. On average, a food truck serves 120 customers per day. Assume the event occurs independently in any given day.
   a. Define the random variable X.
   b. What values does X take on?
   c. What is the probability of getting 150 customers in one day?
   d. What is the probability of getting 35 customers in the first four hours? Assume the store is open 12 hours each day.
   e. What is the probability that the store will have more than 12 customers in the first hour?
   f. What is the probability that the store will have fewer than 12 customers in the first two hours?

2. On average, 8 teenagers in the U.S. die from motor vehicle injuries each day. As a result, states across the country are debating raising the driving age. Assume the event occurs independently in any given day.

   a. Define the random variable X.

   b. What values does X take on?

   c. For the given values of the random variable X, fill in the corresponding probabilities.

   d. Is it likely that there will be no teens killed from motor vehicle injuries on any given day in the U.S.? Justify your answer numerically.

   e. Is it likely that there will be more than 20 teens killed from motor vehicle injuries on any given day in the U.S.? Justify your answer numerically.

3. A law office gets an average of 5.5 incoming phone calls during the noon hour on Mondays. Experience shows that the existing staff can handle up to 6 calls in an hour. Let $X =$ the number of calls received at noon.

   a. Find the mean and standard deviation of X.

   b. What is the probability that the office receives at most 6 calls at noon on Monday?

   c. Find the probability that the law office receives 6 calls at noon. What does this mean to the law office staff who get, on average, 5.5 incoming phone calls at noon?

   d. What is the probability that the office receives more than 8 calls at noon?

4. The maternity ward at Dr. Jose Fabella Memorial Hospital in Manila in the Philippines is one of the busiest in the world, with an average of 60 births per day. Let $X =$ the number of births in an hour.

   a. Find the mean and standard deviation of X.

   b. Find the average number of babies delivered in one hour.

   c. What is the probability that the maternity ward will deliver 3 babies in one hour?

   d. What is the probability that the maternity ward will deliver at most 3 babies in one hour?

   e. What is the probability that the maternity ward will deliver more than 5 babies in one hour?

5. A manufacturer of Christmas tree light bulbs knows that 3% of its bulbs are defective. Find the probability that a string of 100 lights contains at most 4 defective bulbs using both the binomial and Poisson distributions.

6. Fertile female cats produce an average of 3 litters per year. Suppose that 1 fertile female cat is randomly chosen. Using the time frame of 1 year, find the following:

   a. In words, define the random variable X.

   b. List the values that X may take on.

   c. Give the distribution of X.

d. Find the probability that the cat has no litters in 1 year.

e. Find the probability that the cat has at least 2 litters in 1 year.

f. Find the probability that the cat has exactly 3 litters in 1 year.

7. According to a survey, a university professor gets an average of 7 students visiting during office hours per day. Let X = the number of student visitors a professor receives per day. The discrete random variable X takes on the values 0, 1, 2 .... The random variable X has a Poisson distribution: $X \sim P(7)$. The mean is 7 student visits.

a. What is the probability that a professor receives exactly 2 student visitors per day?

b. What is the probability that a professor receives at most 2 student visitors per day?

c. What is the standard deviation?

Try It Solutions

Solution 4.1

This is a binomial problem because there is only a success or a failure, and there are a definite number of trials. The probability of a success stays the same for each trial.

Solution 4.2

1. $X \sim B(80, 0.613)$

2. Mean $= np = 80(0.613) = 49.04$
    Standard Deviation $= \sqrt{npq} = \sqrt{80(0.613)(0.387)} \approx 4.356$

3. $P(x = 60) = 0.0036$

4. $P(x > 50) = 1 - P(x \le 50) = 1 - 0.6282 = 0.3718$

Solution 4.3

1. Let X = the number of airplanes arriving and departing from Hartsfield-Jackson in one hour. The average number of arrivals and departures per hour is 2,500/24 $\approx$ 104.1667.

2. $X \sim P(104.1667)$, so $P(X = 100) = \dfrac{104.1667^{100} \cdot e^{-104.1667}}{100!} = 0.0366$
    A TI-83 or TI-84 graphing calculator can find the same answer using these steps:
    "distr", "poissonpdf(", enter the value for mean and
    x: poissonpdf(104.1667, 100) $\approx$ 0.0366.

3. You can also use a graphing calculator TI-83 or TI-84 to find cumulative probability. For cumulative probability, you will use "poissoncdf(". To find $P(x \leq 100)$, enter on your graphing calculator poissoncdf(104.1667, 100) it returns the value 0.3651. This means that the probability that at most 100 airplanes arriving and departing from the airport is 0.3651. It would take a very long time to calculate this probability using the formula because you will need to add 100 probabilities.

Chapter 5
Continuous Probability Distributions

In Chapter 4, you learned about discrete probability distributions and found probabilities of random variables from a discrete distribution. This chapter will focus on continuous random variables and their probability distributions.

Jyoti is an ecologist. She is studying two neighboring mountain lakes to determine why one has an abundant trout population and why one doesn't. Her team wants to examine whether lake depth and water temperature at different depths have anything to do with trout populations. They take depth measurements and water temperature readings at random locations around the lakes. In this case, lake depth is the random variable, and because Jyoti is measuring depths and could get any value, she is working with continuous random variables. The temperatures are also continuous random variables because the temperature scale has infinite possible values. Continuous random variables are everywhere and used in many ways. Baseball batting averages, the time a computer chip lasts, and investment return rates are examples of continuous random variables.

The values of continuous random variables can be confusing. For example, if X equals the *number of books* in a backpack, X is a discrete random variable because we can count the books. If X is the *weight of a book*, X is a continuous random variable because weights are measured.

After reading this chapter, you will be able to do the following:

1. Understand the probability density function
2. Find probabilities involving continuous random variables
3. Understand the properties of uniform distribution
4. Find probabilities involving uniform distribution
5. Understand the properties of the Normal Distribution and the Standard Normal Distribution
6. Apply the Empirical Rule in problem-solving
7. Estimate the Binomial with the Normal Distribution
8. Understand the central limit theorem
9. Calculate probabilities using the central limit theorem
10. Understand the law of large numbers

5.1 Basic Concepts of Continuous Probability Distributions

Overview

There are many continuous probability distributions. You'll encounter the uniform and normal distributions in this chapter. You've got the concept of a continuous variable down already. In this section, you'll learn about the kinds of continuous probability distributions, how to select the appropriate distribution for a particular situation, and how to calculate the probability of events with that distribution.

Continuous Probability Density Function

A continuous probability density function (pdf) describes the distribution of a continuous random variable. The probability density function graph is called a density curve, which describes the distribution of the random variable. If the density curve is a horizontal line, then the random variable has a uniform distribution. Otherwise, the random variable has a non-uniform distribution. The normal distribution is one of the examples of a non-uniform distribution.

For a continuous random variable X, we use the function symbol f to represent its probability density function. In other words, the values of X, such as x, are the input of the density function, and $f(x)$ is the output of the density function. The density curve is plotted by a set of ordered pairs:

$$\{(x_1, f(x_1)), \quad (x_2, f(x_2)), \quad (x_3, f(x_3)), \ldots, (x_n, f(x_n))\}$$

where each pair represents a point on the density curve.

The properties of probability density function:

1. $f(x) \geq 0$ *for all* x
2. The area between the density curve and horizontal X-axis over the entire domain of X is equal to 1.

The area under a density curve of $f(x)$ describes the probability of the random variable taking on any value over an interval. We use the cumulative distribution function $P(X)$, or cdf, to describe probabilities of X taking on values over different intervals. For example, $P(a < x < b)$ is the probability that the random variable X has any value between a and b. This probability is the area under the curve, above the x-axis, to the right of a, and the left of b.

The properties of cumulative distribution function:

1. $P(X)$ is a non-decreasing function, meaning $P(X)$ is either a constant or increasing function.
2. The minimum value of $P(X)$ is 0, and the maximum value of $P(X)$ is 1.

Figure 5.1 shows how the pdf and cdf interact and relate to each other.

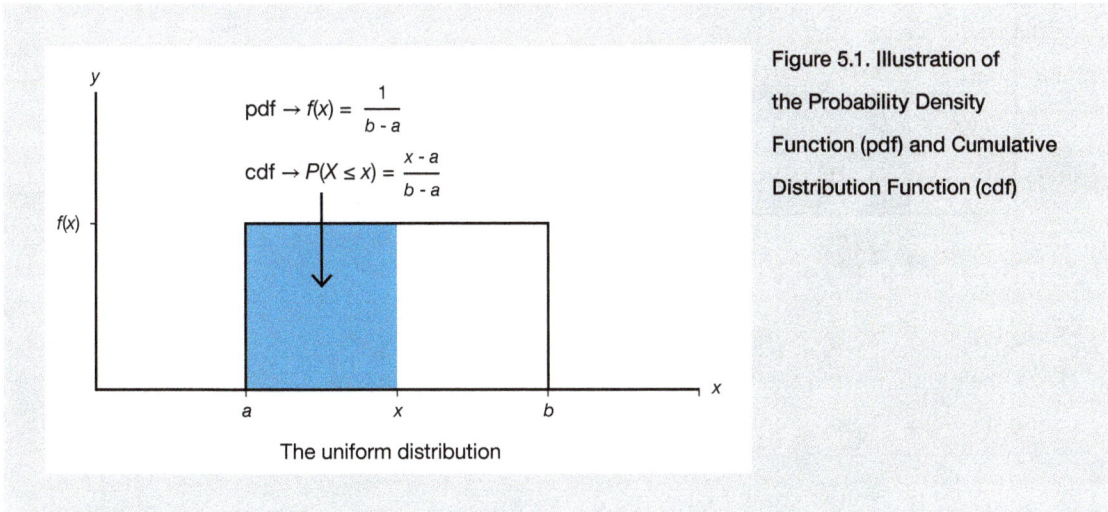

pdf $\rightarrow f(x) = \dfrac{1}{b-a}$

cdf $\rightarrow P(X \leq x) = \dfrac{x-a}{b-a}$

The uniform distribution

Figure 5.1. Illustration of the Probability Density Function (pdf) and Cumulative Distribution Function (cdf)

There are a few unique probabilities of a continuous random variable:

1. The probability that x takes on any individual value is 0 because X describes an interval of values. In other words, the width of a line is zero, and therefore, the area is 0. Since the probability is equal to the area, the probability is also 0.

$$P(x = a) = 0$$

2. $P(a < x < b)$ is the same as $P(a \leq x \leq b)$ because the probability is equal to the area, and there is no area on the boundaries of the interval.

$$P(a < x < b) = P(a \leq x \leq b)$$

Example 1

Find the area between the graph of $f(x) = \frac{1}{20}$ and the x-axis where $4 < x < 15$. Is f a probability density function?

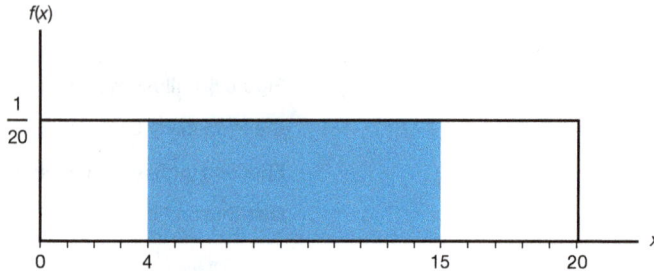

Figure 5.2. Area Representing

$P(4 < x < 15)$

Solution 1

From the graph, we can see the area is the rectangle shaded in blue. The base of the rectangle is $(15 - 4) = 11$, and

$$\text{Area} = (15 - 4)\left(\frac{1}{20}\right) = 0.55$$

Examining the graph, it is true that $f(x) \geq 0$ *for all* x. The entire area under the function and the horizontal X-axis over the interval $(0, 20)$ is:

$$20 \cdot \frac{1}{20} = 1$$

Therefore, f is a probability density function of the random variable X. The area under a density curve corresponds to probability.

$$P(4 < x < 15) = 0.55$$

Exercises 5.1

1. In the distributions below, what does the shaded area represent in terms of probability?

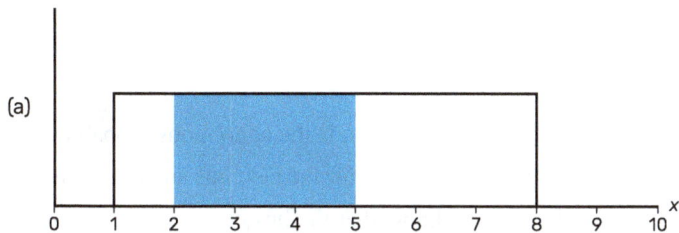

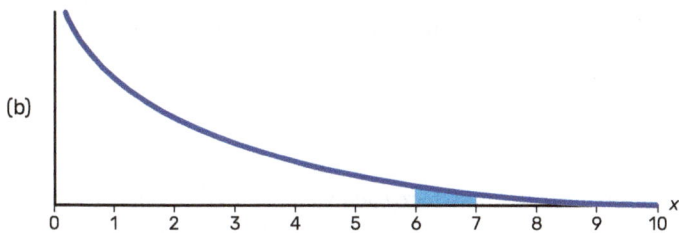

(a)

(b)

2. A continuous probability density function is restricted to the portion between $x = 0$ and 7. What is $P(x = 10)$?

3. $f(x)$ for a continuous probability density function is $\frac{1}{5}$ and the function is restricted to $0 \leq x \leq 5$. What is $P(x < 0)$?

4. $f(x)$, a continuous probability density function, is $\frac{1}{12}$ and the function is restricted to $0 \leq x \leq 12$. What is $P(0 < x < 12)$?

5.2 The Uniform Distribution

Overview

When events are equally likely to occur, the uniform distribution is the continuous probability distribution to use. Think about this scenario: you're at the bus stop waiting for the next bus and don't have a timetable. You don't know when the last bus arrived, but you do know that the bus comes once an hour. What's the probability that the bus will arrive in the next ten minutes? Because you're working with time, and you don't know when the bus last arrived, there's an equal probability that it will arrive at any point in the next hour. For this, you'd use the uniform distribution.

Uniform Distribution Basics

A **uniform distribution** is a probability distribution where each value in a given range is equally likely to occur. The density curve of a uniform distribution is one of the easiest to calculate because it is a horizontal line, as you saw in Figure 5.1. When working out problems that have a uniform distribution, be careful to note if the data is inclusive or exclusive of endpoints.

Probability Density Function

$$f(x) = \frac{1}{b-a} \text{ for } a \leq x \leq b$$

The random variable X has a uniform distribution with parameters a and b, which can be expressed as

$$X \sim U(a, b)$$

U = Uniform Distribution Function, a = the lowest value of X, and b = the highest value of X.

Example 2

Suppose a stock's value varies daily from $16 to $25 with a uniform distribution.

1. Find the probability that the value of the stock is between $19 and $22.
2. Find the probability that the value of the stock is more than $19.

Solution 2

1. The probability density function:

$$f(x) = \frac{1}{b-a} = \frac{1}{25-16} = \frac{1}{9}$$

where $16 \leq x \leq 25$

The probability that the value of the stock is between $19 and $22:

$$P(19 < X < 22) = \frac{1}{9} \cdot (22 - 19) = \frac{1}{3}$$

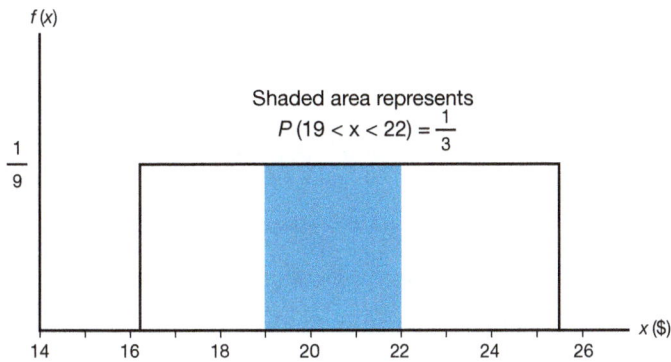

$f(x)$

Shaded area represents
$P(19 < x < 22) = \frac{1}{3}$

$\frac{1}{9}$

14   16   18   20   22   24   26 x ($)

Figure 5.3. Area Representing

$P(19 < X < 22)$

2. The probability that the value of the stock is more than $19:

$$P(X > 19) = \frac{1}{9} \cdot (25 - 19) = \frac{2}{3}$$

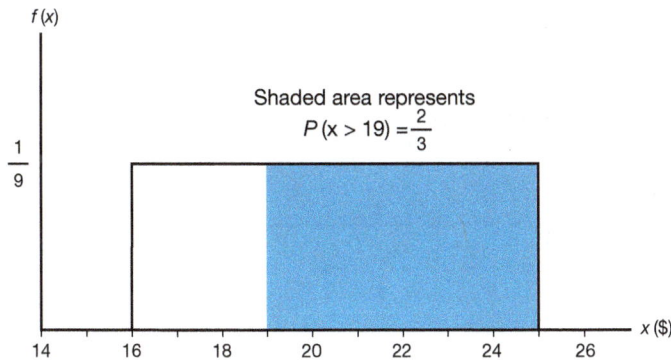

$f(x)$

Shaded area represents
$P(x > 19) = \frac{2}{3}$

$\frac{1}{9}$

14   16   18   20   22   24   26 x ($)

Figure 5.4. Area Representing

$P(X > 19)$

Parameters of Uniform Distribution

Using these formulas, you can find the mean and standard deviation for a uniform distribution.

Mean for Uniform Distribution

$$\mu = \frac{a + b}{2}$$

Standard Deviation for Uniform Distribution

$$\sigma = \sqrt{\frac{(b - a)^2}{12}}$$

Example 3

The amount of time, in minutes, that a person must wait for a bus is uniformly distributed between 0 and 15 minutes, inclusive.

1. What is the probability that a person waits fewer than 12.5 minutes?
2. On average, how long must a person wait (the mean, μ)? Find the standard deviation, σ.
3. Find the 90th percentile for a person's waiting time. In other words, 90 percent of the time, how many minutes must a person wait?

Solution 3

1. Let X = the number of minutes a person must wait for a bus. X can take on any value over the interval (0, 15). The parameters of the uniform distribution a and b are 0 and 15, respectively, with a being the minimum value of X and b the maximum value of X.

$$X \sim U\,(0,\ 15)$$

First, we need to find the pdf of X:

$$f(x) = \frac{1}{b - a} = \frac{1}{15 - 0} = \frac{1}{15}$$

where $16 \leq x \leq 25$

Next, we find the probability that a person waits fewer than 12.5 minutes, $P(x < 12.5)$

$$P(X < 12.5) = (12.5 - 0) \cdot \frac{1}{15} = 0.8333$$

The probability a person waits less than 12.5 minutes is 0.8333. This probability is the area of the blue rectangle.

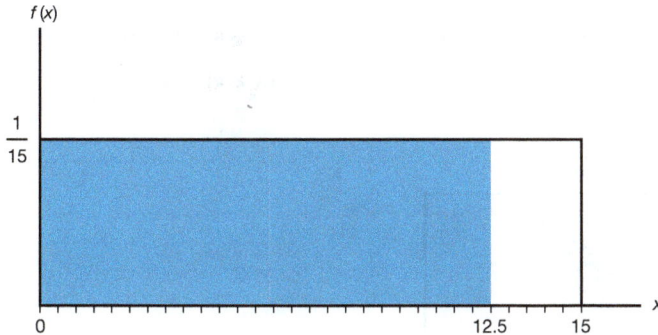

Figure 5.5. Area Representing $P(x < 12.5)$

2. The mean (μ) *and* the standard deviation (σ) of the random variable X:

$$\mu = \frac{15 + 0}{2} = 7.5 \text{ minutes}$$

$$\sigma = \sqrt{\frac{(15 - 0)^2}{12}} = 4.3 \text{ minutes}$$

On average, a person must wait 7.5 minutes. The standard deviation of the wait time is 4.3 minutes.

3. The 90th percentile for a person's waiting time is equivalent to the wait time, k, that is on the 90th percentile. This is interpreted as the probability that a person waits under k minutes, which is 0.9.

$$P(X < k) = 0.9$$

Since X is uniformly distributed, the area of the rectangle on $(0, k)$ is equivalent to the probability as

$$P(X < k) = (k - 0) \cdot \frac{1}{15} = \frac{k}{15}$$

Making the right sides of the two equations equal gives you

$$\frac{k}{15} = 0.9$$

Solving the equation for k, we have

$$k = 0.9 \cdot 15 = 13.5 \text{ minutes}$$

Ninety percent of the time, the wait time for a bus is at most 13.5 minutes.

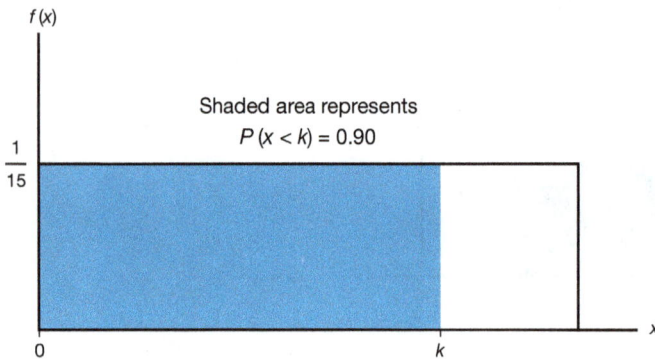

Figure 5.6. Area Representing $P(X < k)$

Shaded area represents
$P(x < k) = 0.90$

Try It 5.1

The total duration of major league baseball games in the 2011 season was uniformly distributed between 447 hours and 521 hours, inclusive.

1. Find a and b and describe what they represent.
2. Write the distribution.
3. Find the mean and the standard deviation.
4. What is the probability that the duration of games for a team for the 2011 season is between 480 and 500 hours?

Exercises 5.2

1. A distribution is given as $X \sim U(0, 12)$.
   a. What is the probability density function?

    b.   What is the mean of the distribution?

    c.   What is the standard deviation of X?

    d.   Draw the graph of the distribution for $P(x > 9)$.

    e.   Find $P(x > 9)$.

    f.   Find the 40th percentile.

2.   The distribution can be written as $X \sim U(1.5, 4.5)$.

    a.   Graph $P(2 < x < 3)$.

    b.   What is $P(2 < x < 3)$?

    c.   What is $P(x = 1.5)$?

3.   The age of cars in the staff parking lot of a college is uniformly distributed from six months to 9.5 years.

    a.   What is being measured here?

    b.   Describe the random variable X.

    c.   Are the data discrete or continuous?

    d.   The interval of values for x is _____.

    e.   The distribution for X is _____.

    f.   Write the probability density function.

    g.   Sketch the graph of the probability distribution.

4.   For the two uniform distributions below, find the probability that x falls in the shaded area.

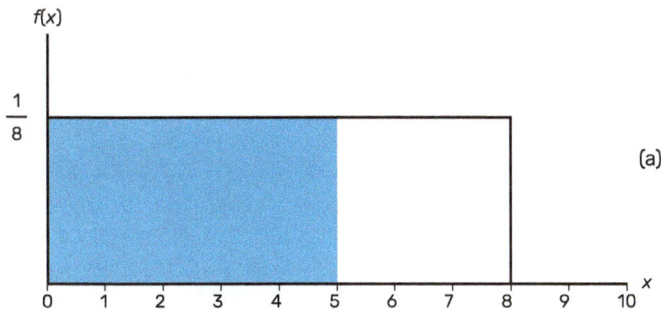

(a)

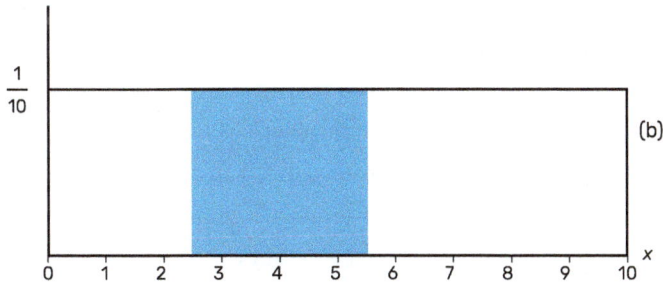

(b)

5.3 The Normal Distribution

Overview

The normal probability distribution is the most important of all the distributions. Its graph is bell-shaped and symmetrical. You'll see the bell curve in almost all subjects and fields, including psychology, business, economics, the sciences, nursing, and statistics. Sometimes, instructors use the normal distribution to help determine your grade. Most IQ scores and home prices have a normal distribution. In this section, you'll discover when and how to use a normal distribution to find probabilities.

Normal Distribution Basics

A **normal distribution** for a continuous random variable is a bell-shaped probability distribution where most data falls near the middle and outliers are rare, as seen in Figure 5.7. While the bell shape of the curve makes calculating the probabilities a bit trickier, the connection between area and probability is the same. Where there is more area under the curve, there is a higher probability of X occurring.

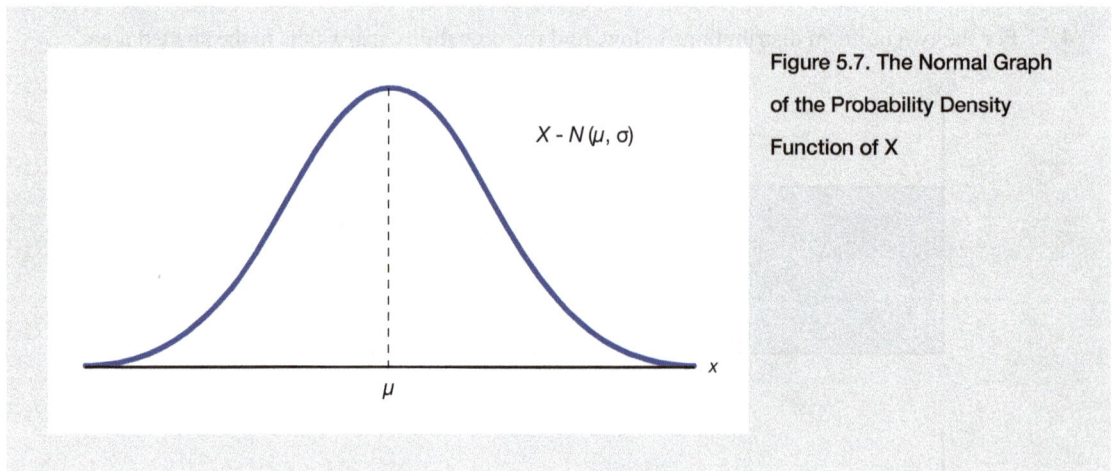

$X - N(\mu, \sigma)$

Figure 5.7. The Normal Graph of the Probability Density Function of X

A normal distribution is described by the mean (μ) and the standard deviation (σ). If the continuous random variable with a normal distribution has these two parameters, you can use the normal probability density function to describe the probability distribution. This formula is complicated, and you don't need to memorize it, but know that the mean and standard deviation are the keys.

Normal Probability Density Function

$$f(x) = \frac{1}{\sigma\sqrt{2\pi}} \, e^{-\frac{1}{2}\left(\frac{x-\mu}{\sigma}\right)^2}$$

The random variable X that has a normal distribution with mean μ and standard deviation σ can be expressed as

$$X \sim N(\mu, \sigma)$$

The curve is symmetrical around a vertical line drawn through the mean μ. The mean is the same as the median, which is the same as the mode. As the notation indicates, the normal distribution depends only on the mean and the standard deviation. Since the area under the curve must equal 1, a change in the standard deviation causes a change in the shape of the normal curve. The curve becomes fatter and wider or skinnier and taller depending on σ. A change in μ causes the graph to shift to the left or right. This means there are an infinite number of normal probability distributions. One of particular interest is called the standard normal distribution.

Standard Normal Distribution

The **standard normal distribution** is a special case of the normal distribution. In the standard normal distribution, the mean (μ) is 0, and the standard deviation (σ) is 1. This means the curve is symmetric around 0, and the standard deviation controls its spread. The total area under the curve is still equal to 1.

The standard normal distribution is useful because it allows us to convert any normal distribution to a standard form. This conversion uses standardization, which involves subtracting the mean and dividing by the standard deviation. The resulting value is called a z-score, which we learned in Chapter 2.

Probability Density Function for the Standard Normal Distribution

$$f(z) = \frac{1}{\sqrt{2\pi}} \, e^{-\frac{z^2}{2}}$$

z-score

$$z = \frac{x-\mu}{\sigma}$$

The random variable Z with a standard normal distribution with mean 0 and standard deviation 1 can be expressed as

$$Z \sim N(0, 1)$$

The z-score represents the number of standard deviations a value x is above (to the right of) or below (to the left of) the mean μ. Values of x larger than the mean have positive z-scores, and values of x smaller than the mean have negative z-scores. If x equals the mean, then x has a z-score of 0. You can find z-score value tables in the Appendix.

Example 4

Suppose $X \sim N(5, 6)$, how many standard deviations is x from the mean if $x = 17$ and $x = 11$?

Solution 4

$X \sim N(5, 6)$ means that X is a normally distributed random variable with the mean $\mu = 5$ and standard deviation $\sigma = 6$. Using the standardizing formula, you can find the z-score:

$$z = \frac{x - \mu}{\sigma} = \frac{17 - 5}{6} = 2$$

The value $x = 17$ is 2 standard deviations above, or to the right, of its mean.

$$z = \frac{x - \mu}{\sigma} = \frac{11 - 5}{6} = 1$$

The value $x = 11$ is 0.67 standard deviations below, or to the left, of its mean.

The empirical rule was introduced in Chapter 2, but it makes even more sense once you've learned about the normal distribution. If the random variable Z has the standard normal distribution, then approximately 68% of the distribution falls between -1 and 1, 95% falls between -2 and 2, and 99.7% falls between -3 and 3.

Example 5

Suppose X has a normal distribution with $\mu = 50$ and $\sigma = 6$. What can you say according to the empirical rule?

Solution 5

For 1 standard deviation from the mean,

$$\mu \pm \sigma = 50 \pm 6 = (44, 56)$$

According to the empirical rule, about 68% of the x values are between 44 and 56. For 2 standard deviations from the mean,

$$\mu \pm 2\sigma = 50 \pm 2(6) = (38, 62)$$

According to the empirical rule, about 95% of the x values are between 38 and 62. For 3 standard deviations from the mean,

$$\mu \pm 3\sigma = 50 \pm 3(6) = (32, 68)$$

According to the empirical rule, about 99.7% of the x values are between 32 and 68.

Calculating Probabilities Using the Normal Distribution

Calculating probabilities with the normal distribution is important for predicting outcomes. We'll use Figure 5.8 to show how you can do this. The shaded region indicates the area to the right of x_1. This area is represented by the probability $P(X > x_1)$.

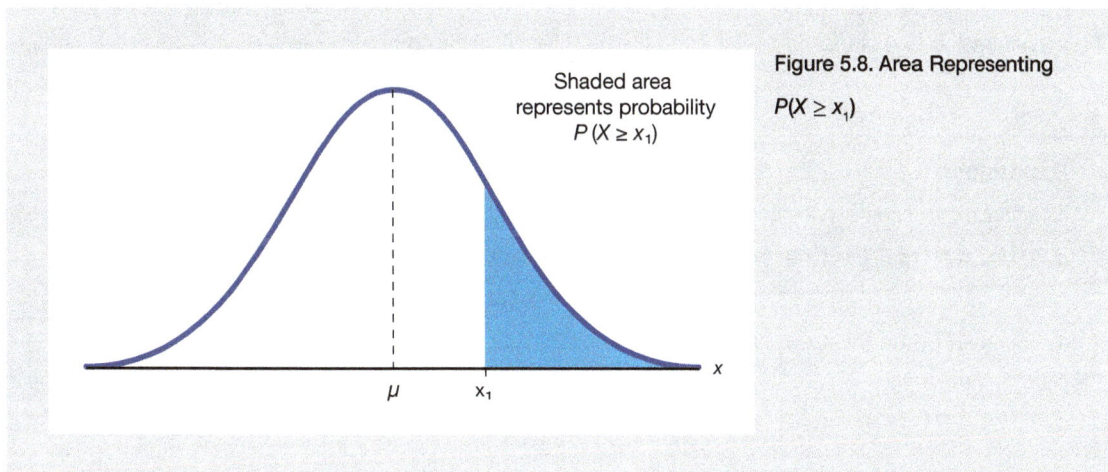

Shaded area represents probability $P(X \geq x_1)$

Figure 5.8. Area Representing $P(X \geq x_1)$

Because the normal distribution is symmetrical, if x_1 were the same distance to the left of the mean, the area in the left tail would be the same as the shaded area in the right tail. Also, because of this distribution's symmetry, half of the probability is to the right of the mean, and half is to the left.

The shaded region in Figure 5.9 shows the area between x_1 and x_2, which is the probability written as $P(x_1 \leq X \leq x_2)$. To compute probabilities for any normal distribution, convert the normal distribution to the standard normal distribution and look up the answer in the Appendix (Table A.1).

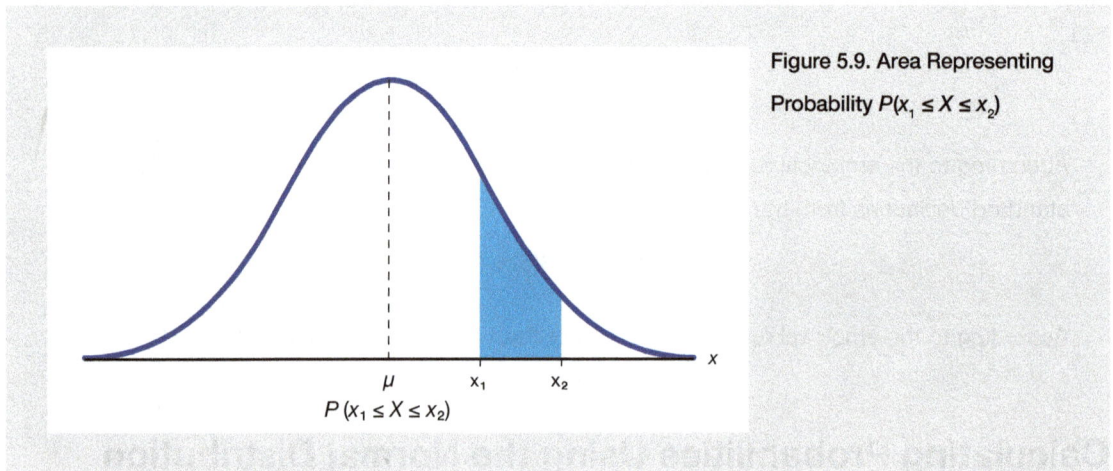

Figure 5.9. Area Representing Probability $P(x_1 \leq X \leq x_2)$

μ x_1 x_2

$P(x_1 \leq X \leq x_2)$

Note

Because the normal distribution is symmetrical, it doesn't matter if the z-score is positive or negative when calculating a probability. One standard deviation to the left (negative z-score) covers the same area as one standard deviation to the right (positive z-score).

Example 6

The final exam scores in a statistics class were normally distributed with a mean of 63 and a standard deviation of 5. Find the probability that a randomly selected student got the following scores:

1. At least 65 on the exam, $P(X \geq 65)$.
2. Less than 85.

Solution 6

1. First, we will define the random variable X:

    Let X = a score on the final exam. X has a normal distribution, so:

    $$X \sim N(63, 5) \text{ with } \mu = 63, \text{ and } \sigma = 5.$$

Next, convert the normal distribution to the standard normal distribution using the z-score formula:

$$z = \frac{X - \mu}{\sigma} = \frac{65 - 63}{5} = 0.4$$

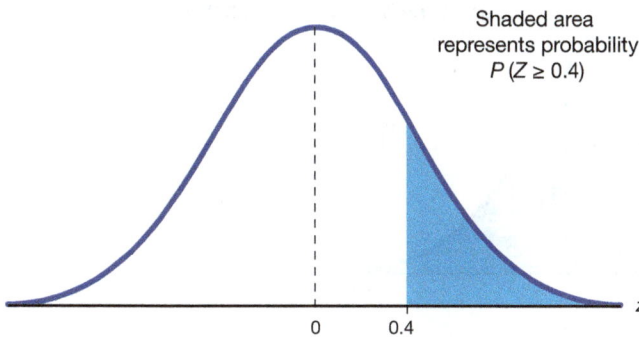

Shaded area represents probability $P(Z \geq 0.4)$

Figure 5.10. Area Representing Probability

We have converted the random variable X to Z, the random variable under the standard normal distribution. This means $X = 65$ is equivalent to $Z = 0.4$, and $P(X \geq 65) = P(Z \geq 0.4)$.

The probability $P(Z \geq 0.4)$ represents the area in the right tail because the z-score is above 0, the mean of the standard normal distribution.

$$P(Z \geq 0.4) = 1 - P(Z < 0.4) = 1 - 0.6554$$
$$= 0.3446$$

The probability that a randomly selected student scored more than 65 on the exam is 0.3446.

2. The probability that a randomly selected student scored less than 85 can be found similarly.

Convert the random variable X to Z:

$$z = \frac{X - \mu}{\sigma} = \frac{85 - 63}{5} = 4.4$$

$$P(X < 85) = P(Z < 4.4) \approx 1$$

Shaded area represents probability $P(z < 4.4)$

Figure 5.11. Area Representing Probability

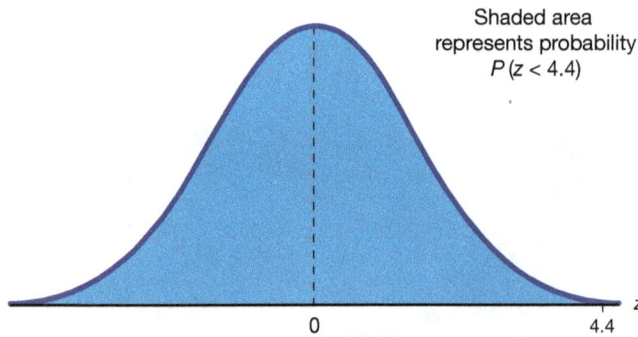

Z = 4.4 is larger than the maximum value in the standard normal table, which means the probability that a student scores less than 85 is approximately 1.

Try It 5.2

The golf scores for a school team were normally distributed with a mean of 68 and a standard deviation of 3. Find the probability that a randomly selected golfer scored less than 65.

Example 7

A household computer is used for work, research, communication, personal finances, education, entertainment, social networking, and more. Suppose the average number of hours a household computer is used for entertainment is 2 hours per day. Assume the times for entertainment are normally distributed, and the standard deviation for the times is half an hour.

1. Find the probability that a household computer is used for entertainment between 1.8 and 2.75 hours per day.
2. Find the maximum number of hours per day that the bottom quartile of households uses a computer for entertainment.
3. Find the probability that a household computer is used for more than 2.5 hours per day.

Solution 7

1. Let X = the time (in hours) a household computer is used for entertainment.
   $X \sim N(2, 0.5)$ where $\mu = 2$ and $\sigma = 0.5$.

The probability we try to find represents the area under the pdf curve *between* $x = 1.8$ and $x = 2.75$. We can accomplish this task in several ways. First, we'll follow the steps from Example 6 and convert the x values to z-scores, then use the standard normal distribution table to find the probability.

$$z_1 = \frac{x - \mu}{\sigma} = \frac{1.8 - 2}{0.5} = -0.4$$

$$z_2 = \frac{x - \mu}{\sigma} = \frac{2.75 - 2}{0.5} = 1.5$$

$$P(-0.4 < z < 1.5) = P(z < 1.5) - P(z < -0.4) = 0.9332 - 0.3446 = 0.5886$$

Now, we will find the probability with TI-84 and R. Table 5.1 has the necessary functions for the calculation.

Table 5.1. Probability Functions for TI-84 and R

Probability	TI – 84	R
Normal Distribution	normcdf (x_1, x_2, μ, σ)	pnorm(x_2, μ, σ) – pnorm(x_1, μ, σ)
Standard Normal Distribution	normcdf $(z_1, z_2, 0, 1)$	pnorm$(z_2, 0, 1)$ – pnorm$(z_1, 0, 1)$
Normal Distribution	normcdf $(-1E99, x, \mu, \sigma)$	pnorm(x, μ, σ)
Normal Distribution	normcdf $(x, 1E99, \mu, \sigma)$	$1 -$ pnorm(x, μ, σ)

As we know, the probability is equivalent to the area under the pdf curve *between* $x = 1.8$ and $x = 2.75$.

Using TI-84:

$$\text{Normcdf } (-0.4,\ 1.5,\ 0,\ 1) = 0.5886$$

or

$$\text{Normcdf } (1.8,\ 2.75,\ 2,\ 0.5) = 0.5886$$

Using R:

$$\text{pnorm}(2.75,\ 2,\ 0.5)- \text{pnorm}(1.8,\ 2,\ 0.5) = 0.5886$$

or

$$\text{pnorm}(1.5,\ 0,\ 1)- \text{pnorm}(-0.4,\ 0,\ 1) = 0.5886$$

The probability that a household computer is used between 1.8 and 2.75 hours per day for entertainment is 0.5886.

2. The bottom quartile of households is the 25th percentile of the population. We denote it as k, and

$$P(X < k) = 0.25$$

where k is the inverse cdf of the normal distribution. We use the appropriate functions to find the 25th percentile.

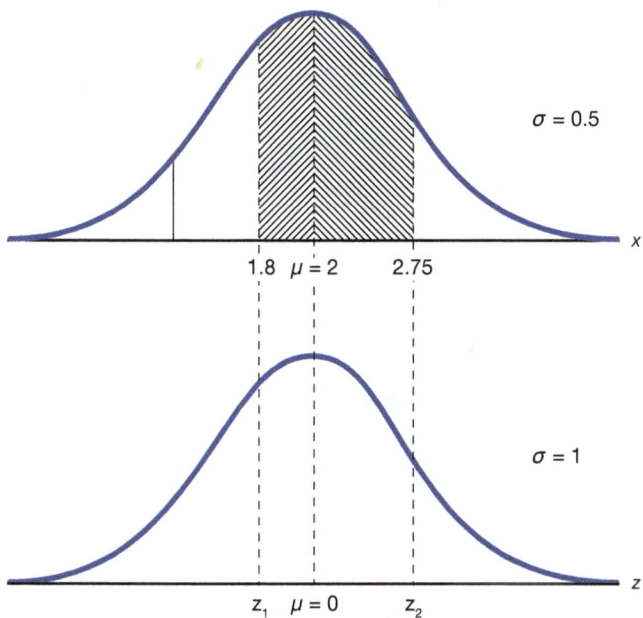

$\sigma = 0.5$

1.8 $\mu = 2$ 2.75

x

$\sigma = 1$

z_1 $\mu = 0$ z_2

z

Using TI-84: $\mathrm{invNorm(P, \mu, \sigma)}$

$k = \mathrm{invNorm(0.25, 2, 0.5)} = 1.66$ hours

Using R: $\mathrm{qnorm(P, \mu, \sigma)}$

$k = \mathrm{qNorm(0.25, 2, 0.5)} = 1.66$ hours

The maximum number of hours per day that the bottom quartile of households uses a computer for entertainment is 1.66 hours.

3.   The probability is equivalent to the area in the right tail of the pdf curve from $x = 2.5$.

Using TI-84: $\mathrm{Normalcdf(2.5, 1E99, 2, 0.5)} \approx 0.1587$

Using R: $P(X > 2.5) = 1 - \mathrm{pnorm(2.5, 2, 0.5)} \approx 0.1587$

The probability that a household computer is used for more than 2.5 hours per day is about 0.1587.

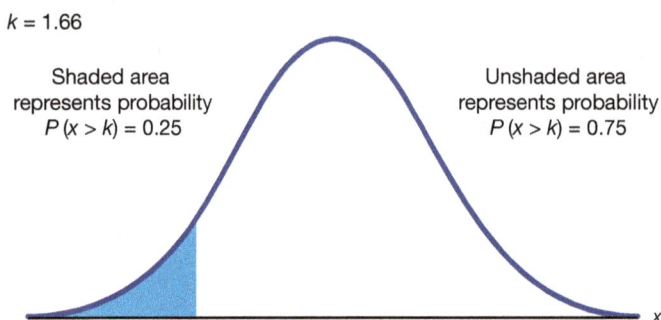

Figure 5.13. Area Representing the 25th Percentile of Households Using a Computer for Entertainment

$k = 1.66$

Shaded area represents probability $P(x > k) = 0.25$

Unshaded area represents probability $P(x > k) = 0.75$

x

Try It 5.3

A school golf team's scores were normally distributed with a mean of 68 and a standard deviation of 3. Find the probability that a golfer scored between 66 and 70.

Estimating the Binomial with the Normal Distribution

The normal distribution can be used to estimate a binomial distribution. If you remember, the binomial distribution formula is complicated. To estimate a binomial distribution using the normal distribution, calculate the mean and standard deviation of the binomial distribution, then use them to standardize the distribution. This allows us to use the normal distribution to estimate the probabilities of certain outcomes.

For a normal distribution to approximate a binomial probability distribution, two requirements must be satisfied:

1. The sample is a simple random sample with sample size n selected from a population where the probability of success is p, and the probability of failure is q.

2. $np \geq 5$ and $nq \geq 5$

The binomial probability distribution can be approximated by the normal distribution with mean $\mu = np$ and standard deviation $\sigma = \sqrt{npq}$. The integer value x of the discrete random variable X is estimated by the interval $(x - 0.5, x + 0.5)$ because the binomial distribution is discrete when the continuous normal distribution approximates it. This process is referred to as **continuity correction**.

Example 8

In a local school district, 53% of the population favors a charter school for grades K–5. A simple random sample of 300 residents is surveyed.

1. Find the probability that 175 residents favor a charter school.
2. Find the probability that more than 155 favor a charter school.
3. Find the probability that at least 150 favor a charter school.

Solution 8

Let X = the number that favors a charter school for grades K–5.

$X \sim B(n, p)$ with $n = 300$ and $p = 0.53$. The requirements $np > 5$ and $nq > 5$ are satisfied because we have

$$np = 300 \cdot 0.53 = 159, \quad \text{and} \quad nq = 300 \cdot 0.47 = 141$$

Using the normal approximation to the binomial, the mean and standard deviation for the normal distribution are $\mu = np$ and $\sigma = \sqrt{npq}$.

$$\mu = np = 159 \quad \text{and} \quad \sigma = \sqrt{npq} = \sqrt{300 \cdot 0.53 \cdot 0.47} \approx 8.645$$

Let Y be the random variable of the normal distribution with a mean of 159 and a standard deviation of 8.6447. We use continuity correction to estimate the integer value of X by the interval $(x - 0.5, x + 0.5)$. That gives us the following answers obtained from TI-84:

1. $P(X = 175)$ has normal approximation $P(174.5 < Y < 175.5) = 0.0083$
2. $P(X > 155)$ has normal approximation $P(Y > 155.5) = 1 - P(Y < 155.5) = 1 - 0.3428 = 0.6572$
3. $P(X \geq 150)$ has normal approximation $P(Y > 149.5) = 0.8641$

The normal distribution makes particularly accurate estimates of a binomial process under certain circumstances. Figure 5.12 is a frequency distribution of a binomial distribution for the experiment of flipping 3 coins where the random variable X is the number of heads. The sample space is listed below the distribution. The experiment assumed that the probability of a success is 0.5. That means the probability of a failure, a tail, is also 0.5. Figure 5.14 shows that the distribution is symmetrical because the probabilities

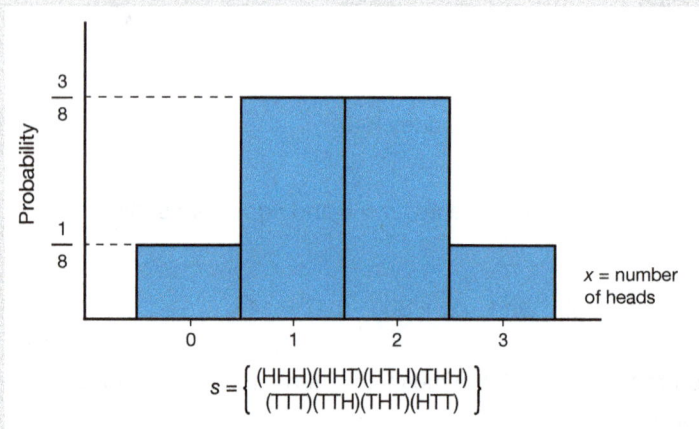

Figure 5.14. Probability Distribution for Tossing Three Coins

of success and failure are the same. If the probability of success were smaller than 0.5, the distribution becomes skewed right. Indeed, as the probability of success diminishes, the degree of skewness increases. If the probability of success increases from 0.5, then the skewness increases in the lower tail, resulting in a left-skewed distribution.

Skewness is important because it affects the accuracy of the distribution estimate. The normal distribution estimates become more accurate as the binomial distribution becomes more symmetric. Look at Figure 5.15. It shows a symmetrical normal distribution laid on top of a graph of a binomial distribution where $p = 0.2$ and $n = 5$. We can see that they don't match. A good guideline to follow is to make sure that both np and $n(1 - p)$ are greater than 5. This is an effective baseline and gives you solid estimates of the binomial probability.

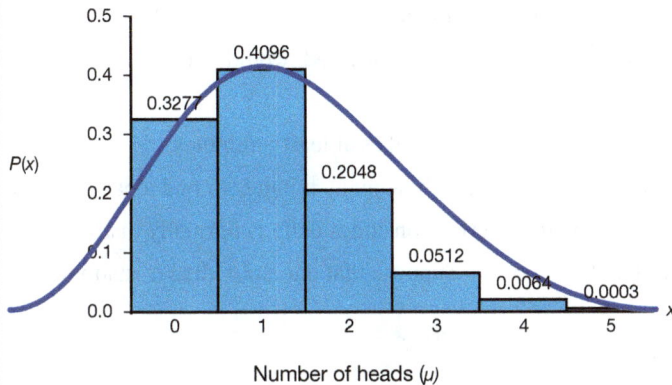

Figure 5.15. Number of Heads While Tossing Three Coins with Skewed Distribution

Exercises 5.3

1. The systolic blood pressure (given in millimeters) of males has an approximately normal distribution with mean $\mu = 125$ and standard deviation $\sigma = 14$. Systolic blood pressure for males follows a normal distribution.

   a. Calculate the z-scores for the male systolic blood pressures 100 and 150 millimeters.

   b. If a male friend of yours said he thought his systolic blood pressure was 2.5 standard deviations below the mean but that he believed his blood pressure was between 100 and 150 millimeters, what would you say to him?

2. Kyle's doctor told him that the z-score for his systolic blood pressure is 1.75. The systolic blood pressure (given in millimeters) of males has an approximately normal distribution with mean $\mu = 125$ and standard deviation $\sigma = 14$. If $X =$ a systolic blood pressure score, then $X \sim N(125, 14)$.

   a. Which answer(s) is/are correct?

      i. Kyle's systolic blood pressure is 175.

      ii. Kyle's systolic blood pressure is 1.75 times the average blood pressure of men his age.

      iii. Kyle's systolic blood pressure is 1.75 above the average systolic blood pressure of men his age.

      iv. Kyles's systolic blood pressure is 1.75 standard deviations above the average systolic blood pressure for men.

   b. Calculate Kyle's blood pressure.

3. The length of time it takes to find a parking space at 9 a.m. follows a normal distribution with a mean of 5 minutes and a standard deviation of 2 minutes.

   a. Based upon the given information and numerically justified, would you be surprised if it took less than 1 minute to find a parking space?

   b. Find the probability that finding a parking space takes at least 8 minutes.

   c. Seventy percent of the time, it takes more than how many minutes to find a parking space?

4. The percent of fat calories that a person in America consumes daily is normally distributed with a mean of about 36 and a standard deviation of 10. Suppose that one individual is randomly chosen. Let X = percent of fat calories.

   a. Describe the distribution of X.

   b. Find the probability that the percent of fat calories a person consumes is more than 40. Graph the situation. Shade in the area to be determined.

   c. Find the maximum number for the lower quarter of percent of fat calories. Sketch the graph and write the probability statement.

5. Facebook provides a variety of statistics that detail the growth and popularity of the site. On average, 28% of 18-to-34-year-olds check their Facebook profiles before getting out of bed in the morning. Suppose this percentage follows a normal distribution with a standard deviation of 5%.

   a. Find the probability that the percent of 18-to-34-year-olds who check Facebook before getting out of bed in the morning is at least 30.

   b. Find the 95th percentile and express it in a sentence.

6. Students in a math course generally have passed the final exam with a probability of 0.9. The exam is given to a group of 70 students.

   a. What is the mean of the binomial distribution?

   b. What is the standard deviation?

   c. Can this binomial distribution be approximate with a normal distribution?

   d. If so, use the normal distribution to find the probability that at least 60 of the students pass the exam.

7. A $1 scratch-off lotto ticket will be a winner 1 out of 5 times. Out of a shipment of n = 190 lotto tickets, find the probability for the lotto tickets that there are

    a. somewhere between 34 and 54 prizes.

    b. somewhere between 54 and 64 prizes.

    c. more than 64 prizes.

8. In a large city, 1 in 10 fire hydrants need repair. If a crew examines 100 fire hydrants in a week, what is the probability they will find 9 or fewer fire hydrants that need repair? Use the normal distribution to approximate the binomial distribution.

5.4 The Central Limit Theorem

Overview

The central limit theorem is among statistics' most useful theorems. It's a concept that can accurately predict a population's mean and standard deviation without surveying the entire population. We'll come across statistics problems that require the central limit theorem all the time, and they are usually easy to spot.

Central Limit Theorem Basics

The **central limit theorem** (CLT) states that if we take many random samples of a sufficiently large size from the same population, the mean of all samples has approximately a normal distribution and is approximately equal to the population's mean. This is true no matter what shape the distributions of the individual samples take.

For the central limit theorem to work, we need a large enough sample (at least 30), or the data should be from a normal distribution. Suppose a random sample of size n (X_1: $x_1, x_2, x_3, ..., x_n$) is selected from a population with mean μ and standard deviation σ. We repeatedly take samples of size n from the population. In this case, we take m random samples of size n and then calculate the mean of each sample. We use the random variable $\overline{X}$ to represent these sample means. The distribution of these m sample means is called the sampling distribution of the mean $\overline{X}$.

$$\overline{X}: \overline{x}_1, \overline{x}_2, \overline{x}_3, ..., \overline{x}_m$$

As the sample sizes increase, the sampling distribution of the mean $\overline{X}$ has an approximately normal distribution with a mean $\mu_{\bar{x}}$, which is the same as the population mean μ, and a standard deviation $\sigma_{\bar{x}} = \frac{\sigma}{\sqrt{n}}$.

Mean of All Sample Means

$$\mu_{\bar{x}} = \mu$$

Standard Deviation of Sample Means

$$\sigma_{\bar{x}} = \frac{\sigma}{\sqrt{n}}$$

If sample sizes are equal to or greater than 30, the sample means have a normal distribution, regardless of the population distribution.

$$\bar{X} \sim N(\mu, \frac{\sigma}{\sqrt{n}})$$

z-Score

$$z = \frac{\bar{x} - \mu}{\frac{\sigma}{\sqrt{n}}}$$

Figure 5.16 illustrates how the central limit theorem works. Notice how the horizontal axis in the top panel is labeled x's. These are the individual observations of the population. This is the unknown distribution of the population values. The graph is purposefully squiggly to show that it does not matter just how odd the shape is. The horizontal axis in the bottom panel is labeled $\bar{x}$'s. This is the sampling distribution of the means. Each observation on this distribution is a sample mean. All these sample means were calculated from individual samples with the same sample size. The sampling distribution contains all the sample mean values from all the possible samples that could have been taken from the population. Of course, no one would ever take all the possible samples, but this is how they would look if they did. The central limit theorem says that they will be normally distributed.

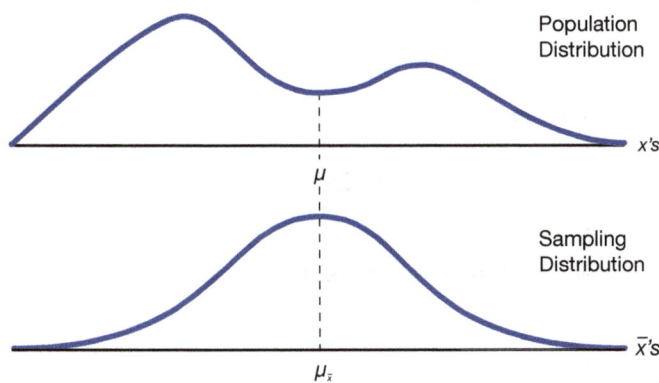

Population
Distribution

x's

Sampling
Distribution

$\bar{x}$'s

μ

$\mu_{\bar{x}}$

Figure 5.16. Sampling
Distribution of the Mean

Example 9

A normal distribution has a mean of 90 and a standard deviation of 15. Samples of size $n = 25$ are drawn randomly from the population.

1. Find the probability that the sample mean is between 85 and 92.
2. Find the mean that is 2 standard deviations above 90, the sample mean.

Solution 9

1. We are asked to find the probability of the sample mean $\bar{X}$ (not X). Let $\bar{X}$ = the mean of a sample of size 25. Since $\mu = 90$, $\sigma = 15$, and $n = 25$, by central limit theorem, we have:

$$\bar{X} \sim N\left(90, \frac{15}{\sqrt{25}}\right)$$

Using TI-84, "normalcdf(lower value, upper value, μ, $\frac{\sigma}{\sqrt{n}}$)" we find the probability $P(85 < \bar{X} < 92) = 0.6997$ because

$$\text{normalcdf}\left(85, 92, 90, \frac{15}{\sqrt{25}}\right) = 0.6997$$

The probability that the sample mean is between 85 and 92 is 0.6997.

2. To find the mean value that is 2 standard deviations above the sample mean 90, we need to use z-score formula:

$$z = \frac{\bar{X} - \mu}{\frac{\sigma}{\sqrt{n}}}$$

After solving for $\bar{x}$ we get

$$\bar{X} = \mu + z\frac{\sigma}{\sqrt{n}}$$

$$= 90 + 2 \cdot \frac{15}{\sqrt{25}} = 96$$

The mean value that is 2 standard deviations above the sample mean is 96.

Try It 5.4

An unknown distribution has a mean of 45 and a standard deviation of 8. Samples of size $n = 30$ are drawn randomly from the population. Find the probability that the sample mean is between 42 and 50.

Example 10

The length of time (in hours) that it takes a group of people over the age of forty to play one game of softball has an unknown distribution with a mean of 2 hours and a standard deviation of 0.5 hours. A sample of size $n = 50$ is drawn randomly from the population. Find the probability that the sample mean is between 1.8 hours and 2.3 hours.

Solution 10

We need to find the probability that the sample mean $\overline{X}$ (not X) time required to play 1 softball game. So we will define $\overline{X}$ = the mean time (in hours) it takes to play one softball game.

Because $\mu = 2$, $\sigma = 0.5$, and $n = 50$, then by the central limit theorem, the sampling distribuion $\overline{X}$ has a normal distribution.

$$\overline{X} \sim N\left(2, \frac{0.5}{\sqrt{50}}\right)$$

Using the TI – 84, we can find

$$P(1.8 < \overline{X} < 2.3) = 0.9977$$

because

$$\text{normalcdf}\left(1.8, 2.3, 2, \frac{0.5}{\sqrt{50}}\right) = 0.9977$$

The central limit theorem can also apply to sums. Say you take a bunch of samples of random variable, $\overline{x}_1, \overline{x}_2, \overline{x}_3, ..., \overline{x}_m$. In our case, we take a total of m samples of size n, and the sums of these random variables, $\sum \overline{x}_1, \sum \overline{x}_2, \sum \overline{x}_3, ..., \sum \overline{x}_m$. The CLT says that the sampling distribution of these sums becomes more and more like a normal distribution as the sample sizes n increase. Note that n is the sample size, not the number of samples.

Central Limit Theorem for the Sums

$$\Sigma X \sim N(n\mu, \sqrt{n}\sigma)$$

Mean of the Sum

$$\mu_{\Sigma X} = n\mu$$

Standard Deviation of the Sum

$$\sigma_{\Sigma X} = \sqrt{n}\sigma$$

z-score for the Random Variable ΣX

$$z = \frac{\Sigma x - n\mu}{\sqrt{n}\sigma}$$

Example 11

An unknown distribution has a mean of 90 and a standard deviation of 15. A sample of size 80 is drawn randomly from the population.

1. Find the probability that the sum of the 80 values is more than 7,500.
2. Find the sum that is 1.5 standard deviations above the mean of the sums.

Solution 11

1. Let X = Any value from the original unknown population. We need to find the probability that the sum of 80 values is more than 7500. In other words, we want to find $P(\Sigma X > 7500)$.

Let's define the random variable ΣX as the sum of the 80 values. Then, according to the CLT, ΣX is normally distributed with mean and standard deviation as

$$\mu_{\Sigma x} = n\mu = 80 \cdot 90 = 7200$$
$$\sigma_{\Sigma x} = \sqrt{n}\sigma = \sqrt{80} \cdot 15 = 134.164$$

We will use the R function to find the probability $P(\Sigma X > 7500)$:

$$P(\Sigma X > 7500) = 1 - P(\Sigma X < 7500)$$

$$= 1 - \text{pnorm}(7500, 7200, 134.164)$$

$$= 1 - 0.9873 \approx 0.0127$$

Note the R function *pnorm* finds the probability in the left tail. The probability or the area we are looking for is in the right tail. So, we find the probability of the complement.

2. We need to find the value of ΣX that is 1.5 above the standard deviation. Using the formula for the z-score, we can change the formula to express ΣX in terms of the z-score because $z = 1.5$.

$$\Sigma x = n\mu + z\sqrt{n}\sigma$$

$$= 7200 + 1.5 \cdot \sqrt{80} \cdot 15$$

$$= 7401.25$$

Try It 5.5

The mean number of minutes for a tablet user is 8.2 minutes. Suppose the standard deviation is 1 minute. Take a sample size of 70. What is the probability that the sum of the sample is between 7 hours and 10 hours? What does this mean in the context of the problem?

Using the Central Limit Theorem

You need to understand when to use the central limit theorem. If we're being asked to find the probability of the mean, use the CLT for the mean. If we're being asked to find the probability of a sum or total, use the CLT for sums.

The law of large numbers says that if we take samples of larger and larger size from any population, then the mean $\bar{x}$ of the sample tends to get closer and closer to μ, the population mean. From the central limit theorem, we know that as n increases, the sample means follow a normal distribution. The larger n gets, the smaller the standard deviation because the standard deviation for $\bar{x}$ is $\frac{\sigma}{\sqrt{n}}$. This means that the sample mean $\bar{x}$ must be closer to the population mean μ. We can say that μ is the value the sample means approach as n increases. The central limit theorem illustrates the law of large numbers.

Several important discoveries come from using the central limit theorem and applying the law of large numbers. First, the probability density curve of the sampling distribution of means is normally distributed. Second, the standard deviation of the sampling distribution decreases as the size of the samples that were used to calculate the means for the sampling distribution increases. The central limit theorem is even included in the formula for standardizing from the sampling distribution to the standard normal distribution.

Figure 5.17 shows a sampling distribution. The mean has been marked on the horizontal axis of the $\bar{x}$'s, and the standard deviation has been written to the right above the distribution.

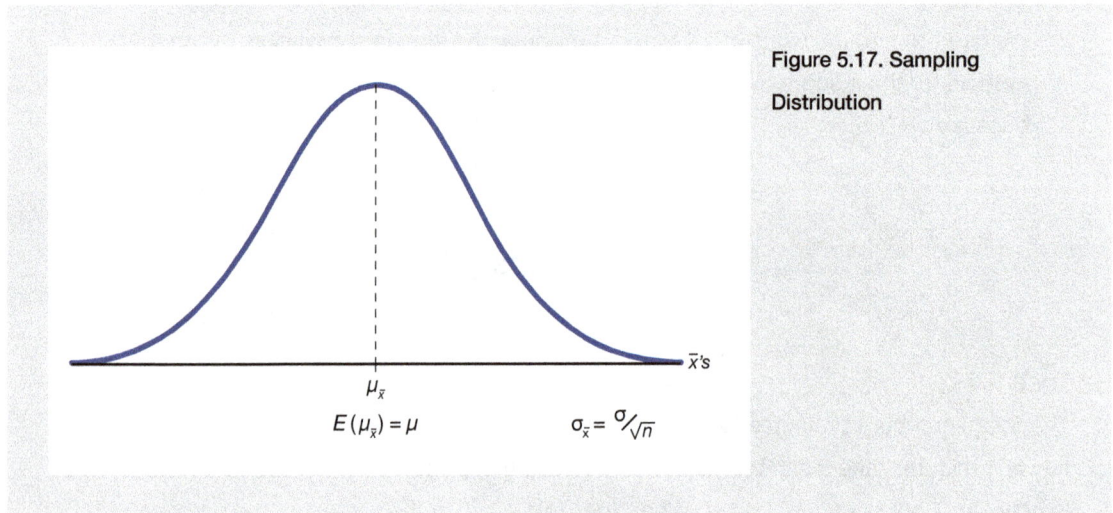

Figure 5.17. Sampling Distribution

$\mu_{\bar{x}}$

$E(\mu_{\bar{x}}) = \mu$

$\sigma_{\bar{x}} = \sigma/\sqrt{n}$

Notice that the sampling distribution's standard deviation is the population's original standard deviation divided by the square root of the sample size. As the sample size increases, the sampling distribution becomes closer and closer to the normal distribution. As this happens, the standard deviation of the sampling distribution changes in another way, too. The standard deviation decreases as n increases. With a very large n, the standard deviation of the sampling distribution becomes very small, and at infinity n, it collapses on top of the population mean. This shows that the expected value of μ_x is the population mean, μ.

Figure 5.18 shows two sampling distributions from the same population. One sampling distribution was created with samples of 10 and the other with samples of 50. All other things are the same.

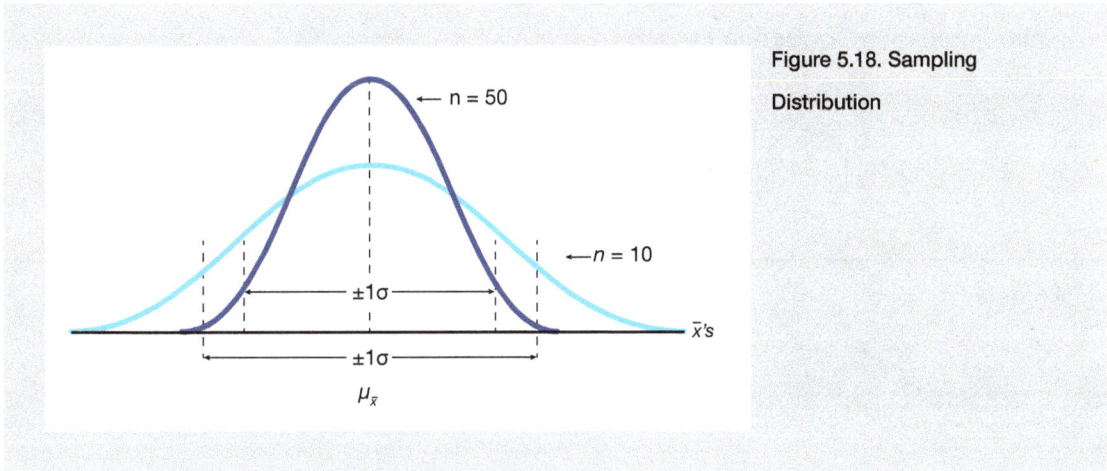

Figure 5.18. Sampling Distribution

The sampling distribution with a sample size of 50 has a smaller standard deviation that causes the graph to be higher and narrower. This distribution has a smaller range of possible values than the other distribution, with the same probability of one standard deviation from the mean. The two arrows mark one standard deviation on the $\bar{x}$ axis for each distribution. If we want to find the probability that the true mean is one standard deviation from the mean, the possible range of values is much greater for the sampling distribution with a sample size of 10.

We want the sample mean to come from the narrower distribution because fewer values are available. That means we're more likely to be closer to the true answer, even when estimating. That's why we should always look at the sample size of a statistic. If the sample size is larger, the sample mean comes from a more compact distribution, and we can be more confident in its accuracy. This concept will be the foundation for confidence intervals in Chapter 6.

Finite Population Correction Factor

We saw that the sample size has an important effect on the standard deviation of the sampling distribution. The proportion of the total population that has been sampled also has an important effect. Up to this point, we've assumed that the population is extremely large and that we've sampled a small part of it. But if the population is smaller and you sample a large part of it, then the samples are not independent of each other.

Here, we need to introduce the **finite population correction factor**. This factor adjusts the variance of the sampling distribution, giving a more accurate estimate of the population parameter when you take a large sample from a smaller, or finite, population. It is useful when more than 5% of the population is being sampled and the population has a known population size.

Finite Population Correction Factor

$$\sqrt{\frac{N-n}{N-1}}$$

where N is the population size, and n is the sample size. Use the correction factor if $n > 0.5N$.

Adjusted z-score

$$z = \frac{\bar{x} - \mu}{\frac{\sigma}{\sqrt{n}} \sqrt{\frac{N-n}{N-1}}}$$

Example 12

The population of white German Shepherds in the United States is 4,000 dogs, and the mean weight for German Shepherds is 75.45 pounds. The population standard deviation is 10.37 pounds.

The sample size is 250 dogs. Find the probability that a sample mean differs from the population mean by less than 2 pounds.

Solution 12

From the description of the question, we know the following:

$$N = 4000, \quad n = 250, \quad \sigma = 10.37, \quad \mu = 75.45$$

For the sample mean to be different from the true probability mean by less than 2 pounds, the following must be true:

$$\bar{x} - \mu = \pm 2$$

This can be interpreted as $\bar{X}$ varies between $75.45 - 2$ and $75.45 + 2$, or $73.45 < \bar{X} < 77.45$.

Now, we are ready to find the probability that the sample mean differs from the population mean by less than 2 pounds. That is written as:

$$P(73.45 < \overline{X} < 77.45)$$

Since $0.5N = 0.5(4000) = 2000$, and $n = 250 < 2000$, there is no need for finite population correction.

$$z = \frac{\overline{X} - \mu}{\frac{\sigma}{\sqrt{n}}} = \frac{\pm 2}{\frac{10.37}{\sqrt{250}}} = \pm 3.05$$

The probability $P(73.45 < \overline{x} < 77.45)$ is the same as the probability $P(-3.15 < z < 3.15)$.

We will use a TI-84 to find $P(-3.15 < z < 3.15)$:

$$\text{normalcdf}(-3.05, 3.05, 0, 1) \approx 0.998$$

The probability that a sample mean differs from the population mean by less than 2 pounds is about 0.998.

Exercises 5.4

1. Determine which of the following is a true statements. Justify your answers in complete sentences.
    a. When the sample size is large, the mean of $\overline{X}$ is approximately equal to the mean of X.
    b. When the sample size is large, $\overline{X}$ is approximately normally distributed.
    c. When the sample size is large, the standard deviation of $\overline{X}$ is approximately the same as the standard deviation of X.
2. Which of the following is false about the distribution for sample means of relatively large samples?
    a. The mean, median, and mode are equal.
    b. The area under the probability density curve is 1.
    c. The probability density curve never touches the x-axis.

    d.   The probability density curve is skewed to the right.

3.   Suppose that a category of world-class runners is known to run a marathon in an average of 145 minutes with a standard deviation of 14 minutes. Consider 49 marathons. Let $\overline{X}$ be the average of the 49 marathons.

    a.   Describe the distribution of $\overline{X}$.

    b.   Find the probability that the runner will average between 142 and 146 minutes in these 49 marathons.

    c.   Find the 80th percentile for the average of these 49 marathons.

    d.   Find the median of the average running times.

4.   The percentage of fat calories that a person in America consumes daily is normally distributed with a mean of about 36 and a standard deviation of about 10. Suppose that 16 individuals are randomly chosen. Let $\overline{X}$ = average percent of fat calories.

    a.   Describe the distribution of $\overline{X}$.

    b.   For the group of 16, find the probability that the average percent of fat calories consumed is more than 40. Graph the situation and shade in the area to be determined.

    c.   Find the first quartile for the average percent of fat calories.

5.   A large population of 5,000 students take a practice test to prepare for a standardized test. The population mean is 140 questions correct, and the standard deviation is 80. What size samples should a researcher take to get a distribution of means of the samples with a standard deviation of 10?

6.   A large population has skewed data, with a mean of 70 and a standard deviation of 6. Samples of size 100 are taken, and the distribution of the means of these samples is analyzed.

    a.   Will the distribution of the means be closer to a normal distribution than the distribution of the population?

    b.   Will the mean of the means of the samples remain close to 70?

    c.   Will the distribution of the means have a smaller standard deviation?

    d.   What is that standard deviation?

7.   A population has a standard deviation of 50. It is sampled with samples of size 100. What is the variance of the means of the samples?

8.   A company has 1,000 employees. The average number of workdays between sick days is 80, with a standard deviation of 11 days. Samples of 90 employees are examined. What is the probability a sample has a mean of workdays with no sick days between 65 and 82 days?

9.   Trucks pass an automatic scale that monitors the 2,000 trucks in a company's fleet. This population of trucks has an average weight of 20 tons with a standard deviation of 2 tons. If a sample of

50 trucks is taken, what is the probability the sample will have an average weight within one-half ton of the population mean?

10. A town keeps weather records. From these records, it has been determined that it rains on an average of 12% of the days each year with a standard deviation of 8 days. If 190 days are selected randomly from one year, what is the probability that the mean number of rainy days is no more than 44 days?

11. A school has 500 students. On average 20 students are absent daily with a standard deviation of 3 students per day. If 30 students are randomly selected on a certain day, what is the probability that on average at least 19 students will be absent?

Try It Solutions

Try It 5.1

1. a is 447, and b is 521. a is the minimum duration of games for a team for the 2011 season, and b is the maximum duration of games for a team for the 2011 season.
2. $X \sim U(447, 521)$.
3. $\mu = 484$, and $\sigma = 21.36$
4. Figure 5.19. Uniform Distribution for Sample
5. $P(480 < x < 500) = 0.2703$

Try It 5.2

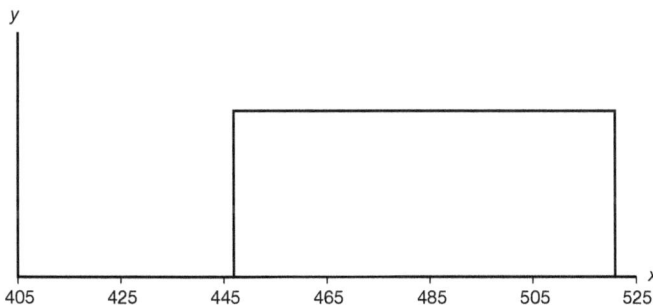

Figure 5.19. Uniform Distribution Sample

normalcdf(0,65,68,3) = 0.1587

Try It 5.3

normalcdf(66,70,68,3) = 0.4950

Try It 5.4

$P(42 < x < 50) = \left(42, 50, 45, \dfrac{8}{\sqrt{30}}\right)$

Try It 5.5

You're given $\mu_x = 8.2$, $\sigma_x = 1$, and $n = 70$

$(n)(\mu_x) = (70)(8.2) = 574$
$(n)(\sigma_x) = (70)(1) = 70$

$\Sigma \sim N(574, 8.367)$

Using TI- 84:

normalcdf (420, 600, 574, 8.367) = 0.999

So, $P(420 < \Sigma X < 600) = 0.999$

The likelihood that the sum of the sample spending between 7 hours and 10 hours on tablet has a very high chance of occurring with a probability of 0.999.

Chapter 6
Confidence Intervals

Melody needs to move to a bigger apartment and is trying to find the average rent of a two-bedroom apartment in her neighborhood. She looks at listings online, writes down the monthly rents of several apartments she likes, and averages the prices together. She found a mean, but it's from just one hour of browsing online. Is it an accurate picture of rent prices?

Austin wants to know the percentage of shots he makes playing basketball with his friends. At the Tuesday night pickup game, he counts the number of shots he makes and divides it by the number of shots he attempts. He found a proportion but was shooting well on Tuesday night, way better than normal. Does that one proportion value indicate he'll always shoot that well, even on an off night? You might think that this data doesn't show that rent is cheap in Melody's neighborhood or that Austin will likely be drafted into the NBA next year. You're right. However, Chapter 5 showed how sample means can help find population means by applying the central limit theorem.

In this chapter, we'll learn about **inferential statistics**, the science of using sample data to generalize about unknown populations. We'll learn how sample data can be used to estimate a population parameter and how to construct confidence intervals. We'll also learn two new distributions: the Student's-t distribution and the chi-square distribution. Using the appropriate distribution, we'll learn how to construct confidence intervals for a population mean, proportion, and variance.

After reading this chapter, you will be able to do the following:

1. Calculate and interpret confidence intervals for estimating a population mean
2. Calculate the sample size required to estimate a population mean given a desired confidence level and margin of error
3. Learn the properties of the Student's-t distribution
4. Calculate and interpret confidence intervals for estimating a population proportion
5. Understand the point estimate of population variance
6. Learn the properties of chi-square distributions
7. Calculate and interpret confidence intervals for estimating a population variance

6.1 Basic Concepts of Confidence Intervals

Overview

Melody and Austin want to know how accurate their data are, but Melody doesn't want to look up the rent for every two-bedroom in her neighborhood, and Austin wants to play basketball, not count shots. Confidence intervals are a tool to help them estimate the true population mean and population proportion with a certain confidence.

Confidence Interval Basics

The sample mean and sample proportion that Melody and Austin took are called point estimates. A **point estimate** is a sample statistic used to estimate a population parameter. A **confidence interval** is a range of values used to estimate a population parameter's true value. The confidence interval is constructed with the sample statistic at the center of the interval. A confidence interval is expected to capture the unknown population parameter with certain degrees of confidence.

The **confidence level (CL)** is the probability, expressed in a percentage, that the confidence interval contains the true population parameter. For example, if the CL = 90%, then in 90 out of 100 samples, the confidence interval estimate will contain the true population parameter. When constructing the confidence interval, choose a CL of 90% or higher because it's important to be certain of the conclusion.

The **significance level (α)** is the probability that the interval does *not* contain the unknown population parameter. It is the complement of the confidence level, so the confidence level is defined as $1 - \alpha$. For example, if the CL = 90%, then $\alpha = 10\%$ or 0.10. The difference between the point estimate and the actual population parameter is called an error, and the largest possible value of that error is called the **margin of error** (E) or the error bound.

Confidence intervals are calculated using the point estimate and the error bound. The sample mean ($\bar{x}$) is the best point estimate for the population mean (μ), The sample proportion ($\hat{p}$) is the best point estimate for the population proportion (p), and the sample standard deviation (s) is the best point estimate for the population standard deviation (σ). The error bound helps us understand how confident we can be in our results. It can be used to determine sample sizes or evaluate the quality of a statistical model. The error bound tells us how much we should expect our estimate or prediction to deviate from the true value.

The error bound for a population mean is the **error bound mean (EBM)**. The confidence interval estimate is $(\bar{x} - \text{EBM}, \bar{x} + \text{EBM})$. The error bound for a population proportion is the **error bound proportion (EBP)**. The confidence interval estimate is $(\hat{p} - \text{EBP}, \hat{p} + \text{EBP})$.

Constructing and Interpreting Confidence Intervals

Confidence intervals deal with probabilities and their distributions. It can be helpful to see the confidence interval to interpret it, so here we'll see how that works. Figure 6.1 shows a normal distribution with the confidence interval.

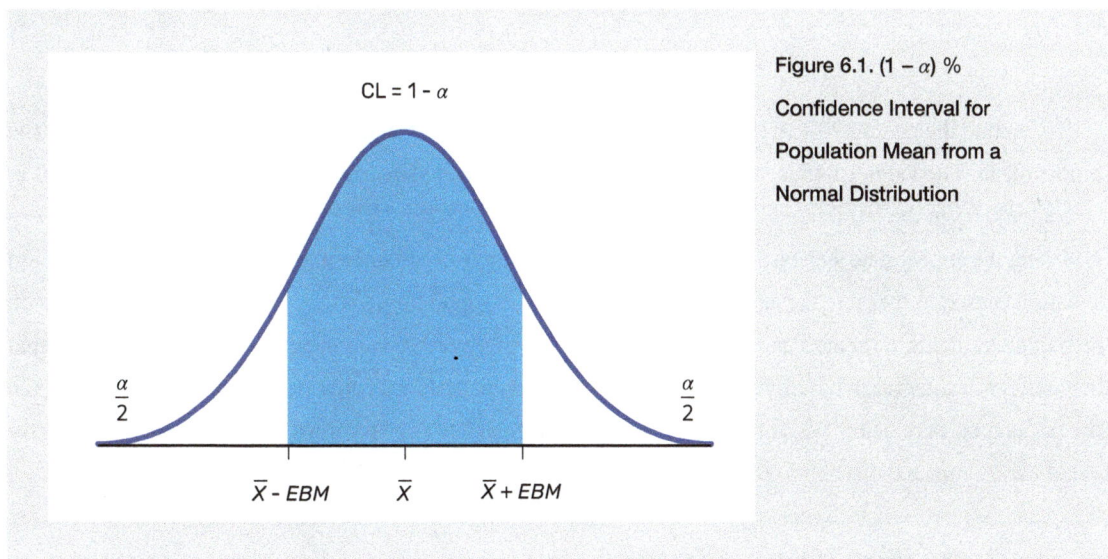

Figure 6.1. $(1 - \alpha)$ % Confidence Interval for Population Mean from a Normal Distribution

A confidence interval for a population mean with a known population standard deviation is based on the central limit theorem. Look at Figure 6.2. Suppose a sample has a mean of $\bar{x} = 10$, and we've constructed the 90% confidence interval (5, 15) where EBM = 5. To get a 90% confidence interval, we must include the central 90% of the probability of the normal distribution. By including that central 90%, we leave out a total of $\alpha = 10\%$ in both tails of the normal distribution, or 5% in each tail.

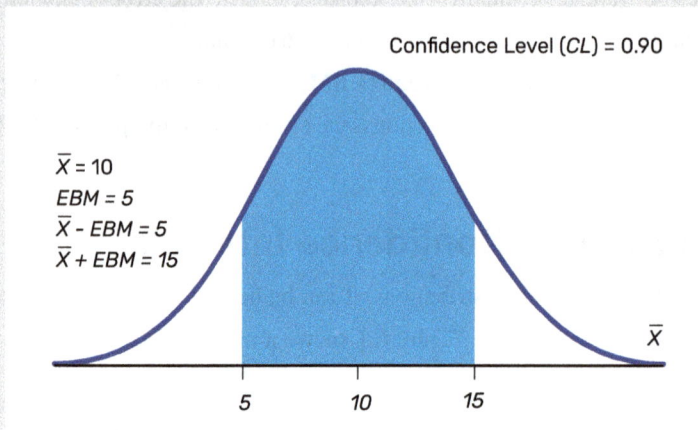

Figure 6.2. 90% Confidence Interval for Population Mean from a Normal Distribution

Confidence Level (CL) = 0.90

$\bar{X} = 10$
$EBM = 5$
$\bar{X} - EBM = 5$
$\bar{X} + EBM = 15$

To capture the central 90%, we must go out a certain number of standard deviations on either side of the sample mean. The value 1.645 is the z-score from a standard normal distribution that puts an area of 0.90 in the center, 0.05 in the left tail, and 0.05 in the right tail.

Using the appropriate standard deviation formula is a key to confidence interval estimates. The standard deviation is used to measure the amount of variability in the population, and the correct formula depends on the parameter being estimated and the type of data being analyzed. Using the correct formula ensures that the confidence interval is based on accurate statistical calculations and provides a reliable estimate of the true population parameter. When the population standard deviation σ is known, we can calculate the error bound using a normal distribution.

Error Bound Mean When σ is Known

$$EBM = z_{\frac{\alpha}{2}}\left(\frac{\sigma}{\sqrt{n}}\right)$$

The fraction $\frac{\sigma}{\sqrt{n}}$ is commonly called the standard error of the mean to distinguish the standard deviation σ.

Confidence Interval

$$\bar{x} - EBM < \mu < \bar{x} + EBM$$

In the error-bound formula, $z_{\frac{\alpha}{2}}$ is called the critical value. A **critical value** is the number that separates significant values from values that aren't. In this case, it's a z-score at the border between a normal distribution and its right-side tail.

The interpretation for confidence intervals should clearly state the confidence level, explain which population parameter is being estimated, and state both endpoints of the confidence interval. "I estimate with __% confidence that the true population mean [include the context of the problem] is between __ and __ [include appropriate units]."

Example 1

Suppose we're interested in the mean scores on a psychology final exam. A random sample of 36 scores gives a sample mean of 68. In this example, we have the unusual knowledge that the population standard deviation is 3.

Find a 90% confidence interval for the true population mean of psychology final exam scores. Interpret the confidence interval.

Solution 1

To find the confidence interval, we need the sample mean $\bar{x}$ and the EBM.

$$\bar{x} = 68$$
$$EBM = z_{\frac{\alpha}{2}}\left(\frac{\sigma}{\sqrt{n}}\right)$$

The confidence level $CL = 0.90$ and

$$\alpha = 1 - CL = 1 - 0.90 = 0.10$$

Find the critical value:

$$\frac{\alpha}{2} = 0.05 \quad \text{and} \quad z_{\frac{\alpha}{2}} = z_{0.05}$$

To find the critical value $z_{0.05}$ we use TI-84:

$$\text{InvNorm(0.05, 0, 1)} = -1.645$$

This is the z-score at the border between the normal distribution and its left-side tail. Likewise, the z-score at the border between a normal distribution and its right-side tail should be 1.645 because normal distribution is symmetric. So, the critical values are:

$$z_{0.05} = -1.645 \text{ and } 1.645$$

We can also find the critical values in the table for the standard normal distribution in Table A.1 in the Appendix. Because the common levels of confidence are 90%, 95%, and 99%, we will use their z-scores often and will probably remember them: 1.645, 1.96, and 2.56.

$$EBM = 1.645\left(\frac{3}{\sqrt{36}}\right) = 0.8225$$
$$\bar{x} - EBM = 68 - 0.8225 = 67.1775$$
$$\bar{x} + EBM = 68 + 0.8225 = 68.8225$$

The 90% confidence interval is (67.1775, 68.8225).

Write the interpretation of the confidence interval: We are 90% confident that the true population mean final exam score for all psychology students is between 67.18 and 68.82.

This means that 90% of all confidence intervals constructed this way contain the true mean psychology exam score. For example, if we constructed 100 of these confidence intervals, we expect 90 of them to contain the true population mean final exam score.

Example 2
Use the information in Example 1, find a 95% confidence interval for the true population mean psychology final exam score.

Solution 2

We know the sample statistics are $\bar{x} = 68$, $\sigma = 3$, and $n = 36$.

The confidence level is 95% ($CL = 0.95$):

$$\alpha = 1 - CL = 1 - 0.95 = 0.05$$

Find the critical value:

$$z_{\frac{\alpha}{2}} = z_{0.025} = 1.96$$

$$EBM = \left(z_{\frac{\alpha}{2}}\right)\left(\frac{\sigma}{\sqrt{n}}\right) = 1.96\left(\frac{3}{\sqrt{36}}\right) = .098$$

$$\mu = \bar{x} \pm 0.98$$

$$67.02 \le \mu \le 68.98$$

We are 95% confident that the true population mean final exam score for all psychology students is between 67.02 and 68.98.

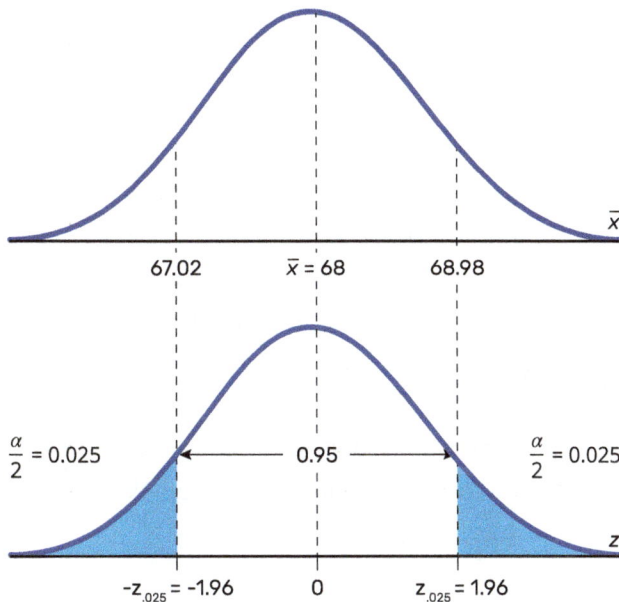

Figure 6.3. The 95% Confidence Interval for Population Mean from a Normal Distribution with $\bar{x} = 68$, $\sigma = 3$, and $n = 36$

Notice that the EBM is larger for a 95% confidence level in Example 2 than a 90% confidence level in Example 1. How should we interpret that information? Figure 6.4 shows the differences. The shape of the distribution is the same, but the unshaded areas in the right and left tails indicate different areas.

Example 1 is illustrated by Figure 6.4a, showing the 90% confidence interval of (67.18, 68.82), or 1.64 points apart. Example 2 is illustrated by Figure 6.4b, showing the 95% confidence interval of (67.02, 68.98), or 1.96 points apart. The 95% confidence interval is wider. The confidence interval necessarily needs to be wider to make sure that it contains the true value of the population mean for all psychology final exam scores.

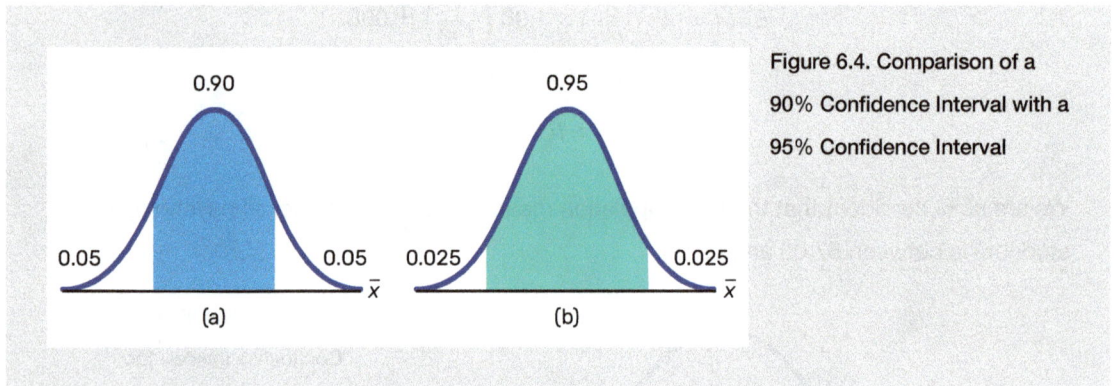

Figure 6.4. Comparison of a 90% Confidence Interval with a 95% Confidence Interval

This demonstrates an important principle of confidence intervals. There is a tradeoff between the level of confidence and the width of the interval. Increasing the confidence level makes the confidence interval wider, but decreasing the confidence level makes the confidence interval narrower. If the confidence interval is wider, we're getting less useful information. But it is also important to have a high level of confidence in the interval, like 95% instead of 90%. This is a decision that statisticians weigh all the time.

Example 3

Change the problem in Example 1 to see what happens to the confidence interval if the *sample size* is changed. Leave everything else the same except the sample size. Use the original 90% confidence level.

1. What happens to the confidence interval if we increase the sample size and use $n = 100$ instead of $n = 36$?
2. What happens if you decrease the sample size to $n = 25$ instead of $n = 36$?

Solution 3

1. Using the new sample size $n = 100$, we find the EBM and calculate the confidence interval:

$$EBM = \left(z_{\frac{a}{2}}\right)\left(\frac{\sigma}{\sqrt{n}}\right) = 1.645\frac{3}{\sqrt{100}} = 0.4935$$

$$\mu = \bar{x} \pm 0.4935$$

$$67.5064 \leq \mu \leq 68.4935$$

If we increase the sample size n to 100, we decrease the width of the confidence interval relative to the original sample size of 36 observations.

2. Using the new sample size $n = 25$, we find the EBM and calculate the confidence interval:

$$EBM = \left(z_{\frac{a}{2}}\right)\left(\frac{\sigma}{\sqrt{n}}\right) = 1.645\frac{3}{\sqrt{25}} = .0.987$$

$$\mu = \bar{x} \pm 0.987$$

$$67.013 \leq \mu \leq 68.987$$

If we decrease the sample size to 25, we increase the width of the confidence interval by comparison to the original sample size of 36 observations.

When constructing a confidence interval we use the sample mean and error bound to calculate the confidence interval. If the confidence interval is known, we can work backward to find both the error bound and the sample mean. There are two ways to do this, and you can choose the method that is easier to use according to the information.

Finding the EBM: 1) Subtract the sample mean from the upper value for the interval. Divide the difference by two. 2) Subtract the lower value from the upper value for the interval. Divide the difference by two.

Finding the Sample Mean: 1) Subtract the error bound from the upper value of the confidence interval. 2) Average the upper and lower endpoints of the confidence interval.

Example 4

We know that a confidence interval estimate for the population mean is (67.18, 68.82). Find the error bound if the sample mean is 68. Find the error bound if the sample mean is unknown.

Solution 4

Calculate the Error Bound

If we know that the sample mean is 68, then:

$$EBM = 68.82 - 68 = 0.82$$

If we don't know the sample mean, then:

$$EBM = \frac{(68.82 - 67.18)}{2} = 0.82$$

Calculate the Sample Mean

If we know the error bound, then:

$$\bar{x} = 68.82 - 0.82 = 68$$

If we don't know the error bound, then:

$$EBM = \frac{(67.18 + 68.82)}{2} = 68$$

Determining Sample Size

In the same way that we can work backward from the EBM or sample mean to find the other, we can use the error-bound formula to calculate the required sample size. A researcher planning a study who wants a specified confidence level and error bound can use this formula to calculate the size of the sample needed for the study.

Sample Size

$$n = \frac{\left(z_{\frac{\alpha}{2}}\right)^2 \sigma^2}{EBM^2}$$

Example 5

The population standard deviation for the age of Foothill College students is 15 years. If we want to be 95% confident that the sample mean age is within 2 years of the true population mean age of Foothill College students, how many randomly selected Foothill College students need to be surveyed?

Solution 5

From the problem, you know that $\sigma = 15$ and EBM = 2 years.

We want a confidence level of 95%. Use the sample size equation:

$$z_{\frac{\alpha}{2}} = z_{0.025} = 1.96$$

$$n = \frac{\left(z_{\frac{\alpha}{2}}\right)^2 \sigma^2}{EBM^2} = \frac{(1.96^2)(15^2)}{2^2} = 216.09$$

Always round the answer up to the next higher integer to ensure the sample size is large enough. Therefore, $n = 217$.

217 Foothill College students should be surveyed to be 95% confident that we are within 2 years of the true population mean age of Foothill College students.

Try It 6.1

The population standard deviation for tuna consumption in the United States is 0.94 pounds per person per year. If you want to be 90% confident that the sample mean tuna consumption per person is within 0.1 pounds per year of the true population mean, how many randomly selected people need to be surveyed?

Exercises 6.1

1. A sample of 16 small bags of the same brand of candies was selected. The weight of each bag was then recorded. The mean weight was two ounces with a standard deviation of 0.12 ounces. The population standard deviation is known to be 0.1 ounce. Assume that the population distribution of bag weights is normal.

   a. In words, define the random variable X.

   b. In words, define the random variable $\overline{X}$.

   c. Which distribution should you use for this problem? Explain your choice.

   d. Construct a 90% confidence interval for the population mean weight of the candies.

2. Announcements for 84 upcoming engineering conferences were randomly picked from a stack of *IEEE Spectrum* magazines. The mean sample length of the conferences was 3.94 days, with the population standard deviation of 1.28 days. Assume the underlying population is normal.

   a. In words, define the random variables X and $\overline{X}$.

   b. Which distribution should you use for this problem? Explain your choice.

   c. Construct a 95% confidence interval for the population mean length of engineering conferences.

      i. State the confidence interval.

      ii. Calculate the error bound.

3. What is meant by the term 90% confident when constructing a confidence interval for a mean?

   a. If we took repeated samples, approximately 90% of the samples would produce the same confidence interval.

   b. If we took repeated samples, approximately 90% of the confidence intervals calculated from those samples would contain the sample mean.

   c. If we took repeated samples, approximately 90% of the confidence intervals calculated from those samples would contain the true value of the population mean.

   d. If we took repeated samples, the sample mean would equal the population mean in approximately 90% of the samples.

4. The American Community Survey (ACS), part of the US Census Bureau, conducts a yearly census similar to the one taken every ten years but with a smaller percentage of participants. The most recent survey estimates with 90% confidence that the mean household income in the United States falls between $69,720 and $69,922. Find the point estimate for mean U.S. household income and the error bound for mean U.S. household income.

5. The average height of young adult males has a normal distribution with a standard deviation of 2.5 inches. You want to estimate the mean height of students at your college or university to within 1 inch with 93% confidence. How many male students must you measure?

6.2 Estimating a Population Mean When σ Is Unknown

Overview

In practice, we rarely know the population standard deviation σ. With a large sample size, an unknown σ isn't a problem for statisticians. We use the sample standard deviation s as an estimate for σ and calculate a confidence interval with close enough results. However, statisticians run into problems when the sample size is small because a small sample size produces inaccuracies in the confidence interval.

The Student's t-Distribution

William S. Gosset of the Guinness Brewery in Dublin, Ireland, ran into this problem in the early 1900s. He was the Head Experimental Brewer at Guinness, and his experiments with hops and barley produced tasty beverages but very few samples. Simply replacing σ with s didn't produce accurate results when he tried to calculate a confidence interval. He realized that he could not use a normal distribution for the calculation and found that the actual distribution depends on the sample size. This led him to "discover" the student's t-distribution. The name comes from the fact that Gosset wrote under the pen name "A Student."

The **student's t-distribution** is a probability function good at estimating population parameters for small sample sizes and when the population standard deviation is unknown. If you draw a simple random sample of size n from a population with mean μ and unknown population standard deviation σ, you can calculate the t-score with the following formula.

Student's t-score

$$t = \frac{\bar{x} - \mu}{\left(\frac{s}{\sqrt{n}}\right)}$$

The t-scores follow a Student's t-distribution with $n - 1$ degrees of freedom where n = the sample size, s = sample standard deviation, $\bar{x}$ = sample mean, and μ = population variance.

The t-score has the same interpretation as the z-score. It measures how far $\bar{x}$ is from its mean μ in terms of the standard deviation. A t-distribution is determined by the degrees of freedom. The **degrees of freedom (df)** is the number of values in a data set that can vary freely without influencing the overall outcome of the statistical analysis. For a t-distribution, $df = n - 1$.

Constructing a Confidence Interval

A *t*-distribution has one of the most desirable properties of a normal distribution: it is symmetrical around μ. Like the normal distributions, there are an infinite number of *t*-distributions, one for each adjustment to the sample size (*n*). As the sample size increases to 30 or beyond, the *t*-distribution becomes more and more like the normal distribution. This relationship between the *t*-distribution and the normal distribution is shown in Figure 6.5. There are two requirements for using the *t*-distribution to construct a confidence interval:

1. The sample is a simple random sample.
2. The population is normally distributed or sample size $n > 30$.

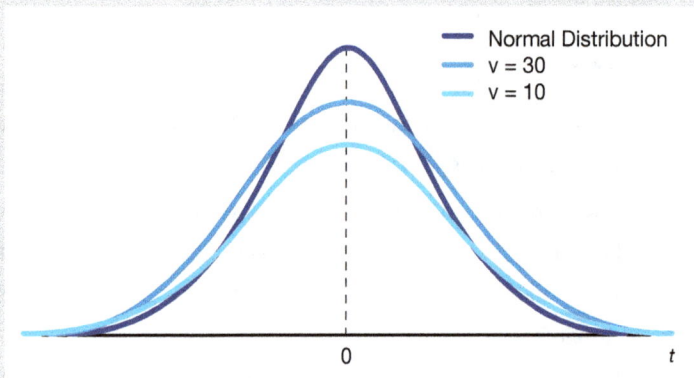

Figure 6.5. Comparing *t*-distributions with Sample Sizes of 10 and 30 to the Standard Normal Distribution

---

Error Bound Mean When σ Is Unknown

$$EBM = \left(t_{\frac{\alpha}{2}}\right)\left(\frac{s}{\sqrt{n}}\right)$$

The critical value $t_{\frac{\alpha}{2}}$ is the *t*-score. s = sample standard deviation, and *n* is the sample size.

Confidence Interval

$$\bar{x} - EBM < \mu < \bar{x} + EBM$$

Requirement

The sample must be a random sample

Example 6

The average earnings per share (EPS) for 10 industrial stocks randomly selected from those listed on the Dow-Jones Industrial Average (DJIA) is $1.85 with a standard deviation of $0.395. Calculate a 99% confidence interval for the average EPS of all the industrial stocks listed on the DJIA. Assume that DJIA industrial stocks have a normal distribution.

Solution 6

We know the sample mean $\bar{x} = 1.85$, the sample standard deviation $s = 0.395$, and $df = 9$. We don't know the population standard deviation. We do know that the population has a normal distribution, so we can use the t-distribution for the confidence interval.

The 99% confidence level corresponds to the significance level $\alpha = 0.01$, meaning the area of each tails is $\frac{\alpha}{2} = 0.005$, as shown in Figure 6.6.

Using TI-84 inverse t-distribution function $invT(\textit{left tail area, df})$ to find the critical value:

$$invT(0.005, 9) = -3.2498$$

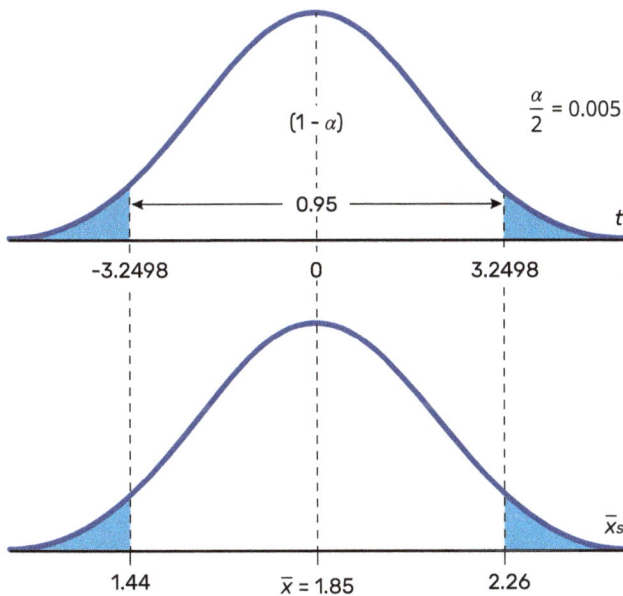

Figure 6.6. 99% Confidence Interval for Population Mean from the t-distribution with $\bar{x} = 1.85$ and $n = 10$

This is the t-score at the border between a t-distribution and its left tail. Likewise, the t-score at the border between a t-distribution and its right tail should be 3.2498 because the t-distribution is symmetric. So, the critical values are:

$$t_{0.005} = -3.2498 \ and \ 3.2498$$

We can also use the t-table in the Appendix to find the t-scores corresponding to the confidence level and degrees of freedom.

$$EBM = \left(t_{\frac{\alpha}{2}}\right)\left(\frac{s}{\sqrt{n}}\right) = (3.248)\left(\frac{0.395}{\sqrt{10}}\right) = 0.4057$$

The confidence interval for μ:

$$\bar{x} - EBM < \mu < \bar{x} + EBM$$
$$1.445 < \mu < 2.257$$

The 99% confidence interval for the population mean is ($1.45, $2.26).

Now, write the interpretation of the confidence interval.

We are 99% confident that the true population mean EPS of all the industrials listed on the DJIA is between $1.45 and $2.26.

Example 7

A study is done to determine how effective hypnotherapy is in increasing the hours of sleep subjects get each night. We measure hours of sleep for 12 participants with the following results.

| 8.2 | 9.1 | 7.7 | 8.6 | 6.9 | 11.2 | 10.1 | 9.9 | 8.9 | 9.2 | 7.5 | 10.5 |

Construct a 95% confidence interval for the mean number of hours slept for the population. Assume a normal distribution.

Solution 7

Since $n = 12$ is less than 30, the population standard deviation is unknown, and the normal distribution for the population is assumed, we will use the t-distribution to construct the confidence interval for μ.

We will demonstrate using both the TI-84 and R.

TI-84:

Enter data in STAT, EDIT, and under L1

Go to TEST, choose Tinterval, select Data

Using R:

Enter data in the vector form, and use R function: t.test(x, Conf.levl =)

```
> Hours <- c(8.2, 9.1, 7.7, 8.6, 6.9, 11.2, 10.1, 9.9, 8.9, 9.2, 7.5, 10.5)
> t.test(Hours, conf.level = 0.95)
```

R Display

```
                       data: Hours

 t = 24.116, df = 11, p-value = 7.119e-11
 alternative hypothesis: true mean is not equal to 0

 95 percent confidence interval:
 8.163448 9.803219

 Sample estimates:
 mean of x
 8.983333
```

The results from both TI-84 and R show that the 95% confidence interval for the population mean is between 8.16 hours and 9.8 hours. We are 95% confident that the true population mean number of hours of sleep is between 8.16 hours and 9.8 hours.

The R function (t.test(X. Conf.levl =) displays the mean of a random variable X and the confidence interval estimate for the population mean based on the t-distribution. It also displays the t-test result for the null hypothesis $\mu = 0$, which we will study in the next chapter.

Try It 6.2

We want to know how much glass garbage American families generate each year. A survey of 25 American families found that they generate an average of 37.2 pounds of glass garbage each year with a standard deviation of 4.5 pounds.

1. Construct a 95% confidence interval for the population's mean number of pounds of glass garbage.
2. Interpret the meaning of the confidence interval.

Exercises 6.2

1. A random survey of enrollment at thirty-five community colleges across the United States yielded the following figures:

6414	1550	2109	9350	21828	4300	5944	5722	2825	2044	5481
5200	5853	2750	10012	6357	27000	9414	7681	3200	17500	9200
7380	18314	6557	13713	17768	7493	2771	2861	1263	7285	28165
5080	11622									

Assume the underlying population is normal.

    a. Define the random variables X and $\overline{X}$ in words.

    b. Calculate s_x.

    c. Which distribution should you use for this problem? Explain your choice.

    d. Construct a 95% confidence interval for the population mean enrollment at community colleges in the United States.

       i. State the confidence interval.

       ii. Sketch the graph.

       iii. Calculate the error bound.

e. What will happen to the error bound and confidence interval if 500 community colleges are surveyed? Why?

2. A pharmaceutical company makes a drug used during surgery. It is assumed that the distribution for the length of time the drug is effective is approximately normal. Researchers in a hospital used the drug on a random sample of 9 patients. The effective period of the drug for each patient, in hours, was as follows:

$$2.7 \quad 2.8 \quad 3.0 \quad 2.3 \quad 2.3 \quad 2.2 \quad 2.8 \quad 2.1 \quad 2.4$$

a. Define the random variables X and $\bar{X}$ in words.

b. Calculate s_x.

c. Which distribution should you use for this problem? Explain your choice.

d. Construct a 95% confidence interval for the population mean length of time.

i. State the confidence interval.

ii. Calculate the error bound.

e. What does it mean to be 95% confident in this problem?

3. Unoccupied seats on flights cause airlines to lose revenue. Suppose a large airline wants to estimate its mean number of unoccupied seats per flight over the past year. To accomplish this, the records of 225 flights are randomly selected, and the number of unoccupied seats is noted for each of the sampled flights. The sample mean is 11.6 seats, and the sample standard deviation is 4.1 seats. Construct a 92% confidence interval for the population mean number of unoccupied seats per flight.

4. In a recent sample of 84 used car sales prices, the sample mean was $6,425, with a standard deviation of $3,156. Assume the underlying distribution is approximately normal.

a. Which distribution should you use for this problem? Explain your choice.

b. Construct a 95% confidence interval for the population mean cost of a used car.

c. Explain what a 95% confidence interval means for this study.

5. Six different national brands of chocolate chip cookies were randomly selected at the supermarket. The grams of fat per serving are as follows: 8, 8, 10, 7, 9, 9. Assume the underlying distribution is approximately normal.

a. Construct a 90% confidence interval for the population mean grams of fat per serving of chocolate chip cookies sold in supermarkets.

b. If you wanted a smaller error bound while keeping the same level of confidence, what should have been changed in the study before it was done?

6.3 Estimating a Population Proportion

Overview

During an election year, you hear about confidence intervals in terms of proportions or percentages all the time. For example, a poll for candidate A might show that 40% of voters say they'll vote for her, with a margin of error of 3 percentage points. Election polls are usually calculated with 95% confidence, so the pollsters would be 95% confident that the true proportion of voters who favored candidate A would be between 0.37 and 0.43. The procedure to find the confidence interval for a population proportion is like that for the population mean, but the formulas are different.

Constructing A Confidence Interval

How do you know you are dealing with a proportion problem? First, the underlying distribution has a binomial distribution. In Chapter 4, we learned that binomial distributions have only two possible outcomes: success or failure. Remember the binomial probability function: if the binomial random variable X represents the number of successes, then $X \sim B(n, p)$ where n is the number of trials and p is the probability of a success.

A **proportion** or percentage is a ratio of the subgroup to the entire group. For example, if there are 24 pets at an animal shelter, and 17 of them are cats, then the proportion of cats at the shelter is $\frac{17}{24}$. To form a sample proportion $\hat{p}$, said as "p-hat," divide the number of successes x by the sample size or the number of trials n. The sample proportion is written as $\hat{p} = \frac{x}{n}$.

When n is large and p is not close to 0 or 1, we can use the normal distribution to approximate the binomial. Recall in section 5.3, $X \sim B(n, p)$ can be approximate by a normal distribution whose mean and standard deviation are the same as the binomial mean np and standard deviation $\sqrt{npq}$ as the following:

$$X \sim N(np, \sqrt{npq})$$

Dividing by n, we get

$$\frac{X}{n} \sim N\left(\frac{np}{n}, \frac{\sqrt{npq}}{n}\right)$$

Where the random variable, $\frac{X}{n}$ is the proportion. After simplifying the above formula, the sample proportion $\hat{p}$ becomes

$$P \sim N\left(\hat{p}, \sqrt{\frac{\hat{p}\hat{q}}{n}}\right)$$

So, P follows a normal distribution where n is the sample size, x is the number of successes, $\hat{p}$ is the sample proportion, and $\hat{q} = 1 - \hat{p}$. The confidence interval estimate for the population proportion p is $(\hat{p} - \text{EBP}, \hat{p} + \text{EBP})$.

Error Bound for Proportion

$$\text{EBP} = \left(z_{\frac{\alpha}{2}}\right)\left(\sqrt{\frac{\hat{p}\hat{q}}{n}}\right)$$

where $\hat{p}$ is the sample statistic and proportion of successes, and $\hat{q}$ is the proportion of failure, $\hat{q} = 1 - \hat{p}$

Confidence Interval

$$\hat{p} - \text{EBP} < p < \hat{p} + \text{EBP}$$

Requirement

1. The sample is a simple random sample.
2. The conditions for binomial distribution are satisfied.
3. $np \geq 5$ and $nq \geq 5$

If a symbol appears over a letter, like $\bar{x}$ or $\hat{p}$, we're dealing with a sample statistic. We're dealing with a population parameter if it's only a letter, like x or p. Therefore, $\hat{p}$ is a sample proportion, and p is the unknown population proportion. We use the sample proportion $\hat{p}$ to construct a confidence interval estimate for the unknown population proportion in which $\hat{p}$ is the point estimate for p. However, in the error bound proportion formula, we use $\sqrt{\frac{\hat{p}\hat{q}}{n}}$ as the standard deviation. In the EBP formula, the sample proportions $\hat{p}$ and $\hat{q}$ are estimates of the *unknown* population proportions p and q.

Example 8

A market research firm is hired to estimate the percentage of adults living in a large city who own smartphones. Five hundred randomly selected adult residents in this city are surveyed to determine whether they own smartphones. Of the 500 people sampled, 421 responded, "Yes, I own a smartphone." Construct a 95% confidence interval estimate for the true proportion of adult residents of this city who own smartphones.

Solution 8

We know that $n = 500$, and the number of successes $x = 421$. We need to find the sample proportion $\hat{p}$.

$$\hat{p} = \frac{x}{n} = \frac{421}{500} = 0.842$$

$$\hat{q} = 1 - \hat{p} = 1 - .0842 = 0.158$$

Checking the requirements, we find that the random sample of 500 satisfies the simple random sample requirement. The condition for binomial distribution is satisfied because there is a fixed number of trials (500), and the trials are independent of each other. The probability of success 0.842 is constant from one trial to another, and both $np = 421$ and $nq = 79$ are greater than 5.

Next, we need to construct the 95% confidence interval estimate for p. Since the requested confidence level is CL = 0.95, then

$$\alpha = 1 - CL = 1 - 0.95 = 0.05$$

$$\frac{\alpha}{2} = 0.025$$

$$z_{\frac{\alpha}{2}} = 1.96$$

$$EBP = \left(z_{\frac{\alpha}{2}}\right)\left(\sqrt{\frac{\hat{p}\hat{q}}{n}}\right) = 1.96\left(\sqrt{\frac{0.842 \cdot 0.158}{500}}\right) = 0.032$$

$$\hat{p} - EBP < p < \hat{p} + EBP$$

$$0.842 - 0.032 < p < 0.842 + 0.032$$

$$0.81 < p < 0.874$$

We are 95% confident that interval from 0.81 to 0.874 captures the true proportion of adult residents who own smartphones. We can also state that we are 95% confident that the population proportion of adult residents of this city who own smartphones is between 0.81 and 0.874.

Both TI-84 and R can calculate the confidence interval estimate for p.

For TI – 84, we use the function 1-PropZInt.

For R: we use the function "prop.test( )"

```
> prop.test(x=421, n=500, p=0.842)
```

R Display

```
data: 421 out of 500, null probability 0.842
X-squared = 2.5563e-30, df = 1, p-value = 1
alternative hypothesis: true p is not equal to 0.842

95 percent confidence interval:
 0.8074376, 0.8713473

sample estimates:
    p
0.842
```

Try It 6.3

Suppose 250 randomly selected people are surveyed to determine if they own a tablet. Of the 250 surveyed, 98 reported owning a tablet.

1. Using a 95% confidence level, compute a confidence interval estimate for the true proportion of people who own tablets.
2. Interpret the meaning of the confidence interval.

Example 9

The Dundee Dog Training School has a larger-than-average proportion of clients who compete in competitive professional events. A confidence interval for the population proportion of dogs that compete in professional events from 150 different training schools is constructed as (0.08, 0.16). Determine the confidence level used to construct the confidence interval of the population proportion.

Solution 9

The information from the problem shows the confidence interval to be $0.08 < p < 0.16$.

Find the sample proportion:

$$\hat{p} = \frac{0.08 + 0.16}{2} = 0.12$$
$$\hat{q} = 1 - 0.012 = 0.088$$

The method used in Section 6.1 for calculating EBM also applies to finding EBP. Just like the sample mean is a point estimate of the population mean, the sample proportion is a point estimate of the population proportion. Use the boundaries of the confidence interval to find the EBP:

$$EBP = \frac{0.16 - 0.08}{2} = 0.04$$

Solve for z:

$$EBP = z_{\frac{\alpha}{2}} \sqrt{\frac{\hat{p}\hat{q}}{n}}$$

$$z_{\frac{\alpha}{2}} \sqrt{\frac{(0.12)(0.88)}{150}} = 0.04$$

Find the critical value by solving for $z_{\frac{\alpha}{2}}$:

$$z_{\frac{\alpha}{2}} \cdot 0.02653 = 0.04$$

$$z_{\frac{\alpha}{2}} = \frac{0.04}{0.02653} = 1.5077$$

This is the z-score at the border between the normal distribution and its right tail of the area $\frac{\alpha}{2}$.

Use TI-84 to find the area:

$$\frac{\alpha}{2} = normalcdf\ (1.5077,\ 1E99,\ 0,\ 1)\ =\ 0.0658$$

$$\alpha = 2 \cdot 0.0658 = 0.1316$$

Since the significance level $\alpha = 0.1316$, the confidence level:

$$CL = 1 - \alpha = 1 - 0.1316 = 0.8684$$

Therefore, the confidence level used to construct the interval of the population proportion of dogs that compete in professional events is 0.8684 or about 86.90%.

Determining Sample Size

Usually, we have no control over the sample size of a data set. However, if we can set the sample size, it's helpful to know how large it should be to provide the most information. This is especially useful when conducting surveys. Sampling can be costly in both time and product. Knowing exactly how many responses we need before starting can save resources.

The formula for determining the sample size is a standardized version of the error bound for proportions formula.

Sample Size

When $\hat{p}$ is known

$$n = \frac{\left(z_{\frac{\alpha}{2}}\right)^2 (\hat{p}\,\hat{q})}{(EBP)^2}$$

When $\hat{p}$ is unknown

$$n = \frac{\left(z_{\frac{\alpha}{2}}\right)^2 (0.5)(0.5)}{(EBP)^2}$$

Because the binomial distribution is a single-parameter distribution, if we know $\hat{p}$ and $\hat{q}$ then the first formula can be used to find n. But since we rarely know the sample proportion $\hat{p}$ before selecting the sample, we chose 0.5 for $\hat{p}$ and 0.5 for $\hat{q}$ to find the largest possible sample size that gives the confidence level needed. In this case, the second formula should be used to calculate the sample size.

Example 10

A cell phone company wants to know the percentage of customers over 70 who text on their cell phones. How many customers over 70 should the company survey to be 90% confident that the sample proportion is within 3 percentage points of the true proportion of customers over 70 who text on their cell phones?

Solution 10

From the problem, we know that the EBP = 3% or 0.03, and the critical value $z_{\frac{\alpha}{2}} = z_{0.05} = 1.645$.

Without information about $\hat{p}$ or $\hat{q}$, we use the second formula. This gives a large enough sample to be 90% confident that we're within 3 percentage points of the true population proportion.

$$n = \frac{(z_{\frac{\alpha}{2}})^2 (0.5)(0.5)}{(EBP)^2} = \frac{1.645^2 \cdot 0.5 \cdot 0.5}{0.03^2} = 751.7 = 752$$

Round the answer to the next higher value. A simple random sample of at least 752 cell phone customers over 70 should be surveyed to be 90% confident that the sample proportion is within 3 percentage points of the true population proportion of all customers over 70 who use text messaging on their cell phones.

Exercises 6.3

1. When designing a study to determine the population proportion, what is the minimum number you would need for the survey to be 90% confident that the population proportion is estimated to within 0.05?

2. In 6 packages of multicolored fruit snacks, there were 5 red snack pieces. The total number of snack pieces in the 6 bags was 68. We wish to calculate a 95% confidence interval for the population proportion of red snack pieces.

    a. Define the random variables X and $\hat{p}$ in words.

    b. Which distribution should you use for this problem? Explain your choice.

    c. Calculate $\hat{p}$.

    d. Construct a 95% confidence interval for the population proportion of red snack pieces per bag.

        i.   State the confidence interval.

        ii. Calculate the error bound.

    e. Do you think that 6 packages of fruit snacks yield enough data to give accurate results? Why or why not?

3. On May 23, 2013, a polling group reported that of the 1,005 people surveyed, 76% of U.S. workers believe that they will continue working past retirement age.

    a. Determine the estimated proportion from the sample.

    b. Identify CL and α.

    c. Calculate the error bound based on the information provided.

    d. Construct a confidence interval estimate for the population proportion.

e. A reporter is covering the release of this study for a local news station. How should she explain the confidence interval to her audience?

4. In the previous question, the confidence level of study was reported at 95% with a ±3% margin of error.

   a. If you want to be 95% confident that your estimate of workers believing that they will continue working past retirement age is 3% within the true population proportion and you don't know or you don't have a value for sample proportion, how large of a sample size is required?

   b. Compare the error bound in part c to the ±3% margin of error reported by Gallup. Explain any differences between the values.

5. You plan to conduct a survey on your college campus to learn about the political awareness of students. You want to estimate the true proportion of college students on your campus who voted in the 2020 presidential election with 95% confidence and a margin of error no greater than five percent. How many students must you interview?

6.4 Estimating a Population Variance or Standard Deviation

Overview

Population variance is an important measure of the spread of population data. Statisticians calculate variance to determine how individual numbers in a data set relate to each other. There are many jobs and fields where variance is important, including manufacturing, finance, and health care. In this section, you'll learn about estimating the population variance and how to interpret a confidence interval.

Point Estimates of Population Variance and Standard Deviation

When estimating the population variance, the best point estimate for the population variance σ^2 is the sample variance (s^2) and the best point estimate for the population standard deviation σ is the sample standard deviation (s). The method for estimating the population variance requires using the chi-square distribution.

Chi-Square Distribution

Have you ever wondered if lottery numbers are evenly distributed or if some numbers occur more frequently? How about if the types of movies people like differ across age groups? What about if an automatic coffee machine dispenses approximately the same amount of coffee each time?

These are all questions of variance. The **chi-square distribution** tests questions about the variance of a normally distributed population based on sample statistics. "Chi" is pronounced "kigh" and is the Greek letter χ. The chi-square distribution is skewed to the right, meaning it has a long tail on the right side. However, as the degrees of freedom increase, the chi-square becomes more of a normal distribution in shape. In Figure 6.7a, the *df* is 2, but in Figure 6.7b, the *df* is 24. You can see how much the shapes change depending on the *df*.

The distribution is commonly used in categorical data analysis, as well as to construct confidence intervals for population variance and standard deviation. If independent samples are randomly selected from a normally distributed population with variance σ^2, the sample variances $s_1^2, s_2^2, s_3^2, \ldots$ have a chi-square distribution. The sample statistic χ^2 follows a chi-square distribution with $n-1$ degrees of freedom.

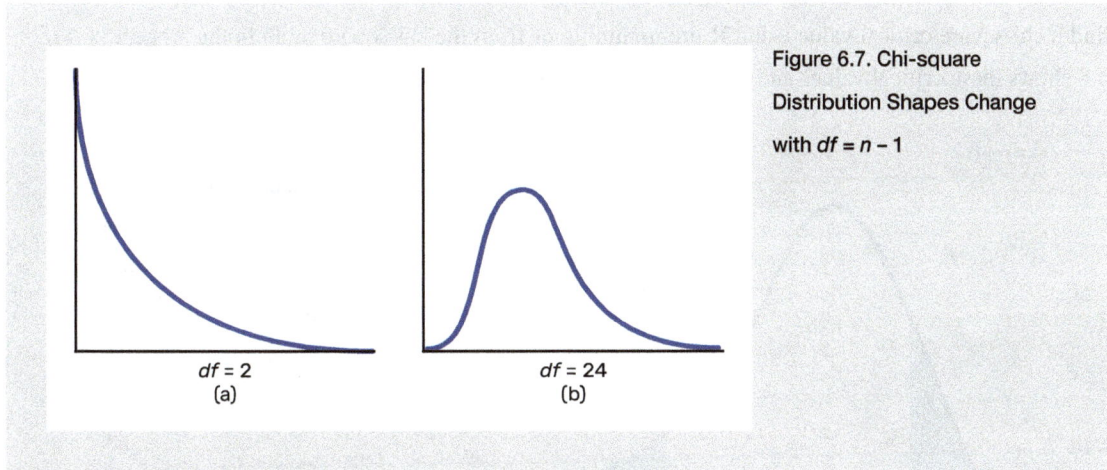

Figure 6.7. Chi-square Distribution Shapes Change with $df = n - 1$

$df = 2$
(a)

$df = 24$
(b)

Chi-Square Distribution

$$\chi^2 = \frac{(n-1)s^2}{\sigma^2}$$

where n = the sample size, s^2 = sample variance, and σ^2 = population variance.

There are several properties of a chi-square distribution to keep in mind:

1. All chi-square values are greater or equal to zero: $\chi^2 \geq 0$.
2. The area under each curve of the chi-square distribution equals 1.
3. Unlike the normal and Student's-t distributions, all chi-square distributions are positively skewed and asymmetric.

Constructing Confidence Intervals of Population Variance and Standard Deviation

Because chi-square distributions are not symmetric and positively skewed, there are two positive critical values for each confidence level. The value χ_L^2 represents the left-tail critical value and χ_R^2 represents the right-tail critical value. For different degrees of freedom and areas, we can use the `qchisq( )` function to find a chi-square critical value using R programming or from the chi-square table in the Appendix. Figure 6.8 shows these critical values and the chi-square distribution.

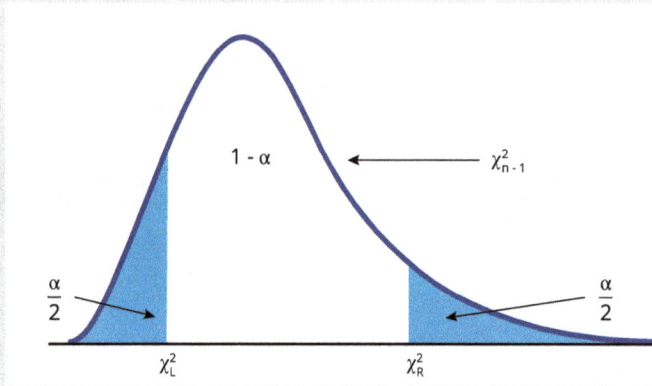

Figure 6.8. Chi-square Distribution with Citical Values χ_L^2, χ_R^2, and $(1 - \alpha)\%$ Confidence Interval

Confidence Interval for Population Variance

$$\frac{(n-1)s^2}{\chi_R^2} \leq \sigma^2 \leq \frac{(n-1)s^2}{\chi_L^2}$$

Confidence Interval for Standard Deviation

$$\sqrt{\frac{(n-1)s^2}{\chi_R^2}} \leq \sigma \leq \sqrt{\frac{(n-1)s^2}{\chi_L^2}}$$

where χ_L^2 = the left-tail critical value, χ_R^2 = the right-tail critical value, n = sample size, σ^2 = population variance, and s^2 = sample variance

Requirements

1. The sample is a simple random sample.
2. The population must be normally distributed.

Example 11

Find the critical values, χ_L^2 and χ_R^2, for a 90% confidence interval estimate of a normally distributed population variance. The sample size is 25.

Solution 11

We are given $n = 25$ and the confidence level = 90%. We have the following:

$$df = n - 1 = 24, \; \alpha = 0.1, \; and \; \tfrac{\alpha}{2} = 0.05$$

Using R we find the critical values χ_L^2 and χ_R^2 as follows:

> qchisq(p = 0.05, df = 24, lower.tail = FALSE)
p = 0.05 is equivalent to $\tfrac{\alpha}{2} = 0.05$

[1] 36.41503 -- $\chi_R^2 = 36.415$

> qchisq(p = 0.05, df = 24, lower.tail = TRUE)
[1] 13.84843 -- $\chi_L^2 = 13.848$

Note: the command lower.tail = FALSE returns the χ_R^2 value, and the command lower.tail = TRUE returns the χ_L^2 value.

Alternatively, using Table A.3, The Chi-Square Distribution table from the appendix, we find the critical values, χ_L^2 and χ_R^2, as follows:

» Locate 24 in the degrees of freedom column.
» From the top row, locate 0.05. we find the right critical value $\chi_R^2 = 36.415$. Again, with the same degrees of freedom column (24), locate from the top row 0.95, we find the left critical value $\chi_L^2 = 13.848$.

Example 12

The manager of a furnace manufacturer wants to assess the temperature variation in the control system. It's known that the temperatures are normally distributed. A random sample of 30 temperatures is taken over ten days. The sample variance s^2 is 15. Find a 95% confidence interval for the population temperature variance.

Solution 12

The question stated that the temperatures are normally distributed. The normal population requirement is satisfied. We know that $n = 30$ and $s^2 = 15$. A 90% confidence interval means the following:

$$\alpha = 0.1 \text{ and } \tfrac{\alpha}{2} = 0.05$$

Using the chi-square distribution table, locate 29 in the degree of freedom column, and from the top row, locate the two χ^2 values corresponding to 0.95 and 0.05. We find that

$$\chi_L^2 = 17.708 \text{ and } \chi_R^2 = 42.557$$

The two χ^2 values can also be found in R using the R function qchisq:

```
> qchisq(p = 0.05, df = 29, lower.tail = TRUE) returns χ² = 17.708
                                                          L
> qchisq(p = 0.05, df = 29, lower.tail = FALSE) returns χ² = 42.557
                                                          R
```

Find the confidence interval estimate for populations variance. The left confidence interval limit:

$$\frac{(n-1)s^2}{\chi_L^2} = \frac{(29)(15)}{17.708} = 24.565$$

The right confidence interval limit:

$$\frac{(n-1)s^2}{\chi_R^2} = \frac{(29)(15)}{42.557} = 10.221$$

The 90% confidence interval for the population variance temperature is (10.221, 24.565). We are 90% confident that population temperature variance is between 10.221 squared degrees and 24.565 squared degrees.

Try It 6.4

A manufacturer of washing machines wants to know the variation in replacement times for washing machines. It is known that the replacement times are normally distributed. A random sample of 28 replacement times were recorded. The sample standard deviation is 3.1 years. Find a 99% confidence interval for the standard deviation of the replacement times for the population of washing machines.

Exercises 6.4

1. Find the critical values χ_L^2 and χ_R^2 for a 95% confidence interval estimate of a normally distributed population variance when the sample size is 30.

2. A random sample of 40 steel ball bearings with a mean diameter of 10mm and standard deviation of 0.48 mm. Assuming that the diameters are normally distributed with unknown variance.
   a. Find a 95% confidence interval for the variance, σ^2.
   b. Find a 95% confidence interval for the standard deviation, σ.

3. The values listed below are waiting times in minutes of customers at a credit union bank. The customers wait in a single line that feeds into two teller windows. Construct a 90% confidence interval for the population variance, σ^2. Assuming that the waiting times are normally distributed with unknown variance.

   4.2   4.7   4.8   5.5   5.8   6.5   6.7   7.7   7.9   8.1   8.5   8.8

4. A statistician chooses 22 randomly selected dates, and when examining the recorded daily temperature of those dates, he finds a standard deviation of 2.56° F. If the degree of temperature is normally distributed, find the 95% confidence interval for the population standard deviation of the temperature. Assuming that the temperatures are normally distributed.

5. The mean replacement time for a random sample of 28 washing machines is 11.8 years and the standard deviation is 3.1 years. Find a 99% confidence interval for the population standard deviation of the replacement times of this type of washing machine. Assuming that replacement times are normally distributed.

Try It Solutions

Solution 6.1

$$z_{\frac{\alpha}{2}} = z_{0.05} = 1.645$$

$$n = \frac{z^2\sigma^2}{EBM^2} = \frac{(1.645^2)(0.94^2)}{0.1^2} = 239.1$$

$$n = 240$$

Solution 6.2

1. $n = 25$

   $df = 24$

$t_{\frac{\alpha}{2}} = t_{0.025} = 2.064$

$EBM = (t_{\frac{\alpha}{2}})(\frac{s}{\sqrt{n}}) = (2.064)(\frac{4.5}{\sqrt{25}}) = 1.8576$

You construct the confidence interval as follows:

$\bar{x} - EBM < \mu < \bar{x} + EBM = 35.342 < \mu < 39.058$

2. You are 95% confident that the true population mean number of pounds of glass garbage is between 35.342 and 39.058.

Solution 6.3

1. $\hat{p} = \frac{x}{n} = \frac{98}{250} = 0.392$ is the sample proportion.

$q = 1 - \hat{p} = 1 - 0.392 = 0.608$

$\alpha = 0.05, \frac{\alpha}{2} = 0.025$

$z_{\frac{\alpha}{2}} = 1.96$

$EBP = (z_{\frac{\alpha}{2}})(\sqrt{\frac{\hat{p}\hat{q}}{n}}) = 1.96\sqrt{\frac{(0.392)(0.608)}{250}} = 0.0605$

$\hat{p} - EBP < p < \hat{p} + EBP$

$0.392 - 0.0605 < p < 0.392 + 0.0605$

$0.3315 < p < 0.4525$

2. You are 95% confident that the true proportion of people who own tablets lies in the interval (0.3315, 0.4525).

Solution 6.4

$n = 28$

$s = 3.1$

$s^2 = 9.61$

$\alpha = 0.01$

You find with the Chi-square distribution that $\chi_L^2 = 11.808$ and $\chi_R^2 = 49.645$.

$\frac{(n-1)s^2}{\chi_L^2} = \frac{(27)(9.61)}{11.808} = 21.974$

$\frac{(n-1)s^2}{\chi_R^2} = \frac{(27)(9.61)}{49.645} = 5.227$

The 95% confidence interval for the population variance replacement times is (5.227, 21.974).

Chapter 7
Hypothesis Testing

Statisticians help us make statistical educated guesses, or inferences, about a population based on samples taken from it. In Chapter 6, we learned about confidence intervals, which are one way to estimate a population parameter. Another way to make a statistical inference is to decide whether a specific parameter is accurate. For instance, a car manufacturer advertises that its new small truck gets at least thirty-five miles per gallon on average. A tutoring service claims that its methods help 90% of its students earn better grades. These claims are hypotheses. A **hypothesis** is a claim made about a property of a population.

Maybe you've seen claims like these and thought, "Really?" A statistician's job is to determine whether these claims are accurate. This process is called hypothesis testing. A **hypothesis test** involves collecting data from a sample, evaluating and analyzing the data, and making a decision about the claim.

In this chapter, we conduct hypothesis tests on claims about a population mean and a population proportion. We learn different methods for a hypothesis test, its associated errors, what they mean, and why they matter.

After reading this chapter, you will be able to do the following:

1. Understand the difference between the null and the alternative hypotheses
2. Differentiate between Type I and Type II Errors
3. Understand the power of a test
4. Understand the basics of the hypothesis test
5. Determine the correct distribution to perform a hypothesis test
6. Perform a hypothesis test using two different approaches
7. Conduct and interpret hypothesis tests for a population mean when the population standard deviation is known
8. Conduct and interpret hypothesis tests for a population mean when the population standard deviation is unknown
9. Conduct and interpret hypothesis tests for a population proportion
10. Conduct and interpret hypothesis tests for a population standard deviation or variance

7.1 Basic Concepts of Hypothesis Testing

Overview

Statisticians use hypothesis testing to determine if data support a claim or idea. A hypothesis begins by writing two hypotheses that contain opposing viewpoints. They are called the null hypothesis and the alternative hypothesis. From there, we evaluate the data, make a decision based on the data, and write a conclusion.

Two Types of Hypotheses

Jamila is an analyst at a tutoring company, Academics R Us. Her boss is convinced that the tutors are doing such a great job that 95% of students are earning better grades —not 90% like the company's website currently claims. The process Jamila goes through to test the hypothesis goes like this: first, Jamila sets up two contradictory hypotheses. Then, she collects sample data and determines the best distribution to perform the hypothesis test. Then, she analyzes sample data by performing the calculations that ultimately will allow her to reject or fail to reject the null hypothesis. Finally, she writes a meaningful conclusion. Figure 7.1 shows this process in a flow chart.

Figure 7.1. Steps for Hypothesis Test

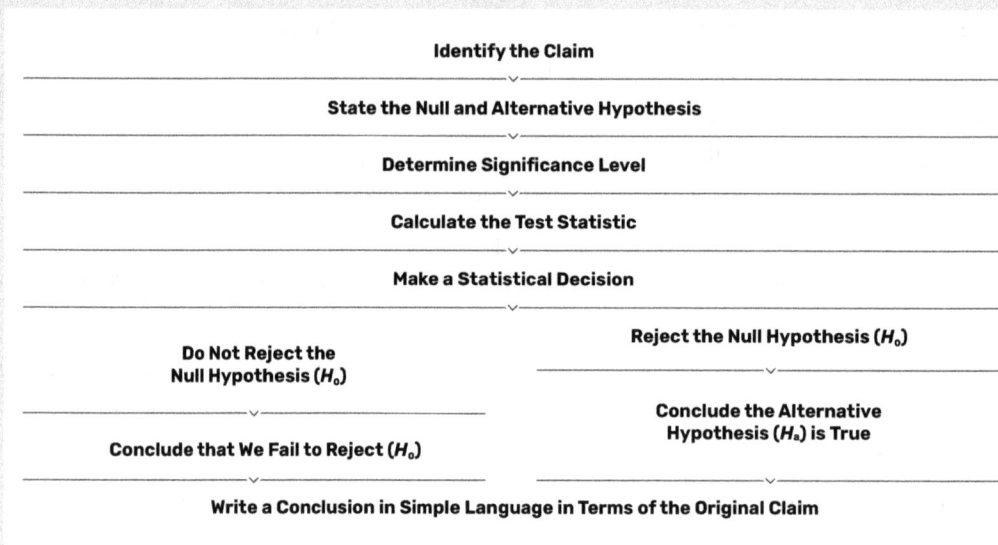

Identify the Claim

State the Null and Alternative Hypothesis

Determine Significance Level

Calculate the Test Statistic

Make a Statistical Decision

Do Not Reject the
Null Hypothesis (H_o)

Reject the Null Hypothesis (H_o)

Conclude that We Fail to Reject (H_o)

Conclude the Alternative
Hypothesis (H_a) is True

Write a Conclusion in Simple Language in Terms of the Original Claim

The **null hypothesis** (H_0) says no difference between groups or variables exists. Hypothesis testing is always about the null hypothesis, not the alternative hypothesis. There are two possible outcomes from a hypothesis test, and both outcomes are about the null hypothesis: either to reject the null hypothesis or fail to reject the null hypothesis. In Jamila's case, the null hypothesis is that Academics R Us's services help 90% of its students earn better grades.

The **alternative hypothesis** (H_a) is a claim about the population that contradicts H_0. It's what we conclude when we reject H_0. For Jamila, the alternative hypothesis is that Academics R Us's services don't help 90% of its students earn better grades.

A claim can either be a null hypothesis or an alternative hypothesis. Since the null and alternative hypotheses are at odds, we must examine the claim and evaluate sample data to reach a conclusion.

As we already know, there are two options for a decision. The first is "reject H_0" if the sample information favors the alternative hypothesis. The second is "don't reject H_0" or "fail to reject H_0" if the sample information is insufficient to reject the null hypothesis. These conclusions are all based upon a level of significance. When we can't reject H_0, it doesn't mean that H_0 is true. It simply means that the sample data have failed to provide enough evidence to doubt the truthfulness of H_0.

Table 7.1 presents the various hypotheses in pairs. For example, if the null hypothesis equals some value, the alternative must not equal that value.

Table 7.1. Null and Alternative Hypotheses Pairs

H_0	H_a
equal ($=$)	not equal ($\neq$)
greater than or equal to ($\geq$)	less than ($<$)
less than or equal to ($\leq$)	more than ($>$)

Note

As a mathematical convention, H_0 always has a symbol with an equal sign ($=$). H_a never has a symbol with an equal sign. The choice of symbol depends on the wording of the hypothesis test.

The statements in Examples 1–3 refer either to the null hypothesis or the alternate hypothesis. State the null hypothesis (H_0), and the alternative hypothesis (H_a), using the appropriate parameter.

Example 1

1. *No more than* 30% of the registered voters in Santa Clara County voted in the primary election.
2. *More than* 30% of the registered voters in Santa Clara County voted in the primary election.

Solution 1

1. "No more than" means "at most." It's also equivalent to "less than or equal to." The symbol for this phrase is ($\leq$). Since the symbol ($\leq$) contains an equal sign (=), the statement is a null hypothesis.

$$H_0: p \leq 0.3$$

2. The symbol for the phrase "more than" is (>). Since the symbol (>) doesn't contain an equal sign (=), the statement is an alternative hypothesis.

$$H_a: p > 0.3$$

Example 2

You want to test whether the mean GPA of students in American colleges is *different from* 2.0 (out of 4.0). Write the null and alternative hypotheses.

Solution 2

"Different from" means "not the same." The symbol for this phrase is ($\neq$), which doesn't contain an equal sign. The statement is an alternative hypothesis.

$$H_0: \mu = 2.0$$
$$H_a: \mu \neq 2.0$$

Example 3

You want to test if, on average, college students take *less than* 5 years to graduate. Write the null and alternative hypotheses.

Solution 3

The symbol for "less than" is (<), which doesn't contain an equal sign, so this is an alternative hypothesis. The null hypothesis must contain the symbol for "greater than or equal to" (≥), as seen in Table 7.1.

$$H_0: \mu \geq 5$$
$$H_a: \mu < 5$$

Significance Levels and Test Statistics

The **significance level** α is the probability value determining when the sample evidence is strong enough to reject the null hypothesis. Common values for α are 0.10, 0.05, and 0.01—which probably look familiar from working with confidence intervals.

A **test statistic** is a value that helps you decide about the null hypothesis. It's like a z-score or t-score. There are specific test statistics for each parameter we test. Table 7.2 is a quick reference for each parameter, sampling distribution, sample size, and corresponding test statistic. We'll learn more about these formulas later in the chapter.

Table 7.2. Test Statistics for Parameters

Parameter	Test Statistic Formula	Sampling Distribution	Requirements
μ (mean)	$z = \dfrac{\bar{x} - \mu}{\frac{\sigma}{\sqrt{n}}}$	Normal (z)	σ is known $n > 30$ or population is normally distributed
μ (mean)	$t = \dfrac{\bar{x} - \mu}{\frac{s}{\sqrt{n}}}$	Student t-distribution	σ is unknown $n > 30$ or population is normally distributed
p (proportion)	$z = \dfrac{\hat{p} - p}{\sqrt{\frac{pq}{n}}}$	Normal (z)	$np \geq 5$ and $nq \geq 5$
σ (standard deviation) σ^2 (variance)	$\chi^2 = \dfrac{(n-1)s^2}{\sigma^2}$	Chi-square (χ^2)	Any n as long as the population is normally distributed

Exercises 7.1

1. Some of the following statements refer to the null hypothesis, and some to the alternate hypothesis. State the null hypothesis, H_0, and the alternative hypothesis, H_a, in terms of the appropriate parameter (μ or p).

    a. The mean number of years Americans work before retiring is 34.

    b. At most, 60% of Americans vote in presidential elections.

    c. Twenty-nine percent of high school seniors get drunk each month.

    d. Fewer than 5% of adults ride the bus to work in Portland, Oregon.

    e. The mean number of cars a person owns in her lifetime is not more than ten.

    f. About half of Americans prefer to live away from cities, given the choice.

    g. Europeans have a mean paid vacation each year of six weeks.

    h. The chance of developing breast cancer is under 11% for women.

2. Over the past few decades, public health officials have examined the link between weight concerns and teen girls' smoking. Researchers surveyed a group of 273 randomly selected girls between 12- and 15-years-old living in Massachusetts. After four years, the girls were surveyed again. Sixty-three said they smoked to stay thin. Is there good evidence that more than 30% of the teen girls smoke to stay thin? Which of the following states the alternative hypothesis?

    a. $p < 0.30$

    b. $p \le 0.30$

    c. $p \ge 0.30$

    d. $p > 0.30$

3. Previously, an organization reported that teenagers spent 4.5 hours per week, on average, on the phone. The organization thinks that, currently, the mean is higher. State the null hypothesis, H_0, and the alternative hypothesis, H_a, in terms of the appropriate parameter.

7.2 Outcomes and Errors

Overview

Whenever we perform a hypothesis test, the decision we make has some chance of resulting in an error. In this section, we will learn the two types of errors associated with a hypothesis test.

Type I and Type II Errors

There are Type I and Type II errors. The type depends on whether the null hypothesis is true or false. Table 7.3 shows each of the errors and the decision outcome under which each of the errors occurs.

Table 7.3. Decision on the Null Hypothesis with the Probabilities of Making the Decision Given the Actual State of H_0

Decision on Null Hypothesis	Null Hypothesis (H_0) Is Actually...	
	True	**False**
Fail to Reject H_0	Correct Decision	Type II error
Reject H_0	Type I error	Correct Decision

A **Type I error** occurs when we reject the null hypothesis H_0 when in fact it is true. The probability of making a type I error is α, which is also the statistical significance level set before the hypothesis test. For example, a hypothesis test with a significance level 0.05 means that there is a 5% chance of making a type I error.

A **Type II error** occurs when we fail to reject the null hypothesis H_0 when in fact it is false. The probability of making a type II error is β. *The* power of a hypothesis test is denoted by $1- \beta$. The **power of a hypothesis test** is the probability that the test correctly rejects a false null hypothesis. Ideally, α and β should be as small as possible because they are probabilities of type I and type II errors, respectively.

The easiest way to see the relationship between the type I error and the level of confidence is by looking at Figure 7.2.

In the center of Figure 7.2 is a normally distributed sampling distribution marked H_0. This is a sampling distribution of $\bar{x}$ and it represents the distribution for the null hypotheses $H_0: \mu = 100$. This is the value that is being tested. The formal statements of the null and alternative hypotheses are listed below the figure.

The distributions on either side of the H_0 distribution represent distributions that could be true if H_0 is false under the alternative hypothesis H_a. An infinite number of distributions could be drawn from the sample data if H_a is true, but only two of them representing all the others are in Figure 7.2.

Figure 7.2. Relationship between the Type I Error and the Level of Confidence

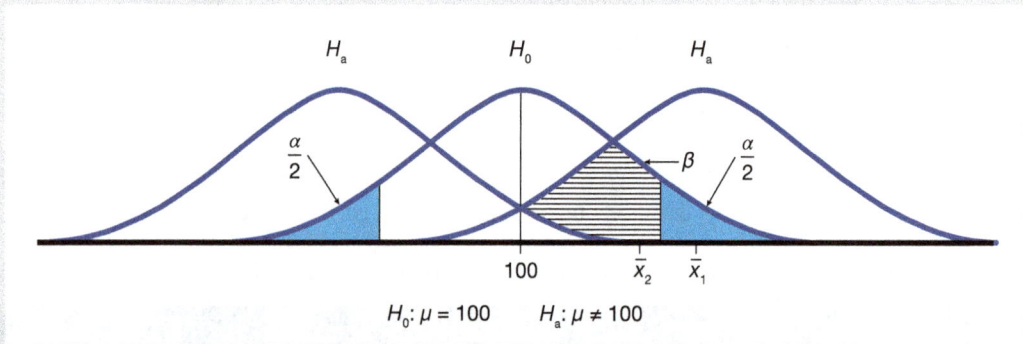

This significance level is marked in Figure 7.2 as the shaded areas in each tail of the H_0 distribution. Each area is actually $\frac{\alpha}{2}$ because the distribution is symmetrical, and the alternative hypothesis allows for the possibility for the value to be either greater than or less than the hypothesized value. This is called a two-tailed test.

If the sample mean marked as $\bar{x}_1$ is in the tail of the distribution of H_0, we conclude that the probability that it could have come from the H_0 distribution is less than α. So, we'd say, "The null hypothesis should be rejected with α level of significance." But it's possible that $\bar{x}_1$ *did* come from the H_0 distribution but falls in the tail. If this is the case, we've falsely rejected a true null hypothesis and made a Type I error.

The sample mean could really be from a H_a distribution but within the boundary set by the α level. Such a case is marked as $\bar{x}_2$. There is a probability that $\bar{x}_2$ came from H_a but shows up in the range of H_0 between the two tails. This probability is β, the probability of accepting a false null hypothesis.

We can only set the alpha error because there are an infinite number of alternative distributions from which the mean could have come that are not equal to H_0. As a result, the statistician places the burden of proof on the alternative hypothesis. Unless there is a probability greater than 0.9, 0.95, or even 0.99 that the null hypothesis is false, the burden of proof lies with the alternative hypothesis.

Example 4

Suppose the null hypothesis H_0 is "Frank's rock-climbing equipment is safe." Define the following: Type I error, Type II error, probability of Type I error, α, and probability of Type II error, β.

Solution 4

Type I error = Frank thinks that his rock-climbing equipment may not be safe when, in fact, it is safe.

Type II error = Frank thinks that his rock-climbing equipment may be safe when, in fact, it isn't safe.

α = Probability that Frank thinks his rock-climbing equipment may not be safe when, in fact, it is safe.

β = Probability that Frank thinks his rock-climbing equipment may be safe when, in fact, it isn't safe.

Notice that, in this case, the error with the greater consequence is the Type II error. If Frank thinks his rock-climbing equipment is safe, he will use it.

Example 5

Suppose the null hypothesis H_0 is "The victim of an automobile accident is alive when he arrives at the emergency room of a hospital." Define the following: Type I error, Type II error, probability of Type I error, α, and probability of Type II error, β.

Solution 5

Type I error = The victim is thought to be dead when, in fact, he is alive.

Type II error = The emergency room staff thinks the victim is alive when, in fact, the victim is dead.

α = Probability that the emergency room staff think the victim is dead when, in fact, he is alive = P (Type I error).

β = Probability that the emergency room staff think the victim is alive when, in fact, the victim is dead = P (Type II error).

Notice that, in this case, the error with the greater consequence is the Type I error. If the emergency crew thinks the victim is dead, they will not treat him.

Try It 7.1

Suppose the null hypothesis, H_0, is "a patient is healthy." Which type of error has the greater consequence, Type I or Type II?

Example 6

A drug company claims that one of its experimental drugs cures males with prostate cancer at least 75% of the time. Describe both the Type I and Type II errors in context. Which error is more serious?

Solution 6

Type I error = The cure rate for the drug is thought to be less than 75% when it is at least 75%.

Type II = The experimental drug has at least a 75% cure rate when, in fact, the cure rate is less than 75%.

In this scenario, the Type II error contains the more severe consequence. If a patient believes the drug works at least 75% of the time, this most likely will influence the patient's (and doctor's) choice about whether to use the drug as a treatment option.

Try It 7.2

Red tide is a bloom of poison-producing algae. When the weather and water conditions cause these blooms, shellfish such as clams develop dangerous levels of a paralysis-inducing toxin. In Massachusetts, the Division of Marine Fisheries (DMF) monitors levels of the toxin in shellfish by regular sampling of shellfish along the coastline. If the mean level of toxin in clams exceeds 800 mg of toxin per kg of clam meat in any area, clam harvesting is banned there until the bloom is over and levels of toxin in clams subside. Describe both a Type I and a Type II error in this context and state which error has the greater consequence.

Exercises 7.2

1. State the Type I and Type II errors in complete sentences given the following statements.

    a. The mean number of years Americans work before retiring is 34.

    b. At most, 60% of Americans vote in presidential elections.

    c. Twenty-nine percent of high school seniors get drunk each month.

    d. Fewer than 5% of adults ride the bus to work in Portland, Oregon.

    e. The mean number of cars a person owns in their lifetime is not more than ten.

    f. About half of Americans prefer to live away from cities, given the choice.

    g. Europeans have a mean paid vacation each year of six weeks.

    h. The chance of developing breast cancer is under 11% for women.

2. When a new drug is created, the pharmaceutical company must subject it to testing before receiving the necessary permission from the Food and Drug Administration (FDA) to market the drug. Suppose the null hypothesis is "the drug is unsafe." What is the Type II Error?

3. It is believed that students in Intermediate Algebra get less than 7 hours of sleep per night, on average. A survey of 22 Intermediate Algebra students generated a mean of 7.24 hours with a standard deviation of 1.93 hours. At a level of significance of 5%, do the Intermediate Algebra students get less than 7 hours of sleep per night on average?

    a. State the null hypothesis, H_0, and the alternative hypothesis, H_a, in terms of the appropriate parameter.

    b. State the Type I and Type II errors.

4. Previously, an organization reported that teenagers spent 4.5 hours per week, on average, on the phone. The organization thinks that, currently, the mean is higher. Fifteen randomly chosen teenagers were asked how many hours they spend on the phone per week. The sample mean was 4.75 hours with a sample standard deviation of 2.0. The Type I error is:

a. to conclude that the current mean hours per week is higher than 4.5, when in fact, it is higher.

b. to conclude that the current mean hours per week is higher than 4.5, when in fact, it is the same.

c. to conclude that the mean hours per week currently is 4.5, when in fact, it is higher.

d. to conclude that the mean hours per week currently is no higher than 4.5, when in fact, it is not higher.

7.3 Hypothesis Testing

Overview

Certain distributions are associated with hypothesis testing. In this section, we learn the concepts of hypothesis tests about claims of a population parameter, such as the population mean.

Critical Value Approach

The critical value approach in hypothesis testing helps us decide whether to reject a null hypothesis by comparing the critical value to the test statistic value. Since we work with a standard normal distribution, think back to Chapter 5 when we learned the z-scores formula. Below is the z-score formula for the z-test statistic.

Testing Claim About a Population Mean When σ is Known

$$z = \frac{\bar{x} - \mu_0}{\frac{\sigma}{\sqrt{n}}}$$

Where μ_0 is the hypothesized value of mean

Requirements

1. The sample is a simple random sample.
2. Population is normally distributed or $n > 30$.

In a hypothesis test for claims about population parameters (μ, σ, or p), there are three types of alternative hypotheses. Table 7.4 shows these three types of tests and decision rule based on the z-test statistic. The information from this table also applies to other test statistics that we learn about later.

Table 7.4. Critical Value Decision Rule for a Hypothesis Test

Alternative hypothesis (H_a)	The test statistic z	Then...
Two-tailed test	$-z_{\frac{\alpha}{2}} < z < z_{\frac{\alpha}{2}}$	Fail to reject H_0
Right-tailed test	$z > z_{\frac{\alpha}{2}}$	Reject H_0
Left-tailed test	$z < -z_{\frac{\alpha}{2}}$	Reject H_0

Example 7

In a hypothesis test for a claim about the population mean, the significance level α is 0.1.

Find the critical value for the following cases:

1. The hypothesis test is two-tailed.
2. The hypothesis test is one-tailed.

Solution 7

1. There are two critical values for the two-tailed test. The two corresponding critical values: $z_{\frac{\alpha}{2}}$ and $-z_{\frac{\alpha}{2}}$. Using the normal distribution table, we find

$$z_{\frac{\alpha}{2}} = z_{0.05} = 1.645$$
$$-z_{\frac{\alpha}{2}} = -z_{0.05} = -1.645$$

2. There is only one critical vale for the one-tailed test. If the one tail is on the left, the corresponding critical value is

$$-z_{\alpha} = -z_{0.1} = -1.28$$

If the one tail is on the right, the corresponding critical value is

$$z_{\alpha} = z_{0.1} = 1.28$$

P-value Approach

Another hypothesis testing method compares the p-value and the significance level to make a decision. The p-value is the probability that a test statistic is at least as extreme as the sample test statistic under the null hypothesis. A large p-value calculated from the data indicates we shouldn't reject the null hypothesis. A very small p-value means there is strong evidence that we should reject the null. Figure 7.3 shows how the p-value approach compares with the calculated z-test statistic.

The test statistic value z is marked on the bottom graph of the standard normal distribution. In this case, z is in the tail, so you would reject the null hypothesis. The sample mean $\bar{x}$ is just too unusually large to believe that it came from the distribution with a mean of μ_0 at the significance level of α.

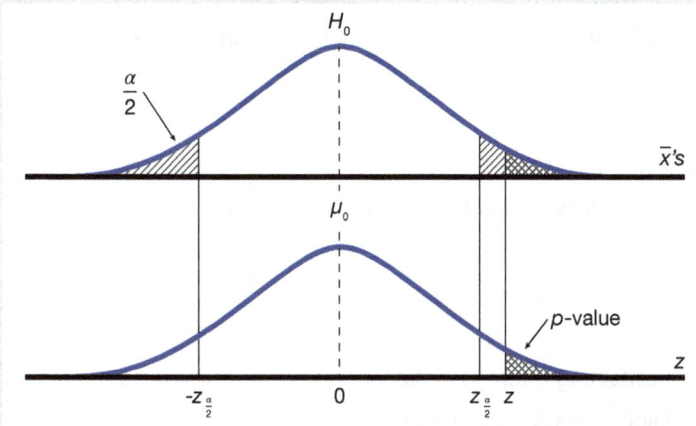

Figure 7.3. Comparing Test Statistic z, the Critical Value $z_{\frac{\alpha}{2}}$, and the p-value

The final step in the p-value approach is to find the p-value, which is the probability associated with the test statistic *value*. Then, we compare it to the significance level α and make a decision about the null hypothesis (H_0).

In Figure 7.3, the p-value is the shaded area to the right of z in the lower graph. When comparing this area to the corresponding area $\frac{\alpha}{2}$, we can see the p-value is less than $\frac{\alpha}{2}$, so we reject H_0. Remember that two researchers, each selecting a random sample from the same population, may find two different p-values from their samples. This happens because the p-value is the probability in the tail farther from the sample mean, assuming that the null hypothesis is correct. Because the sample means will likely be different, this will create two different p-values. But even if the p-values are different, the significance level is the only thing that matters when deciding about the null hypothesis.

Table 7.5 shows you the decision rules for a null hypothesis using the p-value and α.

Table 7.5. p-value Decision Rules

If...	Then...
p-value $< \alpha$	Reject H_0
p-value $\geq \alpha$	Fail to reject H_0

Example 8

Suppose you are performing a test for the claim about a population mean. You are given the test statistic $z = 2.15$.

1. What is the p-value if this is a one-tailed test?
2. What is the p-value if this is a two-tailed test?
3. Based on the two-tailed p-value, would you reject the null hypothesis if the significance level $\alpha = 0.05$?

Solution 8

1. A p-value is the probability of obtaining a test statistic value at least as large as the test statistic z from the sample. Using Table A.1 in the appendix:

$$p\text{-value} = P(z \geq 2.15) = 0.0158$$

2. Since this is a two-tailed test, the p-value could be in either left or the right tail:

$$p\text{-value} = 2P(z \geq 2.15) = 0.0316$$

3. Since the two-tailed p-value $= 0.0316$ is less than $\alpha = 0.05$, you reject the null hypothesis.

One and Two-Tailed Hypothesis Tests

In a one-tailed test, the significance lever is placed in just one tail. A one-tailed test can be either a left-tailed test or a right-tailed test. This can be seen in Figure 7.4, where μ_0 is the hypothesized value of the population mean. However, in a two-tailed test, the significance level is placed in both left and tails, as shown in the top graph of Figure 7.3.

Look at the alternative hypothesis's symbol to determine whether we should use a one- or two-tailed test. Greater than (>) and less than (<) in the alternative hypothesis indicates a one-tailed test. Not equal ($\neq$) indicates a two-tailed test.

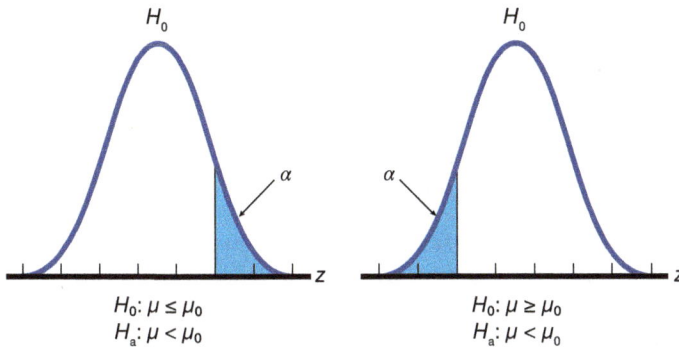

Figure 7.4. One-tailed Test

$H_0: \mu \leq \mu_0$
$H_a: \mu < \mu_0$

$H_0: \mu \geq \mu_0$
$H_a: \mu < \mu_0$

Example 9

A car manufacturer claims that their Model 17B provides gas mileage of greater than 25 miles per gallon. If you test the car manufacturer's claim, is your hypothesis one-tailed or two-tailed?

Solution 9

Since the alternative hypothesis uses a greater-than symbol, this is a one-tailed test.

The null hypothesis is

$$H_0: \mu \leq 25$$

The alternative hypothesis is

$$H_a: \mu > 25$$

Effects of Sample Size on Test Statistics

When we developed the confidence intervals for the population mean from a sample, we usually didn't have the population standard deviation σ. If the sample size is larger than 30, we can use the sample standard deviation, s, as the point estimate for σ and use the t-distribution to correct for this lack of information.

When testing hypotheses, you're faced with the same problem. If σ is unknown, and the sample size is at least 30, substitute s in the formula for the test statistic and use the t-distribution. Remember that the t-distribution can only be computed by knowing the degrees of freedom for the problem.

Exercises 7.3

1. Name the two distributions you can use for hypothesis testing for this chapter.

2. Which distribution do you use when testing a population mean and the population standard deviation is known? Assume the sample size is large. Assume a normal distribution with $n \geq 30$.

3. Which distribution do you use when the standard deviation is not known and you are testing one population mean? Assume a normal distribution with $n \geq 30$.

4. A population mean is 13. The sample mean is 12.8, and the sample standard deviation is 2. The sample size is 20. What distribution should you use to perform a hypothesis test? Assume the underlying population is normal.

5. A population has a mean of 25 and a standard deviation of 5. The sample mean is 24, and the sample size is 108. What distribution should you use to perform a hypothesis test?

6. You are performing a hypothesis test of a single population mean using a Student's t-distribution. What must you assume about the distribution of the data?

7. You are performing a hypothesis test of a single population mean using a Student's t-distribution. The data are not from a simple random sample. Can you accurately perform the hypothesis test?

8. You are performing a hypothesis test of a single population proportion. What must be true about the quantities of np and nq?

9. You are performing a hypothesis test of a single population proportion. You find out that np is less than 5. What must you do to be able to perform a valid hypothesis test?

10. You perform a hypothesis test of a single population proportion. The data comes from which distribution?

11. Assume H_0: $\mu = 9$ and H_a: $\mu < 9$. Is this a left-tailed, right-tailed, or two-tailed test?

12. Assume H_0: $p = 0.25$ and H_a: $p \neq 0.25$. Is this a left-tailed, right-tailed, or two-tailed test?

13. A bottle of water is labeled as containing 16 fluid ounces of water. You believe it is less than that. What type of test would you use?

14. You flip a coin and record whether it shows heads or tails. You know the probability of getting heads is 50%, but you think it is less for this particular coin. What type of test would you use?

15. If the alternative hypothesis has a not equals ($\neq$) symbol, you know to use which type of test?

16. Assume the null hypothesis states that the mean is at least 18. Is this a left-tailed, right-tailed, or two-tailed test?

17. Assume the null hypothesis states that the mean is at most 12. Is this a left-tailed, right-tailed, or two-tailed test?

18. Assume the null hypothesis states that the mean is equal to 88. The alternative hypothesis states that the mean is not equal to 88. Is this a left-tailed, right-tailed, or two-tailed test?

7.4 A Systematic Approach for Hypothesis Testing

Overview

Like the scientific method, hypothesis testing follows a systematic process. This is great because this process will work for all hypotheses we test. By learning these steps, we can test any claim we come across.

Steps for Hypothesis Testing

1. **State the hypotheses.** This is typically the most challenging part of the process. Review the question being asked. What parameter is being tested: a mean, a proportion, or differences in means? Is this a one-tailed test or a two-tailed test?

2. **Decide the significance level (α) required.** Look at the question and determine the significance level. Typical significance levels are 0.1, 0.05, and 0.01. However, the significance level should be based on the risk of making a Type I error and rejecting a true null hypothesis.

3. **Select the appropriate test statistic and find the relevant critical value.** The hypotheses and sample size will determine these selections. Drawing the relevant probability distribution and marking the critical value can be helpful. Be sure to match the graph with the hypothesis, especially if it is a one-tailed test.

4. **Take a sample(s) and calculate the relevant test statistic.** Calculate the test statistic value using the formula for the test statistic from Table 7.2. For the p-value approach, find the p-value using the calculated test statistic value.

5. **For the critical value approach, compare the calculated test statistic and the critical value.** Marking these on the graph will give a good picture of the situation. There are two possible outcomes:

   a. **The test statistic is in the tail:** Reject the null. The probability that this sample came from the hypothesized distribution is too small to believe that it is the real home of these sample data.

   b. **The test statistic isn't in the tail:** Fail to reject the null. The sample data are compatible with the hypothesized population parameter.

6. **For _p_-value approach: Compare the _p_-value and the significance level (α).** There are two possible outcomes:

    a. If _p_-value $< \alpha$: Reject the null.

    b. If _p_-value $\geq \alpha$: Fail to reject the null.

7. **Reach a conclusion.** Write the conclusion in two different ways. First, a formal statistical conclusion such as "With a 95% level of significance, we reject the null hypothesis that the population mean is equal to XX (units of measurement)." Second, a less formal conclusion states whether action is required. For example, "The machine is broken, and you need to shut it down and call for repairs."

Testing a Claim About a Population Mean

Testing a claim about a population mean is one of the most used concepts in statistics. A normal distribution or a Student _t_-distribution can be utilized in a hypothesis test about claims of population mean. Remember, the _t_-distribution is used when the population standard deviation is unknown, and the sample size is at least thirty.

Notation

n = sample size

$\bar{x}$ = sample mean

s = sample standard deviation

μ_0 = population mean from the null hypothesis H_0

Testing a Claim About a Mean When σ is Known

$$z = \frac{\bar{x} - \mu_0}{\frac{\sigma}{\sqrt{n}}}$$

Testing a Claim About a Mean When σ is Unknown

$$t = \frac{\bar{x} - \mu_0}{\frac{s}{\sqrt{n}}}$$

Requirements

The _t_-test statistic formula requires:

1. Sample is a simple random sample

2. The population is normally distributed or $n > 30$

3. The t-distribution is used with degrees of freedom $df = n - 1$

Example 10

Jeffrey swims the 25-yard freestyle. His mean time is 16.43 seconds with a standard deviation of 0.8 seconds. His dad, Frank, bought Jeffrey a new pair of expensive goggles, hoping they would help Jeffrey swim faster. Frank timed 15 of Jeffrey's 25-yard freestyle swims. For the 15 swims, Jeffrey's mean time was 16.0 seconds. Frank believes the goggles helped Jeffrey to swim faster. Conduct a hypothesis test using $\alpha = 0.05$. Assume the population has a normal distribution.

Solution 10

State the Hypotheses

Since the problem is about a mean, this is a test of a single population mean.

The null hypothesis is Jeffrey's original mean time without goggles.

$$H_0: \mu \geq 16.43$$

The alternative hypothesis is that Jeffrey swims faster with goggles.

$$H_a: \mu < 16.43$$

For Jeffrey to swim faster than H_0, his time has to be less than 16.43 seconds. The less-than symbol (<) indicates this is a left-tailed test.

Decide the Significance Level α

The problem states $\alpha = 0.05$.

Calculate the Test Statistic and Find the Critical Value

The random variable $\bar{x}$ = the mean time to swim the 25-yard freestyle. Since the population has a normal distribution and the population standard deviation is known, this is a z-test with $\mu_0 = 16.43$.

We know the following:

$$\bar{x} = 16$$

$$\sigma = 0.8$$

$$n = 15$$

The test statistic is:

$$z = \frac{\bar{x} - \mu_0}{\frac{\sigma}{\sqrt{n}}} = \frac{16 - 16.43}{\frac{0.8}{\sqrt{15}}} = -2.08$$

The *p*-value is the probability of obtaining another test statistic z that is equal to or smaller than -2.08 because this is a left-tailed test.

Therefore, the *p*-value is the probability, $P(z < -2.08)$. Enter on TI-84 "normcdf (-1E99, -2.08, 0, 1)".

The *p*-value is 0.0188.

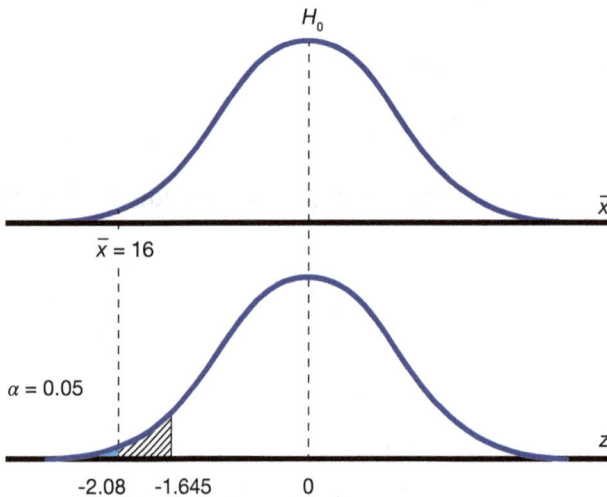

H_0

Figure 7.5. Hypothesis Test for Swim Times

$\bar{x} = 16$

$\bar{x}$

$\alpha = 0.05$

z

-2.08   -1.645   0

The critical value can be found from the z-table, $z_{0.05} = -1.645$, as shown in Figure 7.5.

Comparing the test statistic -2.08 and the critical value -1.645, the test statistic -2.08 falls in the critical region because -2.08 < -1.645. We reject H_0.

We reach the same conclusion when comparing the p-value of 0.0188 to the significance level of $\alpha = 0.05$ because the p-value is the shaded area to the left of -2.08 and is smaller than the striped area, which is the significance level α of 0.05, we come to the same conclusion of rejecting the null hypothesis.

Formal Conclusion

At the significance level of $\alpha = 0.05$, we reject the null hypothesis that Jeffery's mean swimming time is at least 16.43 seconds. There is evidence that expensive goggles help Jeffery swim faster.

Example 11

The National Institute of Standards and Technology provides exact data on the conductivity properties of materials. The following list contains conductivity measurements (in W/mK) for 11 randomly selected pieces of glass.

1.11   1.07   1.18   1.09   1.12   1.08   0.98   0.92   1.02   0.95   0.99

Is there convincing evidence that the average conductivity of this type of glass is greater than 1? Use a significance level of 0.05.

Solution 11

Check the Requirements

First, the sample is a simple random sample. Second, the sample must be from a normal population, or $n > 30$. The sample size $n = 11$, which doesn't meet the sample size, so we need to check for normality. A **normal quantile plot** is a scatter plot with the quantiles from the sample data versus quantiles from the normal distribution. It is the best tool to check for normality, and this can be accomplished in R using the function "qqnorm( )" and "qqline( )."

```
>Glass<-c(1.11,1.07,1.18,1.09,1.12,1.08,0.98,0.92,1.02,0.95,0.99)
>qqnorm(Glass)
>qq(Glass)
```

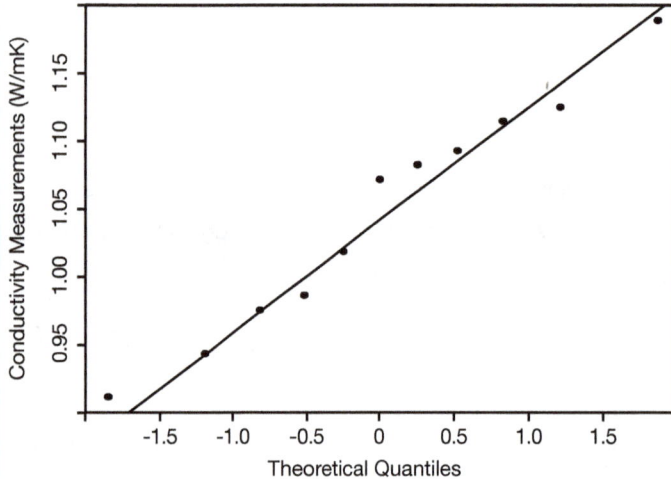

Figure 7.6. Normal Quantile Plot for Glass Data

Examining the Scatter Plot, we can see that the points scattered along the 45° line, $y = x$ indicating data has come from a normally distributed population, so we can use the t-test for the hypothesis testing.

State the Hypotheses

We are interested in whether the average conductivity is greater than 1. This statement doesn't contain the equal sign, so it becomes the alternative hypothesis. The null hypothesis is that the average conductivity is less than or equal to 1 W/mK.

$$H_0: \mu \leq 1$$
$$H_a: \mu > 1$$

Decide the Significance Level α

The significance level is specified as $\alpha = 0.05$.

Use R To Perform The Hypothesis Test

After entering data in data set "Glass", run the R function "t.test(name, mu = , alternative =)" as follows:

```
> t.test (Glass, mu = 1, alternative ="greater")
```

R Display from the T-Test

```
                    One Sample t-test
data: Glass
t = 1.9153, df = 10, p-value = 0.04223
alternative hypothesis: true mean is greater than 1
95 percent confidence interval:
1.00249 Inf
sample estimates:
mean of x
1.046364
```

We will reject the null hypothesis since the p-value = 0.0422 is less than the significance level $\alpha = 0.05$.

Formal Conclusion

At the significance level of $\alpha = 0.05$, we conclude that the population average conductivity level is greater than 1 W/mK.

Try It 7.3

Marco is a quarterback on his high school football team. His mean throwing distance of a football is 40 yards with a standard deviation of two yards. The coach tells Marco to adjust his grip to get more distance on each throw. The coach records the distances for 20 throws. For the 20 throws, Marco's mean distance was 45 yards. The coach believes the new grip helped Marco throw more than 40 yards. Conduct a hypothesis test using a significance level

$\alpha = 0.05$. Assume the throw distances for footballs are normal.

First, determine what type of test this is, set up the hypothesis test, find the p-value, sketch the graph, and state your conclusion.

Example 12

A manufacturer of salad dressings uses machines to dispense liquid ingredients into bottles that move along a filling line. Max is the production manager and is concerned that the machine needs maintenance. The machine should dispense 8 ounces into each bottle. Max takes a sample of 35 bottles and finds that the average amount dispensed is 7.91 ounces with a variance of $s^2 = 0.03$ square ounce. Is there evidence that Max should stop the machine and call the maintenance team? The lost production from a shutdown is so great that management wants the significance level of the analysis to be 0.01.

Solution 12

State the null H_0 and alternative hypothesis H_a

We're concerned that the machine isn't filling properly. From what we're told, whether the machine is overfilling or underfilling doesn't matter. Both seem to be equally bad errors. This tells us that this is a two-tailed test.

$$H_0: \mu = 8$$
$$H_a: \mu \neq 8$$

Determine the Significance Level α

This problem has already set the level of significance at $\alpha = 0.01$.

Calculate the Test Statistic and Find the Critical Values

This is a continuous random variable, and we're interested in the mean. The sample size is greater than 30. That means the appropriate distribution is the normal distribution, and the relevant critical value is 2.575 from the normal distribution table in Appendix A.1.

We know the following:

$$\bar{x} = 7.91$$
$$s^2 = 0.03$$
$$s = \sqrt{0.03} = 0.173$$
$$n = 35$$

The test statistic is:

$$t = \frac{\bar{x} - \mu_0}{\frac{s}{\sqrt{n}}} = \frac{7.91 - 8}{\frac{0.173}{\sqrt{35}}} = -3.08$$

The p-value is the probability of obtaining another test statistic t at least or more extreme than -3.08.

So p-value = $P(t < -3.08) + P(t > 3.08)$.

Enter in TI-84 `tcdf (3.08, 1E99, 34)`, and `tcdf (-1E99, -3.08, 34)` to find the p-value

$$p\text{-value} = 0.002 + 0.002 = 0.004$$

Because the hypothesis is two-tailed, there are two critical values. The two critical values can be found in TI-84 `invT(0.005, 34)`, and `invT(0.995, 34)`

$$t = \pm 2.728$$

Since the p-value of 0.042 is less than the significance level of $\alpha = 0.05$, we will reject the null hypothesis.

Compare the Calculated Test Statistic and the Critical Value

The critical value approach leads to the same conclusion of rejecting the null hypothesis. The test statistic $t = -3.08$ falls in the critical region because $-3.08 < -2.728$. So, we reject the null hypothesis.

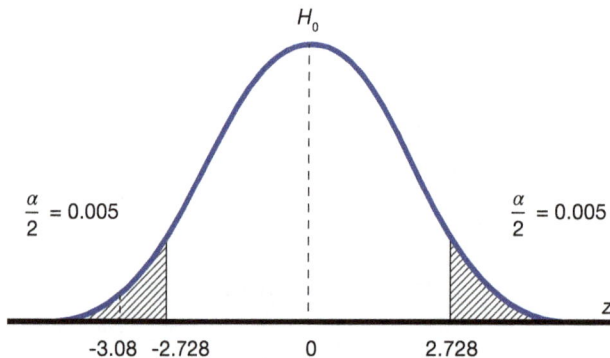

H_0

$\frac{\alpha}{2} = 0.005$ $\frac{\alpha}{2} = 0.005$

z

-3.08 -2.728   0   2.728

Figure 7.7. Hypothesis Test for Mean Bottle Fill Rate

Formal Conclusion

At the significance level $\alpha = 0.01$, we conclude that the machine is underfilling the bottles and needs repair.

Testing a Claim About a Population Proportion

Just as we can construct confidence intervals for population proportions, we can test hypotheses about population proportions. When we perform a hypothesis test of a population proportion p, we take a simple random sample from the population and use the normal distribution to test claims about a population proportion.

Notation

n = sample size or number of trials

$\hat{p}$ = sample proportion and $\hat{p} = \frac{x}{n}$

p = population proportion

$q = 1 - p$

Test Statistic

$$z = \frac{\hat{p} - p}{\sqrt{\frac{pq}{n}}}$$

Requirement

1. Simple random sample
2. Binomial distribution conditions must be met:
   - » Fixed number of independent trials
   - » Outcomes are binary: success or failure
   - » Probability of success is the same in all trials $np \geq 5$ and $nq \geq 5$

The shape of the binomial distribution needs to be like the shape of the normal distribution. To make sure this is true, $np \geq 5$ and $nq \geq 5$. We can approximate the binomial distribution of a hypothesized sample proportion with $\mu = np$ and $\sigma = \sqrt{npq}$. When testing a claim about the population proportion, we can't correct for a small sample size, and so the test statistic can't be used for small samples.

Example 13

A bank's mortgage department is interested in the loans of first-time borrowers. This information will be used to tailor their marketing strategy. They believe at least 60% of first-time borrowers take out smaller loans than other borrowers. They sample 100 first-time borrowers and find 53 of these loans are smaller than other borrowers. For the hypothesis test, use $\alpha = 0.05$.

Solution 13

State the Hypotheses

The words "is the same or different from" implies a two-tailed test. They also tell you that this problem satisfies the binomial distribution requirements of binary outcomes: success or failure.

$H_0: p \geq 0.60$

$H_a: p < 0.60$

Determine the Significance Level α

The level of significance is $\alpha = 0.05$.

Calculate the Test Statistic and Find Critical Values

Because this is a two-tailed test, half of the α value will be in the right tail and half in the left tail, as shown in Figure 7.8. The critical value for the normal distribution at the 95% confidence level is 1.96.

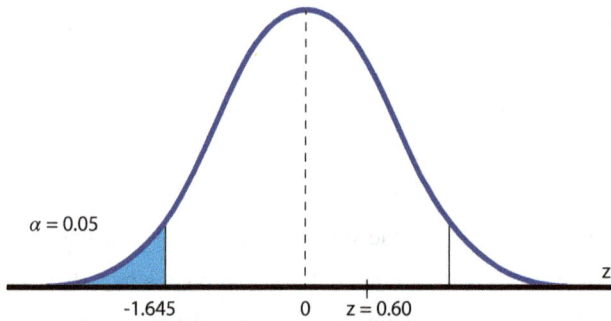

Figure 7.8. Hypothesis Test for the Proportion of First-Time Loan Borrower

Calculate the Test Statistic

$n = 100$

The sample proportion $\hat{p} = 0.53$.

The claimed proportion $p = 0.6$ and $q = 1 - p = 0.4$

The test statistic is:

$$z = \frac{\hat{p} - p}{\sqrt{\frac{pq}{n}}} = \frac{0.53 - 0.6}{\sqrt{\frac{(0.5)(0.5)}{100}}} = -1.429$$

This is a one-tailed test, the p-value is the probability of obtaining another test statistic z that is at least or more extreme than 0.6.

$$p\text{-value} = P(z < -1.429) = 0.0764$$

Because the p-value of 0.0764 is greater than the significance level of $\alpha = 0.05$, we fail to reject the null hypothesis.

The critical value can be found from Table A.1: Normal Distribution Table in the Appendix, z = -1.645. Because the test statistic value is greater than the critical values of -1.645, we fail to reject H_0. The critical value approach leads to the same conclusion as the p-value approach.

Formal Conclusion

At the significance level of $\alpha = 0.05$. We conclude that at least 60% of first-time borrowers take out smaller loans than other borrowers.

Try It 7.4

A second-grade teacher believes that 85% of his students will want to go on a field trip to the local zoo. He performs a hypothesis test to determine if the sample percentage is the same or different from 85%. The teacher samples 50 students and 39 reply that they would want to go to the zoo. For the hypothesis test, use a 1% level of significance.

Testing a Claim About a Population Variance or Standard Deviation

In addition to test claims about population mean and population proportion, there are occasions where we'll want to determine if the population variance is a specified value or range of values. Assessing the amount of variation in a process, natural phenomenon, or experiment is important in many fields. The procedures for testing the population variance σ^2 are based on a point estimate of the sample variance s^2. The techniques of testing claims about population variance or standard deviation use the chi-square distribution. Remember that a chi-square distribution isn't symmetrical, so critical values in a two-tailed test will not be the opposite of each other.

Notation

n = sample size

s = sample standard deviation

σ = population standard deviation

s^2 = sample variance

σ^2 = population variance

Test Statistic

$$\chi^2 = \frac{(n-1)s^2}{\sigma^2}$$

Note: Critical values can be found in Table A.3: Chi-Square Distribution in the appendix with degrees of freedom $df = n - 1$

Requirements

1. The sample is a simple random sample.
2. Samples must come from a normally distributed population.

Example 14

A forester wants to control an invasive species that threatens the health of native plants by using a mist blower to apply an herbicide. She wants to keep a consistent herbicide application rate so that the variability does not exceed 2 liters/acre. She chooses a random sample of 24 types of mist blowers from a normally distributed population. The sample has a variance of 5 liters/acre. Use a 0.05 level of significance test the claim that the variance is significantly different from 2 liters/acre.

Solution 14

State the Hypotheses

This is a two-tailed test. We know the population variance σ^2, the population variance will serve as null hypothesis. The alternative hypothesis would claim that the variance is different from 2 liters/acre.

$$H_0: \sigma^2 = 2$$
$$H_a: \sigma^2 \neq 2$$

Determine the Significance Level α

The significance level $\alpha = 0.05$.

We know the following:

$$n = 24$$
$$s^2 = 5$$
$$\sigma^2 = 2$$

Calculate the Test Statistic and the Critical Values

The test statistic is:

$$\chi^2 = \frac{(n-1)s^2}{\sigma^2} = \frac{(24-1)(5)}{2} = 57.5$$

The critical values are found from Table A.3: Chi-Square Distribution in the appendix with degrees of freedom $df = 23$.

$$\chi_L^2 = 11.69 \ \text{and} \ \chi_R^2 = 38.08$$

Compare the Calculated Test Statistic and the Critical Value

Since $\chi^2 = 57.5$ is significantly greater than the right chi-square critical value, the test statistic $\chi^2 = 57.5$ falls in the right critical region. We reject the null hypothesis.

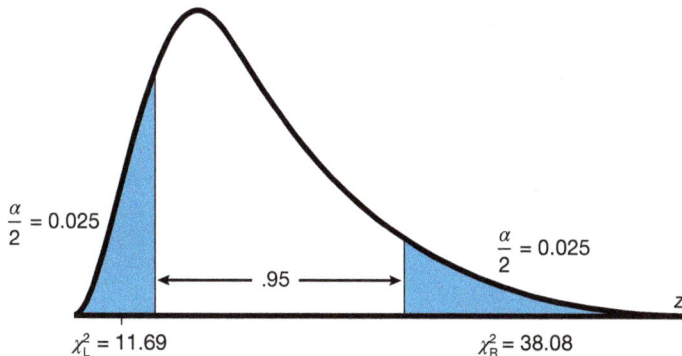

Figure 7.9. Hypothesis Test for Population Variance

$\frac{\alpha}{2} = 0.025$

$\frac{\alpha}{2} = 0.025$

.95

z

$\chi_L^2 = 11.69$

$\chi_R^2 = 38.08$

Formal Conclusion

With the significance level $\alpha = 0.05$, we conclude that the forester has enough evidence to support the claim that the variance significantly differs from 2 liters/acre.

Example 15

A post office lobby has multiple service windows, and each window has a separate line. The post office manager finds that the standard deviation for customer wait times on Friday afternoon is 7.2 minutes. The manager wants to see if a single line decreases wait times. One Friday afternoon, she reorganizes the lobby into one main line. She took a random sample of 25 customers and found that the standard deviation for customer wait times on a Friday afternoon is 3.5 minutes.

With a significance level $\alpha = 0.05$, test the claim that a single line has lower variation among customer waiting times.

Solution 15

State the Hypotheses

Since the claim is that a single line has less variation, this is a test of a single variance. The parameter is the population variance, σ^2. The word "less" implies a left-tailed test.

$H_0: \sigma \geq 7.2$
$H_a: \sigma < 7.2$

Determine the Significance Level α

The problem gives you the significance level: $\alpha = 0.05$

Calculate the Test Statistic and Find the Critical Value

We know the following:

$$n = 25$$
$$s = 3.5$$
$$\sigma = 7.2$$

The Test Statistic Is:

$$\chi^2 = \frac{(n-1)s^2}{\sigma^2} = \frac{(25-1)(3.5)^2}{7.2^2} = 5.67$$

The critical value

$$\chi_L^2 = 13.85$$

Because it's a left-tailed test, we only need to find the left chi-square critical value with 24 degrees of freedom from Table A.3 in the appendix.

Compare the Calculated Test Statistic and the Critical Value.

Since $\chi^2 = 5.67$ is less than the critical value $\chi_L^2 = 13.85$, the test statistic $\chi^2 = 5.67$ falls in the critical region. So, we reject the null hypothesis.

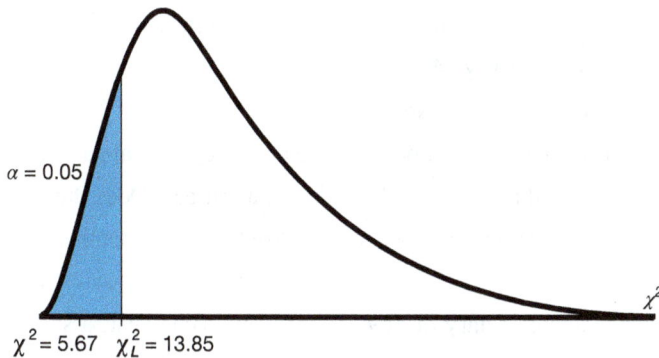

Figure 7.10. Hypothesis Test for Population Variance

$\alpha = 0.05$

$\chi^2 = 5.67$ $\chi_L^2 = 13.85$

Formal Conclusion

At the significance level $\alpha = 0.05$, there is sufficient evidence to conclude that a single line has lower variation among wait times. It seems that wait times from one main line vary less than wait times of customers in separate lines.

Exercises 7.4

1. A particular brand of tires claims that its deluxe tire averages at least 50,000 miles before it needs to be replaced. From past studies of this tire, the standard deviation is known to be 8,000. A survey of owners of that tire design is conducted. From the 28 tires surveyed, the mean lifespan was 46,500 miles with a standard deviation of 9,800 miles. Using $\alpha = 0.05$, is the data highly inconsistent with the claim?

2. From generation to generation, the mean age when smokers first start to smoke varies. However, the standard deviation of that age remains constant at around 2.1 years. A survey of 40 millennials who smoke was done to see if the mean starting age is at least 19. The sample mean was 18.1 with a sample standard deviation of 1.3. Do the data support the claim at the 5% level?

3. The mean number of sick days an employee takes per year is believed to be about 10. Members of a personnel department do not believe this figure. They randomly survey 8 employees. The number of sick days they took for the past year are as follows: 12; 4; 15; 3; 11; 8; 6; 8. Let x = the number of sick days they took for the past year. Should the personnel team believe that the mean number of sick days taken is 10?

4. Your statistics instructor claims that 60% of the students who take her Elementary Statistics class go through life feeling more enriched. For some reason, most people don't believe her. You decide to check this out on your own. You randomly survey 64 of her past Elementary Statistics students and find that 34 feel more enriched because of her class. Now, what do you think?

5. The US Department of Energy reported that 51.7% of homes were heated by natural gas. A random sample of 221 homes in Kentucky found that 115 were heated by natural gas. Does the evidence support the claim for Kentucky at the $\alpha = 0.05$ level in Kentucky? Are the results applicable across the country? Why?

6. Registered nurses earned an average annual salary of $69,110. For that same year, a survey was conducted of 61 Oregon registered nurses to determine if the annual salary is higher than $69,110 for Oregon nurses. The sample average was $70,210 with a sample standard deviation of $7,185. Conduct a hypothesis test.

7. Previously, an organization reported that teenagers spent 4.5 hours per week, on average, on the phone. The organization thinks that, currently, the mean is higher. Fifteen randomly chosen teen-agers were asked how many hours they spend on the phone per week. The sample mean was 4.75 hours with a sample standard deviation of 2.0. Conduct a hypothesis test. At a significance level of $\alpha = 0.05$, what is the correct conclusion?

   a. There is enough evidence to conclude that the mean number of hours is more than 4.75.

   b. There is enough evidence to conclude that the mean number of hours is more than 4.5.

   c. There is not enough evidence to conclude that the mean number of hours is more than 4.5.

   d. There is not enough evidence to conclude that the mean number of hours is more than 4.75.

8. A recent survey indicated that 48.8% of families own stock. A broker wanted to determine if this

survey could be valid. They surveyed a random sample of 250 families and found that 142 owned some type of stock. At the 0.05 significance level, can the survey be accurate?

9. Driver error can be listed as the cause of approximately 54% of all fatal auto accidents, according to the American Automobile Association (AAA). Thirty randomly selected fatal accidents are examined, and it is determined that 14 were caused by driver error. Using $\alpha = 0.05$, is the AAA proportion accurate?

10. The Weather Underground reported that the mean amount of summer rainfall for the northeastern United States is at least 11.52 inches. Ten cities in the northeast are randomly selected and the mean rainfall amount is calculated to be 7.42 inches with a standard deviation of 1.3 inches. At the $\alpha = 0.05$ level, can it be concluded that the mean rainfall was below the reported average? What if $\alpha = 0.01$? Assume the amount of summer rainfall follows a normal distribution.

11. The student academic group on a college campus claims that first-year students study at least 2.5 hours per day, on average. One Elementary Statistics class was skeptical. The class took a random sample of 30 first-year students and found a mean study time of 137 minutes with a standard deviation of 45 minutes. At $\alpha = 0.01$ level, is the student academic group's claim correct?

Try It Solutions

Try It 7.1

The Type II error has a greater consequence. The patient will be thought well when, in fact, he is sick, so he will not get treatment.

Try It 7.2

In this scenario, an appropriate null hypothesis would be that the mean level of toxins is at most 800µg.

$H_0: \mu_0 \leq 800\mu g$

Type I error: The DMF believes that toxin levels are still too high when, in fact, toxin levels are at most 800µg. The DMF continues the harvesting ban.

Type II error: The DMF believes that toxin levels are within acceptable levels (are at least 800μg) when, in fact, toxin levels are still too high (more than 800μg). The DMF lifts the harvesting ban. This error could be the most serious. If the ban is lifted and clams are still toxic, consumers could possibly eat tainted food.

In summary, the more dangerous error would be to commit a Type II error because this error involves the availability of tainted clams for consumption.

Try It 7.3

Since the problem is about a mean, this is a test of a single population mean. Use t -test.

$H_0: \mu \leq 40$
$H_a: \mu > 40$
$p = 4.23 \times 10^{-10}$

Because $p < \alpha$, you reject the null hypothesis. There is sufficient evidence to suggest that the change in grip improved Marco's throwing distance.

Try It 7.4

Since the problem is about percentages, this is a test of single population proportions.

$H_0: p = 0.85$
$H_a: p \neq 0.85$
$p = 0.7554$

Because $p > \alpha$, you fail to reject the null hypothesis. There isn't sufficient evidence to suggest that the proportion of students who want to go to the zoo isn't 85%.

Chapter 8
Correlation and Linear Regression

We often want to know how two or more variables are related. Professor Hartley wants to understand if there is a relationship between a student's grade on the second math exam of the term and their grade on the final exam. If there is a relationship, what is the relationship, and how strong is it? Matteo is a chef making a new menu for his restaurant. He wants to know if there's a relationship between the season and the price of fresh broccoli. If there is, what is it, and how strong is it?

In this chapter, we will learn about correlation, which is a way to describe relationships between variables. Correlation measures how strongly these variables move together, whether in the same or opposite direction. We will also learn about regression analysis, which allows us to determine which variables matter most and how variables influence each other. By creating models and conducting tests, you can decide if these relationships are statistically significant and use them to make predictions. Understanding correlation and regression analysis can help you make better decisions in the real world.

After reading this chapter, you will be able to do the following:

1. Understand the correlation coefficient
2. Describe the relationship between two variables using a scatter plot
3. Perform the hypothesis test of the population correlation
4. Draw conclusions from the test of the population correlation
5. Understand the slope of a least-squares regression line
6. Find residuals and identify outliers
7. Compute the prediction interval for regression equations
8. Find explained and unexplained variation
9. Understand the concept of multiple regression
10. State hypothesis testing for multiple regression
11. Identify categorical variables and procedures of model selection
12. Understand multicollinearity and dummy variables used in linear regression

8.1 The Correlation Coefficient

Overview

Professor Hartley wants to know if a student's grade on the second math exam of the term and their grade on the final exam are related in any way. She begins with a data set with two independent variables and then asks: are these variables related? One way to visually answer this question is to create a scatter plot of the data. Professor Hartley creates a scatter plot of the two data sets of exam grades and looks for relationships. A scatter plot is useful when working with two variables because it can identify a possible relationship between them (the explanatory variable and the response variable). Once Professor Hartley plots the data, she must determine what the scatter plot means. That's where the correlation coefficient comes into play.

Correlation Basics

A **correlation** is a relationship between two variables that exists when the values of one are somehow connected to the values of the other. A **linear correlation** describes a linear relationship between the explanatory and response variables. A straight line can estimate a linear correlation.

The **correlation coefficient** measures the degree of correlation between the two variables. The sample correlation coefficient r is the statistic that measures the strength of a linear relationship between two variables. It is the point estimate for the population correlation coefficient ρ (pronounced rho), which is the parameter that measures the strength of a linear correlation between two variables in a population.

Linear Correlation Coefficient r

$$r = \frac{n\Sigma(xy) - (\Sigma x)(\Sigma y)}{\sqrt{[n\Sigma x^2 - (\Sigma x)^2][n\Sigma y^2 - (\Sigma y)^2]}}$$

Where n is the number of ordered pairs of sample data.

As scary as the formula looks, it is just a measure of relative variances. The sample correlation coefficient r's value ranges from -1 to 1, or $-1 \le r \le 1$. In practice, all correlation analysis can be carried out with statistical software. We can do some of these calculations with the TI calculator or in R programming.

Visualization of Linear Relationship

Linear relationships come in a few general forms. Figure 8.1 shows several scatter plots and the calculated value of *r*. In Figure 8.1, the data generally trend together. Figure 8.1(a) moves upward, and (b) moves downward. Figure 8.1(a) is an example of a positive correlation, and (b) is an example of a negative correlation. If $n > 0$, then the correlation is positive. If $n < 0$, the correlation is negative. Remember, all the correlation coefficient tells us whether the data are *linearly* related. In (c), the variables have no relationship, and $r = 0$. In (d), the variables have a specific relationship, but $r = 0$ still, indicating no linear relationship exists.

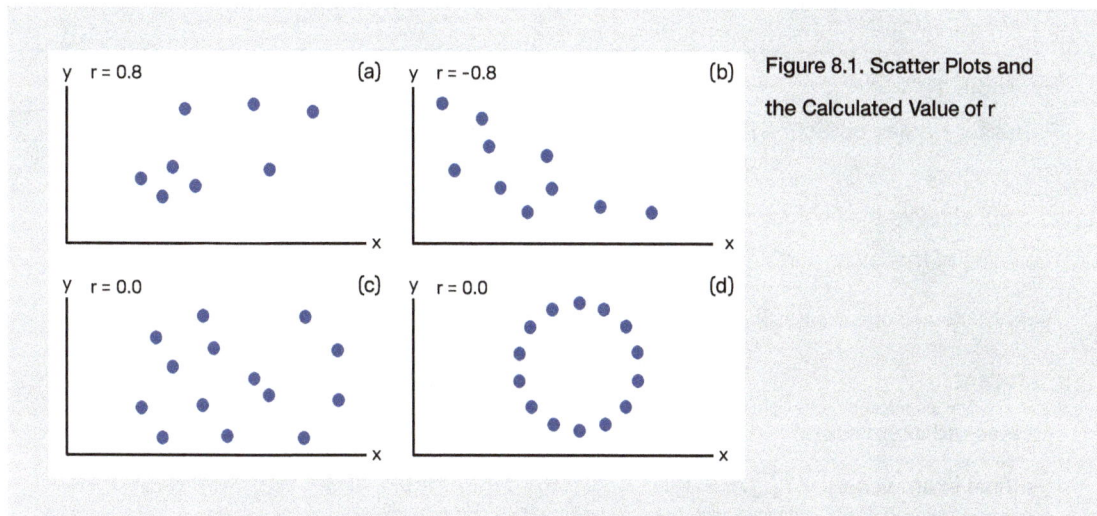

Figure 8.1. Scatter Plots and the Calculated Value of r

The sample correlation coefficient *r* measures the strength of the linear relationship between the explanatory variable *x* and the response variable *y*. If all the *x* and *y* values are on a perfectly straight line, the correlation coefficient will be either 1 or -1, depending on whether the line has a positive or negative slope. The closer *r* is to 1 or -1, the stronger the linear relationship between the two variables. If $r = 1$, there is a perfect positive correlation. If $r = -1$, there is a perfect negative correlation. In both cases, all the original data points lie on a straight line, regardless of the slope. If $r = 0$, there is no linear correlation between *x* and *y*. Figure 8.2 shows the differences between positive, negative, and zero correlation.

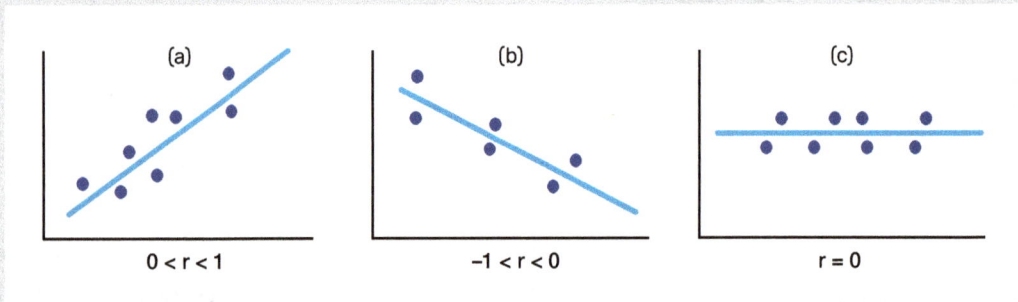

Figure 8.2. Various Forms of Correlation. (a): a scatter plot showing data with a positive correlation. (b): a scatter plot showing data with a negative correlation. (c): a scatter plot showing data with zero correlation.

Of course, we won't usually see a perfect correlation in the real world, but we can see trends. A positive value of r means that when x increases or decreases, y tends to do the same. A negative value of r means that when x increases, y tends to decrease, and when x decreases, y tends to increase.

Example 1

Professor Hartley collects a random sample of 11 students' second and final exam scores. She produces the following data, where x is the second exam score (out of 80 possible points), and y is the final exam score (out of 200 possible points), and creates Table 8.1. Find the value of the linear correlation coefficient r.

Table 8.1. Scores from the Second and Final Exam

student	1	2	3	4	5	6	7	8	9	10	11
x (second exam score)	65	67	71	71	66	75	67	70	71	69	69
y (final exam score)	175	133	185	163	126	198	153	163	159	151	159

Solution 1

To calculate the linear correlation coefficient r, we find the following statistics first

$\Sigma x = 65 + 67 + \ldots + 69 = 761$

$\Sigma x^2 = 65^2 + 67^2 + \ldots + 69^2 = 65^2 + 67^2 + \ldots + 69^2 = 52729$

$(\Sigma x)^2 = 761^2 = 579121$

$\Sigma xy = (65 \times 175) + (67 \times 133) + \dots + (69 \times 159) = 122500$

$\Sigma y = 175 + 133 + \dots + 159 = 1765$

$\Sigma y^2 = 175^2 + 133^2 + \dots + 159^2 = 287529$

$(\Sigma y)^2 = 1765^2 = 3115225$

We use the linear correlation coefficient formula to calculate r.

$$r = \frac{n\Sigma xy - \Sigma x \cdot \Sigma y}{\sqrt{(n\Sigma x^2 - (\Sigma x)^2)(n\Sigma y^2 - (\Sigma y)^2)}}$$

$$= \frac{(11)(122500) - (761)(1765)}{\sqrt{(11\cdot52729 - 579121)(11\cdot287529 - 3115225)}}$$

$$= \frac{4335}{6537.539} = 0.6631$$

The positive correlation coefficient indicates that the second exam score (explanatory variable) and the final exam score (response variable) are positively correlated.

Exercises 8.1

1. In order to have a correlation coefficient between traits A and B, it is necessary to have:
    a. one group of subjects, some of whom possess characteristics of trait A, the remainder possessing those of trait B.
    b. measures of trait A on one group of subjects and of trait B on another group.
    c. two groups of subjects, one which could be classified as A or not A, the other as B or not B.
    d. two groups of subjects, one which could be classified as A or not A, the other as B or not B.
2. Define the Correlation Coefficient and give a unique example of its use.
3. If the correlation between age of a car and money spent on it for repairs is 0.90, then which of the following is true?
    a. 81% of the variation in the money spent for repairs is explained by the age of the car.
    b. 81% of money spent for repairs is unexplained by the age of the car.
    c. 90% of the money spent for repairs is explained by the age of the car.
    d. None of the above.

4. Suppose that college grade-point average and scores on the verbal portion of an IQ test had a correlation of .40. What percentage of the variance do these two have in common?

    a. 20
    b. 16
    c. 40
    d. 80

5. True or false? If false, explain why: The coefficient of determination can have values between -1 and +1.

6. True or False: Whenever r is calculated on the basis of a sample, the value which we obtain for r is only an estimate of the true correlation coefficient which we would obtain if we calculated it for the entire population.

7. Under a scatter diagram, there is a notation that the correlation coefficient is .10. What does this mean?

    a. Plus and minus 10% from the means includes about 68% of the cases.
    b. One-tenth of the variance of one variable is shared with the other variable.
    c. One-tenth of one variable is caused by the other variable.
    d. On a scale from -1 to +1, the degree of linear relationship between the two variables is 0.10.

8. The correlation coefficient for X and Y is known to be zero. We then can conclude that:

    a. X and Y have standard distributions.
    b. The variances of X and Y are equal.
    c. No relationship exists between X and Y.
    d. No linear relationship exists between X and Y.
    e. None of these.

8.2 Testing the Significance of the Correlation Coefficient

Overview

The sample correlation coefficient r accesses the strength and indicates the direction of the linear relationship between *the variables* in a sample. We use r as the point estimate for the unknown population correlation coefficient ρ. We rely on statistical hypothesis testing to determine how accurate r is as an estimate of ρ.

Setting Up the Hypothesis Test

The hypothesis test determines whether the population correlation coefficient ρ is "close to zero" under a specified significance level. If the test concludes that the correlation coefficient is significantly different from zero, we say that the linear correlation is significant.

Testing the significance of the linear correlation requires that certain assumptions about the data are true. The foundation of this test is that the data are a sample of observed points taken from a larger population. In figure 8.3 panel (a) shows how for each x value, the mean of the y values lies on the line. Panel (a) also illustrates how more y values lie near the line than are scattered further away from the line. Panel (b) shows how the y values for each x value are normally distributed about the line with the same standard deviation.

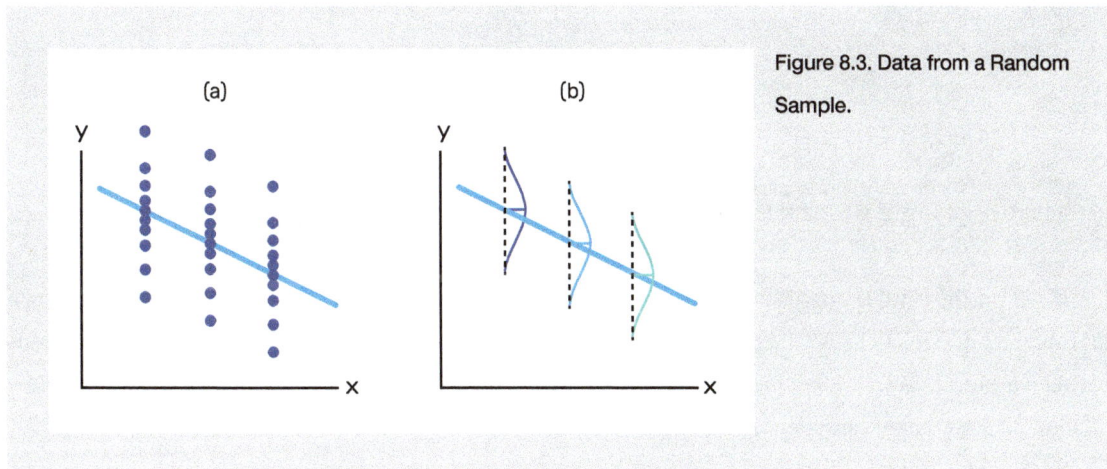

Figure 8.3. Data from a Random Sample.

Here are five things to know about the data that are illustrated in Figure 8.3.

1. The expected value of y for each x value lies on a straight line in the population, as we see in Panel A.
2. The y values for any x value are normally distributed around the line, as we see in Panel B.
3. Each of these normal distributions of y values has the same shape and spread around the line.
4. There are no other patterns or shapes in the differences between the actual y values and the y values on the line.
5. The data are from a simple random sample or randomized experiment.

Performing the Test and Drawing a Conclusion

The hypothesis test in Chapter 7 is the same one that will be used for testing a claim about the population correlation coefficient. The null hypothesis states that there *is not* a significant linear correlation between x and y in the population, or H_0: $\rho = 0$. The alternate hypothesis states that there *is* a significant linear correlation between x and y in the population, or H_a: $\rho \neq 0$.

We can perform a hypothesis test with either the critical value or the p-value approach. We calculate the t-test statistic for the correlation coefficient with the critical value approach. The t-test statistic is then compared with the critical value. If the test statistic is in the tail, then we reject H_0 and conclude that there is a linear relationship between the two variables. Otherwise, we fail to reject H_0 and conclude the opposite.

Test Statistic for Correlation Coefficient r

$$t = \frac{r}{\sqrt{\frac{1 - r^2}{n - 2}}}$$

$$= r\sqrt{\frac{n - 2}{1 - r^2}}$$

where n = sample size and the degrees of freedom are $df = n - 2$

If we use the p-value approach, the two possible outcomes and the corresponding decisions are shown in Table 8.2. Correlations are helpful in identifying the strength and direction of the linear relationships between a pair of variables, but they shouldn't be used to explain a relationship between them because correlation does not imply causation.

Table 8.2. p-value Decisions for Testing the Correlation Coefficient

Hypothesis	If...	Then...	Conclusion
$H_0: \rho = 0$	p-value $< \alpha$	Reject H_0	Significant linear correlation between the variables
$H_a: \rho \neq 0$	p-value $\geq \alpha$	Fail to reject H_0	No evidence for significant linear correlation between the variables

Example 2

Professor Hartley claims that there is a linear correlation between the second exam and final exam scores among her students. State the hypothesis and test her claim at the significance level of 0.05.

Solution 2

We know from Solution 1 that the value of the sample linear correlation coefficient is $r = 0.6631$ and $n = 11$ because there are 11 pairs of data values. We use the following steps to test the professor's claim.

State the Hypotheses

$H_0: \rho = 0$ *(There is no linear correlation between the second exam and final exam scores.)*

$H_a: \rho \neq 0$ *(There is a linear correlation between the second exam and final exam scores.)*

Calculate the Test Statistic

$$t = r\sqrt{\frac{n-2}{1-r^2}} = 0.6631\sqrt{\frac{11-2}{1-0.6631^2}} = 2.66$$

Draw a Conclusion Using Either the *p*-value or Critical Value Approach

The p-value is the probability of obtaining another test statistic t that is at least or more extreme than 2.66.

Because the correlation coefficient t-test is a two-tailed test, the p-value is twice the value of a one-tailed test.

P-value $= 2P(t > 2.66) = 2 \cdot 0.013 = 0.026$ or using TI-84:

```
tcdf (-1E99, -2.66, 9) + tcdf (2.66, 1E99, 9) = 0.026
```

The p-value can be calculated using R function:

```
pt(q = test statstic, df = ,lower tail = )
> pt(q = -2.66, df = 9, lower.tail = TRUE)
```

                                            `[1] 0.013`   `# This is the left tailed p-value`

```
> pt(q = 2.66, df = 9, lower.tail = FALSE)
```

           `[1] 0.013`   `# This is the right tailed p-value`

p-value $= 0.013 + 0.013 = 0.026$

We can also double the one-tailed p-value in R:

```
> 2*pt (q = -2.66, df = 9, lower.tail = TRUE)
```

           `[1] 0.026`   `# This is the two-tailed p-value`

Since the p-value < 0.05, we reject H_0.

The critical value approach yields the same conclusion as that of p-value approach.

Using the Correlation Coefficient r values table from the Appendix (Table A.5), we find the correlation coefficient critical value $r = 0.602$. Follow the instruction at the end of Table A.5, we reject H_0 because $0.602 < 0.6631$.

Conclusion: At the significance level of 0.05, we support Professor Harley's claim that there is a linear correlation between the second exam and final exam scores among her students.

Note: This hypothesis test can also be performed in TI-84 using `LinRegTTest.`

Note

In R programming

q = t test statistic

df = degrees of freedom (n – 2)

If "lower.tail = TRUE," then R calculates the probability to the left of q for a left-tailed test. If "lower.tail = FALSE," then R calculates the probability to the right of q for a right-tailed test.

Exercises 8.2

1. Define a t-test of a regression coefficient and give a unique example of its use.

2. The correlation between the number of hours spent studying and grades received in a class is high and positive; therefore:

   a. Studying causes better grades.

   b. Those who study less tend to have lower grades.

   c. Those who study more tend to have lower grades.

   d. No prediction from hours spent studying to grades received can be meaningfully made.

3. The following data are based on information from a study by a health insurance company. Let X be the average number of employees in a group health insurance plan, and y be the average administrative cost as a percentage of claims. At a 0.05 significance level, is there a linear correlation between x and y?

x	3	7	15	35	75
y	45	33	30	23	21

8.3 Basic Concepts of Linear Regression

Overview

Go back to Professor Hartley's question: is a student's grade on the second math exam of the term related to their grade on the final exam? She can use the correlation coefficient to see if they are related, but what's next when she finds out that they are?

Linear Regression Basics

Regression analysis is a statistical method that tests the hypothesis that a variable is dependent upon one or more other variables. Regression analysis can also estimate the strength of the impact of a change in the explanatory variable on the response variable. This last feature is important in predicting future values of the response variable. For Professor Hartley, doing a regression analysis on her sample data is the next step. Regression analysis examines how variables relate to each other in a functional way. Simple linear regression only looks at one independent variable rather than many.

Linear Regression Equation from a Sample

$$\hat{y} = b_0 + b_1 x$$

where b_0 is the y-intercept and b_1 is the slope.

Linear Regression Equation from a Population

$$y = \beta_0 + \beta_1 x$$

where β_0 and β_1 are parameters estimated by b_0 and b_1.

In the linear regression equation, the variable x is the independent (explanatory) variable, and $\hat{y}$ is the dependent (response) variable. Another way to think about the sample linear regression equation is a statement of input and output. The x variable is the input (observed), and the $\hat{y}$ variable is the output (predicted). The population parameter y is estimated by $\hat{y}$.

The Regression Line

The linear regression equation maps to a graph of the independent and dependent variables, just like the correlation coefficient. In the linear regression equation, b_1 is the slope, and b_0 is the y-intercept of the line. The **slope** is a number that describes the steepness of a line, and the **y-intercept** is the y coordinate of the point $(0, b_0)$ where the line crosses the y-axis.

Slope b_1

$$b_1 = r\frac{s_y}{s_x}$$

Where r is the sample linear correlation coefficient, s_y is the standard deviation of the y values, and s_x is the standard deviation of the x values.

---

y-intercept b_0

$$b_0 = \bar{y} - b_1\bar{x}$$

In a linear function, the slope is where we can read the mathematical expression as "the change in y due to a change in x." Figure 8.4 illustrates three possible graphs of the linear regression equation with different values of b_1.

Figure 8.4. Possible Graphs of the Linear Regression Equation. (a): If $b_1 > 0$, the line slopes upward to the right. (b): If $b_1 = 0$, the line is horizontal. (c): If $b_1 < 0$, the line slopes downward to the right.

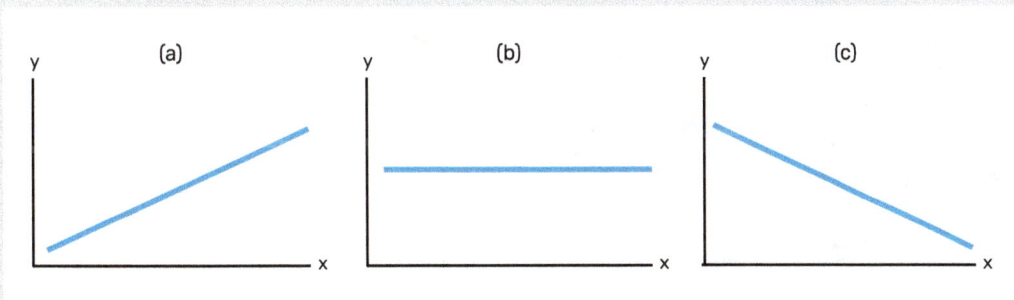

The slope of the regression line tells us how the dependent variable y changes for every one unit increase or decrease in the independent variable x, on average. It's important to interpret the slope of the line in the context of the situation represented by the data. We should be able to write a sentence interpreting the slope in plain language. Let's look at an example and write a clear interpretation of the data.

Let's revisit Professor Hartley's class test scores in Table 8.3. She knows there's a correlation between the second exam score and the final exam score. Now, she will test the linear regression of that data to measure the strength of the relationship between the two exam scores. She wants to know if she can predict the final exam score of a random student if she knows the student's second exam score.

Table 8.3. Scores from Second and Final Exam

x (Second exam score)	65	67	71	71	66	75	67	70	71	69	69
y (Final exam score)	175	133	185	163	126	198	153	163	159	151	159

The second exam score, x, is the independent variable and the final exam score, y, is the dependent variable. First, take each ordered pair of x and y values (x, y) and plot one point per pair on the graph. Then, draw a line through the approximate center of the data. In practice, if each student reading this book were to plot the data and draw a freehand line, we'd all draw different lines. Instead, we use the linear regression to be more precise in drawing the line of best fit.

Because it requires more advanced mathematical techniques to find the equation of the line of best fit, we use software to find it. Below are the steps of using a graphing calculator or R programming for finding the regression line for the second exam and final exam scores as seen in Figure 8.5.

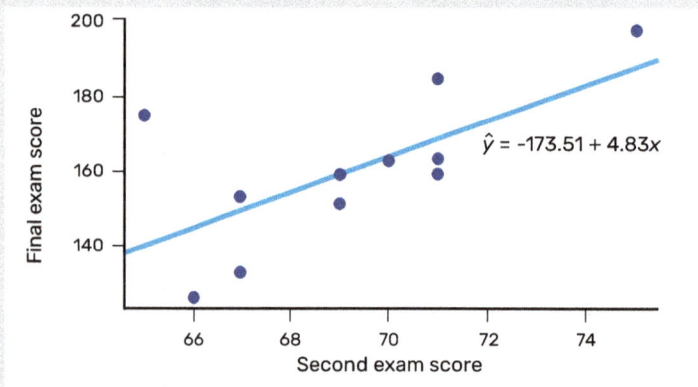

$\hat{y} = -173.51 + 4.83x$

Figure 8.5. Scatter Plot for Exam Scores

TI- 84:

Enter the second and final exam scores under L1 and L2 from STAT and EDIT then 1
Obtain the linear regression equation from STAT, CALC, 4 to select LinReg(ax+b) enter
on CALCULATE.

The linear regression equation is $\hat{y} = -173.51 + 4.83x$

R Programming:

```
> x <- c(65,67,71,71,66,75,67,70,71,69,69)
> y <- c(175,133,185,163,126,198,153,163,159,151,159)
> plot(x,y, main="Scatter Plot for Exam", xlab="Second Exam
  Score",ylab="Final Exam Score")
> abline(lm(y~x), col="blue")
```
                    # The abline( ) function draws the regression line in Figure 8.6
```
> lm(y~x)
```
        # The lm( ) function in R finds the formula of the linear regression equation shown in Figure 8.6

R Display for the Equation

```
Call:
lm(formula = y~x)
Coefficients:
(Intercept) x
-173.513    4.827
```

Note: in the above R display, the value -173.513 is the y-intercept and the value 4.827 is the slope.

Residuals and the Least-Squares Property

Residuals are the differences between the observed y values in a sample and the predicted $\hat{y}$ values from the regression equation. Take a look at Figure 8.6. Each data point is written as (x, y) and is plotted in blue. Each point of the regression line has the form $(x, \hat{y})$. The $\hat{y}$ is the **estimated value of** y and is the predicted value of y found using the regression line. It usually is different than any observed y from the sample data. The residual measures the difference between the values of y and $\hat{y}$. In other words, it measures the vertical distance between the actual or observed data point and the predicted point on the line.

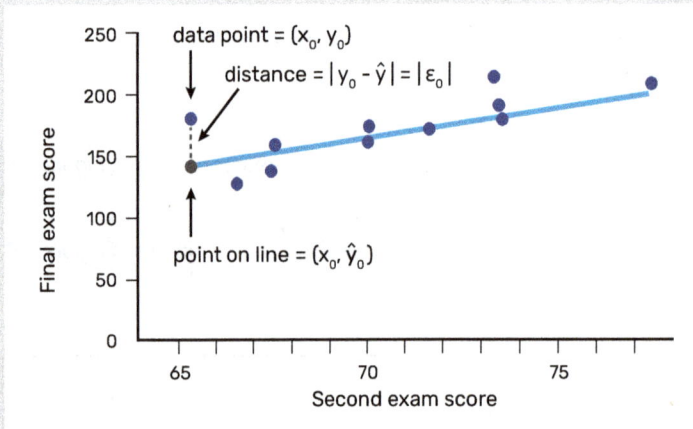

Figure 8.6. Residuals

Residuals ε

$$\varepsilon = y - \hat{y}$$

where y is the observed value in the sample and y is the predicted value from the linear regression equation

If the observed data point lies above the line, the residual is positive, and the line underestimates the actual data value for y. If the observed data point lies below the line, the residual is negative, and the line overestimates that actual data value for y. In Professor Hartley's data set, there are 11 data points. Therefore, there are 11 residuals. If we square each residual and add them together, you get the **sum of squared errors SSE**.

Sum of Squared Errors SSE

$$SSE = \sum(y - \hat{y})^2$$

The **least-squares property** says that a regression line is the line of best fit with the SSE being the smallest sum possible. It simply means that if we drew the regression line anywhere else, the SSE would have a higher number. It's the exact place on the graph where the difference between each y and $\hat{y}$ is the least.

Outliers

Outliers are observed data points that are far from the regression line. They have large residuals and need to be examined closely. Sometimes, they shouldn't be included in the data analysis because an outlier may result from erroneous data. Other times, an outlier may hold valuable information about the population under study and should remain included in the data. The key is to examine what causes a data point to be an outlier. As a rough principle, we can flag any point that exceeds two standard deviations above or below the best-fit line as an outlier. The standard deviation used is the standard deviation of the residuals, as defined in the formula below.

Standard Deviation of Residuals s_e

$$s_e = \sqrt{\frac{SSE}{n-2}}$$

Where n is the total number of data points, and the degrees of freedom is $df = n - 2$

We can do this visually in the scatter plot by drawing an extra pair of lines that are two standard deviations above and below the best-fit line. Any data points outside this extra pair of parallel lines are flagged as potential outliers. Or we can do this numerically by calculating each residual and comparing it to twice the standard deviation.

Besides outliers, a sample may contain one or several influential points. **Influential points** are observed data points far from the other observed data points in the horizontal direction. These points may significantly affect the slope of the regression line. To identify an influential point, remove it from the data set, and if the slope of the regression line changes significantly, then it is an influential point.

Computers and calculators can be used to identify outliers from the data. Computer output for regression analysis will often identify both outliers and influential points so that we can examine them further.

Example 3

For Professor Hartley's data set in Table 8.1, determine if there is an outlier.

Solution 3

Using the values of the variables: x (second exam scores) and y (final exam scores) in Table 8.4, we calculate the predicted $\hat{y}$ values, the residuals, ε, and the squared standard deviation of residuals, s_e below.

Table 8.4. Professor Hartley's Predicted y Values and the Standard Deviation of Residuals

x	y	$\hat{y}$	$\varepsilon = y - \hat{y}$	$(y - \hat{y})^2 = \varepsilon^2$
65	175	140	175 − 140 = 35	1225
67	133	150	133 − 150= −17	289
71	185	169	185 − 169 = 16	256
71	163	169	163 − 169 = −6	36
66	126	145	126 − 145 = −19	361
75	198	189	198 − 189 = 9	81
67	153	150	153 − 150 = 3	9
70	163	164	163 − 164 = −1	1
71	159	169	159 − 169 = −10	100
69	151	160	151 − 160 = −9	81
69	159	160	159 − 160 = −1	1

Entering x and y values in TI – 84 under L1 and L2, we find these statistics under "1-Var stats":

$$\bar{x} = 69.18 \quad \Sigma x = 761 \quad \Sigma x^2 = 52729 \quad (\Sigma x)^2 = 579121$$

The sum of squared errors SSE

$$SSE = \Sigma(y - \hat{y})^2 = 1225 + 289 + \dots + 1 = 2440$$

and the standard deviation of residuals s_e

$$s_e = \sqrt{\frac{SSE}{n-2}} = \sqrt{\frac{2440}{11-2}} = 16.46$$

We're looking for all data points that are more than two standard deviations (2s) from the regression line.

$$2s = 2(16.4) = 32.8$$

Looking for data points that are either greater than 32.8 or less than –32.8 and compare them to the residuals in the fourth column of Table 8.4. The only such data point is the student who had a grade of 65 on the second exam and 175 on the final exam; the residual for this student is 35. The two lines that are 2 standard deviations above and below the line of best fit are shown in Figure 8.7.

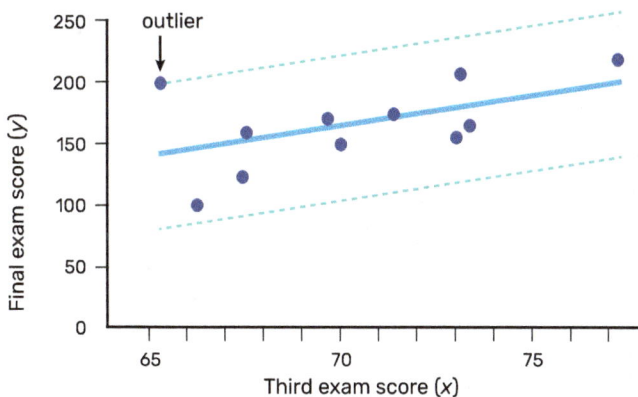

Figure 8.7. Outliers in Professor Hartley's Data Set

Numerically and graphically, we've identified the point (65, 175) as an outlier. Re-examine the data for this point to see if there are any problems with the data. If there is an error, fix it if possible, but if the data is correct, leave it in the data set.

Try It 8.1

Identify the potential outlier in the scatter plot in Figure 8.8. The standard deviation of the residuals or errors is approximately 8.5.

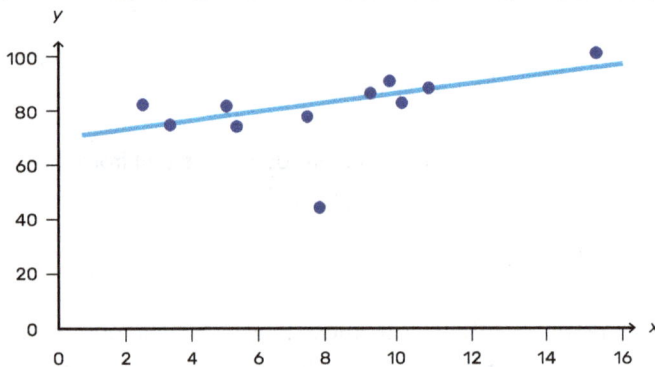

Figure 8.8. Identify Potential Outlier

Exercises 8.3

1. True or False? If false, correct it: Suppose a 95% confidence interval for the slope β of the straight-line regression of Y on X is given by $-3.5 < \beta < -0.5$. Then a two-sided test of the hypothesis would result in rejection of H0 at the 1% level of significance.

2. True or False: It is safer to interpret correlation coefficients as measures of association rather than causation because of the possibility of false correlation.

3. We are interested in finding the linear relation between the number of widgets purchased at one time and the cost per widget. The following data has been obtained:

   X = Number of widgets purchased: 1, 3, 6, 10, 15

   Y = Cost per widget (in dollars): 55, 52, 46, 32, 25

   Suppose the regression line is $\hat{y} = -2.5x + 60$. We compute the average price per widget if 30 are purchased and observe which of the following?

   a. $\hat{y} = 15$ dollars; obviously, we are mistaken; the prediction $\hat{y}$ is 15 dollars.

b. $\hat{y} = 15$ dollars; which seems reasonable judging by the data.

c. $\hat{y} = -15$ dollars; which is obvious nonsense. The regression line must be incorrect.

d. $\hat{y} = -15$ dollars; which is obvious nonsense. This reminds us that predicting Y outside the range of X values in our data is a very poor practice.

4. Briefly discuss the distinction between correlation and causality.

5. True or False: If r is close to +1 or -1, we say there is a strong correlation, with the understanding that we are referring to a linear relationship and nothing else.

6. Suppose that you know the information in the list below about each of 30 drivers. Propose a model (including a very brief indication of symbols used to represent independent variables) to explain how miles per gallon vary from driver to driver based on these factors:

» Miles driven per day.

» Weight of car.

» Number of cylinders in car.

» Average speed.

» Miles per gallon.

» Number of passengers.

7. Consider a sample least squares regression analysis between a dependent variable (Y) and an independent variable (X). A sample correlation coefficient of -1 (minus one) tells us that:

a. there is no relationship between Y and X in the sample.

b. there is no relationship between Y and X in the population.

c. there is a perfect negative relationship between Y and X in the population.

d. there is a perfect negative relationship between Y and X in the sample.

8. In correlational analysis, when the points scatter widely about the regression line, this means that the correlation is:

a. negative.

b. low.

c. heterogeneous.

d. between two measures that are unreliable.

9. The following table shows data on average per capita coffee consumption and heart disease rate in a random sample of 10 countries.

Yearly coffee consumption in liters	2.5	3.9	2.9	2.4	2.9	0.8	9.1	2.7	0.8	0.7
Death from heart diseases	211	167	131	191	220	297	71	172	211	300

a. Enter the data into your calculator and make a scatter plot.

b. Use your calculator's regression function to find the equation of the least-squares regression line. Add this to your scatter plot from part a.

c. Explain in words what the slope and y-intercept of the regression line tell us.

d. How well does the regression line fit the data? Explain your response.

e. Which point has the largest residual? Explain what the residual means in context. Is this point an outlier? An influential point? Explain.

f. Do the data provide convincing evidence that there is a linear relationship between the amount of coffee consumed and the heart disease death rate? Carry out an appropriate test at a significance level of 0.05 to help answer this question.

10. The following table consists of one student athlete's time (in minutes) to swim 2000 yards and the student's heart rate (beats per minute) after swimming on a random sample of 10 days:

Swim Time	34.12	35.72	34.72	34.05	34.13	35.73	36.17	35.57	35.37	35.57
Heart Rate	144	152	124	140	152	146	128	136	144	148

a. Enter the data into your calculator and make a scatter plot.

b. Use your calculator's regression function to find the equation of the least-squares regression line. Add this to your scatter plot from part a.

c. Explain in words what the slope and y-intercept of the regression line tell us.

d. How well does the regression line fit the data? Explain your response.

e. Which point has the largest residual? Explain what the residual means in context. Is this point an outlier? An influential point? Explain.

8.4 Predicting with a Regression Equation

Overview

A useful application of the regression equation is that it predicts what will happen to the response variable if we change the explanatory variable. Professor Hartley could use this information to predict students' final exam scores based on their previous exam scores. If a student was unhappy with that prediction, Professor Hartley could suggest that they visit the math tutoring center or explore other study strategies.

Prediction Intervals

While Professor Hartley was able to use regression analysis to predict a student's score on the final exam based on their score on the second exam, she can't tell how accurate those predictions are. In this section, we'll learn about the **prediction interval**. A prediction interval is an interval estimate for a predicted y value given an input x_0. A prediction interval isn't the same as a confidence interval. Prediction intervals deal with the estimated value of a response variable, while a confidence interval is a range of values used to estimate a population parameter.

Prediction Interval for a Single y at x_0

$$\hat{y} \pm t_{\frac{a}{2}} s_e \sqrt{1 + \frac{1}{n} + \frac{n(x_0 - \bar{x})^2}{n(\Sigma x^2) - (\Sigma x)^2}}$$

Computers and calculators have built-in regression functions to calculate predicted values of y given various x_0 values. But it's important to know whether we're constructing a prediction or confidence interval because the difference in the size of the standard deviations will change the size of the interval estimated. Figure 8.9 shows the difference the standard deviation makes in the size of the estimated intervals. The confidence interval, measuring the expected value of the dependent variable, is smaller than the prediction interval for the same level of confidence.

All regression equations go through the **point of means**, which is the point on the graph where the mean, $\bar{y}$ of y values, and the mean, $\bar{x}$ of x values meet, as shown in Figure 8.10. When the value of x is farther from the point of means, the estimate for y is less precise. If we use x values outside of the data range, the estimates become less reliable.

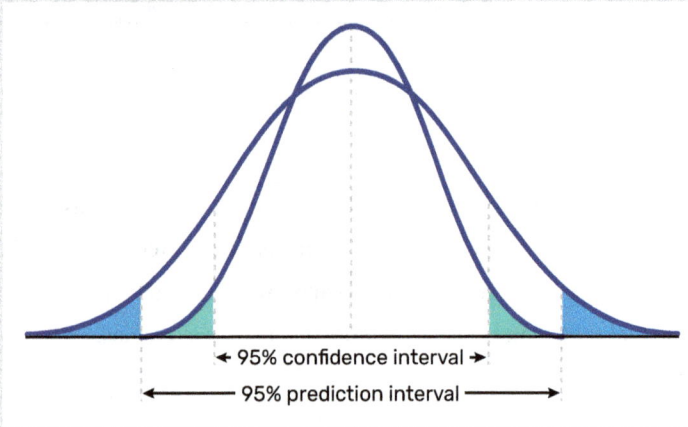

Figure 8.9. Prediction and Confidence Intervals for Regression Equation

← 95% confidence interval →

← 95% prediction interval →

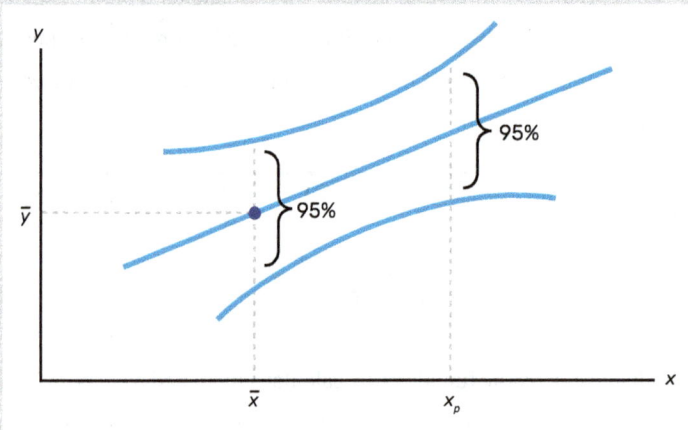

Figure 8.10. Prediction Interval for an Individual Value x_p at 95% Confidence Level

Figure 8.9 shows how the quality of the estimated interval can be different whether it is a prediction interval or a confidence interval. In figure 8.10, the value chosen to predict y is x_p, which is further from the sample mean $\bar{x}$ of the data. We can see the farther a chosen value of x is from $\bar{x}$, the less accurate its prediction will be.

Example 4

Let's go back to Professor Hartley again and use the least-squares regression line for prediction. Assume the coefficient for x was determined to be significantly different from 0.

Suppose we want to predict the mean final exam score of Professor Hartley's students who
earned a 73 on the second exam. The exam scores (x-values) range from 65 to 75. Since 73
falls between the x-values, substitute $x = 73$ into the equation as follows:

$$\hat{y} = -173.51 + 4.83(73) = 179.08$$

We predict that on average students who earn a grade of 73 on the second exam will earn a
grade of 179.08 on the final exam.

1. Construct a 95% prediction interval for the final exam score 179.08.
2. What would you predict the final exam score to be for a student who scored a 90 on the
   second exam?

Solution 4

1. To construct the 95% prediction interval, we need to find key information.

First, find the standard deviation of residuals:

$$s_e = \sqrt{\frac{\Sigma(y_i - \hat{y})^2}{n - 2}} = \sqrt{\frac{35^2 + (-17)^2 + 16^2 + \ldots + (-1)^2}{11 - 2}} = 16.47$$

From Example 3, we know the following:

$$\bar{x} = 69.18 \quad \Sigma x = 761 \quad \Sigma x^2 = 52729 \quad (\Sigma x)^2 = 579121$$

The critical value from the t-table with $df = 9$ and area of 0.05 in two tails is:

$$t_{\frac{\alpha}{2}} = 2.262$$

Use the prediction interval formula to find $\hat{y}$ for a final exam score of 179.08 when the second
exam score $x_0 = 73$.

$$\hat{y} \pm t_{\frac{a}{2}}s_e\sqrt{1 + \frac{1}{n} + \frac{n(x_0 - \bar{x})^2}{n(\Sigma x^2) - (\Sigma x)^2}}$$

$$= 179.08 \pm 2.262 \times 16.47\sqrt{1 + \frac{1}{11} + \frac{(11)(73 - 69.18)2}{11 \times 52729 + 579127}}$$

$$= 179.08 \pm 41.999$$

The 95% prediction interval is $137.08 < y < 221.08$.

2. The x values in the data are between 65 and 75. Since 90 is outside of the range of the observed x value, we can't reliably predict the final exam score for this student. Even though we can enter 90 into the equation for x and calculate a corresponding y value, the y value will have a confidence interval that may not be meaningful. To understand how unreliable the prediction can be, try making the substitution $x = 90$.

$$\hat{y} = -173.51 + 4.83(90) = 261.19$$

The final exam score is predicted to be 261.19, but there are only 200 points possible.

The standard deviation of residuals s_e can be also calculated from TI – 84 under the "Lin-RegTest" function.

Try It 8.2
Use the data from Professor Hartley's class in Table 8.1.
1. What would you predict the final exam score to be for a student who scored a 68 on the second exam?
2. Construct a 95% prediction interval for the final exam score found above.

Explained and Unexplained Variation

Data variation happens all the time. We first learned about it way back in Chapter 1. But in a linear regression, a main goal is to be able to predict future data with some accuracy. Understanding variation is critical to making accurate predictions. The **total variation** is the sum of the distance squared between the point (x, y) and the horizontal line that passes through the sample mean $\bar{y}$. Total variation is the Total Sum of Squares SST. It is the sum of explained and unexplained variation.

Total Variation

$$\Sigma(y - \bar{y})^2 = \Sigma(\hat{y} - \bar{y})^2 + \Sigma(y - \hat{y})^2$$

Explained variation $\sum(\hat{y} - \bar{y})^2$ is the total squared distance between the predicted y value and the horizontal line that passes through the sample mean $\bar{y}$. Explained variation is the Regression Sum of Squares SSR. This is part of the total variation that might be explained by the relationship between the two variables x and y. **Unexplained variation** $\sum(y - \hat{y})^2$ is the total squared distance between the point (x, y) and the predicted y value or the regression line. The unexplained variation is the Sum of Squared Residuals SSE.

Earlier in this chapter we learned about the correlation coefficient r. Here, we'll learn about the **coefficient of determination r^2**, which is the proportion of total variation in the response variable y explained by the regression line. The coefficient of determination is usually expressed as a percent, rather than a decimal. We can also express the proportion of variation in y that isn't explained by variation in x as $(1 - r^2)$.

Coefficient of Determination r^2

$$r^2 = \frac{\Sigma(\hat{y} - \bar{y})^2}{\Sigma(y - \bar{y})^2}$$

To illustrate all these concepts, look at Figure 8.11. It shows how the total deviation of the dependent variable y is broken up into these two pieces.

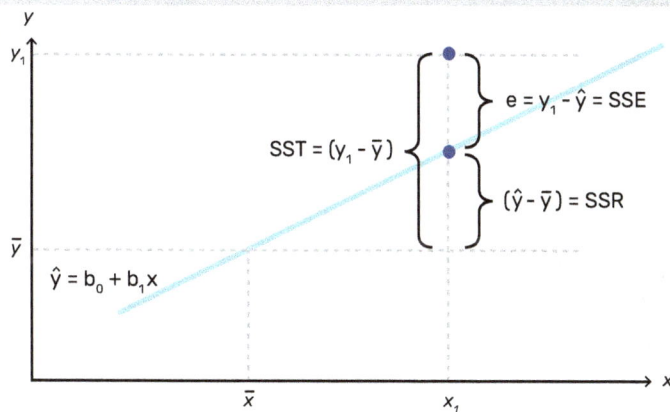

Figure 8.11. The Sum of Squared Total:

SST = SSR + SSE

Figure 8.12 shows the estimated regression line and a single observation x_1. The value of y at observation x_1 varies from $\hat{y}$ by the difference $(y_1 - \hat{y})$. The actual value of y at x_1 deviates from the estimated value, $\hat{y}$ by the difference between the estimated value and the actual value $(y - \hat{y})$. This is also the residual, ε.

Example 5

Let's return to Professor Hartley's second exam and final exam scores again. We know the following:

The best-fit regression line

$$\hat{y} = -173.51 + 4.83x$$

The correlation coefficient

$$r = 0.6631$$

The coefficient of determination

$$r^2 = 0.6631^2 = 0.4397$$

Based on this information, interpret and find the explained and unexplained variations.

Solution 5

Since r^2 is the proportion of total variation explained by the regression line, we conclude that approximately 44% of the variation in the final exam scores can be explained by the variation in the scores on the second exam using the regression line.

Approximately 56% of the variation $(1 - 0.44 = 0.56)$ in the final exam scores *can't* be explained by the variation in the scores on the second exam using the regression line.

The explained variation is 0.44, and the unexplained variation is 0.56.

Exercises 8.4

1. The hypothesis test of a simple linear regression of Y on X tests that the slope β is zero against a two-sided alternative. The computed test (t) statistic is 2.6, and the p-value is between .01 and .02. At the significance level $\alpha = .02$, what is your conclusion?

2. An economist is interested in the possible influence of Miracle Wheat fertilizer on the average yield of wheat in a district. To do so, she fits a linear regression of average yield per year against year after introduction of Miracle Wheat for a ten-year period.

The line of best fit is

$$\hat{y}_j = 80 + 1.5x_j$$

Where y_j = Average yield in j year after introduction, and $x_j = j$ year after introduction.

    a.   What is the estimated average yield for the fourth year after introduction?

    b.   Do you want to use this trend line to estimate yield for, say, 20 years after introduction? Why? What would your estimate be?

3.   An interpretation of $r = 0.5$ is that the following part of the Y-variation is associated with which variation in X:

    a.   most

    b.   half

    c.   very little

    d.   one quarter

    e.   none of these

4.   Which of the following values of r indicates the most accurate prediction of one variable from another?

    a. $r = 1.18$

    b. $r = -.77$

    c. $r = .68$

5.   Recently, the annual number of driver deaths per 100,000 for the selected age groups was as follows:

Age	Number of Driver Deaths per 100,000
16–19	38
20–24	36
25–34	24
35–54	20
55–74	18
75+	28

    a.   For each age group, pick the midpoint of the interval for the x value. (For the 75+ group, use 80.) Make a scatter plot.

b. Using "ages" as the independent variable and "number of driver deaths per 100,000" as the dependent variable, make a scatter plot of the data.

c. Calculate the least squares (best–fit) line. Put the equation in the form of: $\hat{y} = a + b_x$

d. Find the correlation coefficient.

e. Predict the number of deaths for ages 40 and 60.

f. What is the slope of the least squares (best-fit) line? Interpret the slope.

6. A researcher is investigating whether population impacts homicide rate. He uses demographic data from Detroit, Michigan, to compare homicide rates and the number of the population.

Population Size	Homicide rate per 100,000 people
558,724	8.6
538,584	8.9
519,171	8.52
500,457	8.89
482,418	13.07
465,029	14.57
448,267	21.36
432,109	28.03
416,533	31.49
401,518	37.39
387,046	46.26
373,095	47.24
359,647	52.33

a. Use your calculator to construct a scatter plot of the data. What should the independent variable be? Why?

b. Use your calculator's regression function to find the equation of the least-squares regression line. Add this to your scatter plot.

c. Interpret the meaning of each of the following in the context of the question.

    i.   The slope of the regression equation.

    ii.   The y-intercept of the regression equation.

    iii.   The correlation coefficient, r.

    iv.   The coefficient of determination r^2.

d. Do the data provide convincing evidence that there is a linear relationship between population size and homicide rate? Carry out an appropriate test at a significance level of 0.05 to help answer this question.

8.5 Multiple Regression

Overview

Sometimes, we want to know how multiple variables interact with each other. Up to this point, we've learned about the relationships between two variables: the independent and dependent variables. Once we start looking at the relationships between the dependent variable with two or more independent variables, we're into an area called multiple regression.

Multiple Regression Basics

Even though multiple regression looks at multiple independent variables, we're still working with linear relationships. The multiple regression equation works only if the dependent variable y is normally distributed, the residuals are also normally distributed with a mean of 0 and a constant standard deviation, and the residuals are independent of the size of x and independent of each other. It's not practical to calculate this equation without using statistical software, but we'll explore the concepts here.

Multiple Regression Equation from a Sample

$$\hat{y} = b_0 + b_1x_1 + b_2x_2 + \ldots + b_kx_k$$

Multiple Regression Equation from a Population

$$y = \beta_0 + \beta_1x_1 + \beta_2x_2 + \ldots + \beta_kx_k$$

Note

k is the number of independent variables. All coefficients β are parameters and are estimated by the values of sample statistic b.

Adjusted Multiple Correlation Coefficient R^2

$$\bar{R}^2 = 1 - \frac{(n-1)}{[n-(k+1)]}(1-R^2)$$

Where $R^2 = \frac{SSR}{SST}$

The multiple correlation coefficient R^2 ranges from 0 to 1. When R^2 is 0, the equation does not explain any variation in y, and when R^2 is 1, it explains all variation in y. The value of R^2 is found using the same formula as for r^2, the simple correlation coefficient. When there are more independent variables, R^2 increases, but to get a proper measure, we use the adjusted $\bar{R}^2$, which accounts for degrees of freedom.

Hypothesis Test for Multiple Regression

A way to test the general quality of the overall relationships is to test the coefficients as a group, rather than independently. Because this is multiple regression (more than one predictor x), the F-test is used to determine whether the coefficients collectively affect y. In other words, we test whether all predictors (independent variables) jointly can explain a significant part of the variance of the dependent (response) variable.

The null and alternative hypotheses.

$$H_o: \ \beta_1 = \beta_2 = ... = \beta_i = 0$$

$$H_a: \ at \ least \ one \ of \ the \ \beta_i \neq 0$$

If the null hypothesis can't be rejected, we conclude that none of the independent variables x contribute to explaining the variation in y.

F statistic

$$F = \left(\frac{n-k-1}{k}\right)\left(\frac{R^2}{1-R^2}\right)$$

Where degrees of freedom $df_1 = k$, $df_2 = n - k - 1$.

We can determine whether the variation is valid by comparing the calculated F-test statistic with the critical value at the desired confidence level. If F is in the tail of the distribution, we can conclude that this variation is valid because at least one of the estimated coefficients is significantly different from 0. We can also use the p-value method to reach same conclusion.

Example 6

A college administrator wanted to predict student success in college. She decided to use various scores as the indicator of college success. A random sample of 55 students was selected and data from four predictors were collected: college GPA, high school GPA, SAT scores, and AP math scores. The multiple regression analysis was performed in R using the lm( ) function. The output shows the following multiple linear regression equation and the value of multiple correlation coefficient.

$$College\_GPA = -0.215 + 0.324 * (high\ school\ GPA) + 0.0031 * (SAT) + 0.048 * (AP)$$

$$R^2 = 0.387$$

Perform the hypothesis test and interpret the results using $\alpha = 0.05$.

Solution 6

There are three eplanatory variables: hs GPA, SAT, and AP. The response variable is college GPA. The hypothesis test for overall model fitting is:

$H_0: \beta_1 = \beta_2 = \beta_3$
$H_a:$ At least one of the $\beta_i \neq 0$

With $n = 55$ and $k = 3$, the test statistic F is:

$$F = \left(\frac{n - k - 1}{k}\right)\left(\frac{R^2}{1 - R^2}\right) = \left(\frac{55 - 3 - 1}{3}\right)\left(\frac{0.387}{1 - 0.387}\right) = 10.732$$

With the degrees of freedom 3 and 51, the p-value can be calculated using R function 1 − pf(10.732, 3, 51)

$$p\text{-value} = 0.000014$$

Since p-value < 0.05, we can conclude that at least one of the regression coefficients is significantly different from 0. This indicates that the three predictors explained a significant portion of the variance in college GPA.

The regression coefficient for the variable "high school GPA" is 0.324. This means the predicted college GPA would increase 0.324 points with each point increase in high school GPA while keeping SAT and AP scores unchanged.

Similarly, keeping high school GPA and AP scores constant, the predicted college GPA would increase 0.00312 points with each point increase in SAT scores.

Also, the predicted college GPA would increase 0.048 points with each point increase in AP math score while keeping high school GPA and SAT constant.

The adjusted R^2:

$$\bar{R}^2 = 1 - \left(\frac{n-1}{n-k-1}\right)(1 - R^2) = 1 - \left(\frac{55-1}{55-3-1}\right)(1 - 0.387) = 1 - 0.649 = 0.351$$

From the R programming output, $R^2 = 0.387$. This means that about 39% of the variation in college GPA can be explain by the multiple linear regression with high school GPA, SAT scores, and AP exam scores. The adjusted R^2 is slightly smaller because multiple predictors were considered in the regression model.

Dummy Variables

The concepts in this chapter to this point have assumed that the independent variables were continuous random variables. However, the regression model has no restrictions on the type of independent variables. We can test hypotheses with categorical variables that take on only two values: 1 and 0. These are the same as success or failure from the binomial probability distribution. In statistics, these categorical variables are sometimes called **dummy variables**. In this case, the regression equation changes a bit.

Linear Regression Equation with Categorical Values

$$\hat{y} = b_0 + b_1 x_1 + b_2 x_2$$

where x_1 is a continuos random variable, x_2 is a categorical variable

---

To understand how this changes the regression line, look at Figure 8.12 where the categorical variable is x_2 = {0, 1} and x_1 is some continuous random variable. The y-intercept b_0 is the value where the line crosses the y-axis. When the value of $x_2 = 0$, the estimated line crosses at b_0. When the value of $x_2 = 1$, the estimated line

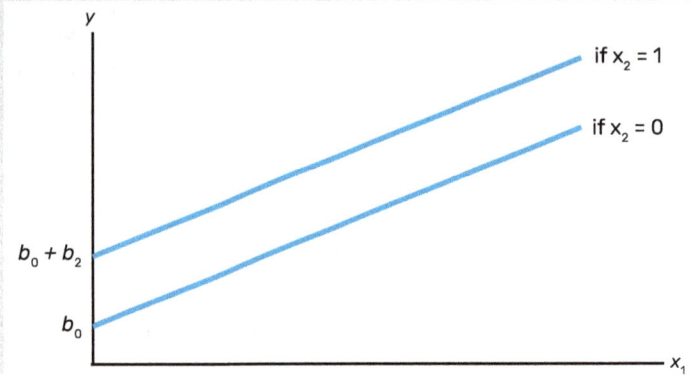

Figure 8.12. Regression Model with Categorical Variables

crosses at $b_0 + b_2$. The categorical variable causes the estimated line to shift either up or down. This is a simple vertical shift and doesn't affect the impact of the other independent variable x_1, a continuous random variable.

As an example, in a study about pay gap between men and women, a categorical variable like gender can be used. Suppose we compare the salaries of male and female elementary and secondary school teachers in a specific state. Many factors influence salaries, like education level and experience, but we're interested in the effect of gender. We use a binary variable where male is 1 and female is 0. We also look at district's average revenue per average daily attendance (ADA) as a measure of their ability to pay. The results of the regression analysis using data on 24,916 schoolteachers are presented in Table 8.5.

Table 8.5. Earnings Estimate for Elementary and Secondary School Teachers

Variable	Regression Coefficients (b)	Standard Errors of the Estimates for Teacher's Earnings Function (s_b)
Intercept	4269.9	
Gender (male = 1)	632.38	13.39
Total Years of Experience	52.32	1.10
Years of Experience in Current District	29.97	1.52
Education	629.33	13.16
Total Revenue per ADA	90.24	3.76
$\bar{R}^2$	.725	
n	24,916	

The coefficients for all the independent variables are significantly different from 0 because each t-test yields a t-statistic greater than the critical value 1.96, based on 0.05 significance level. Therefore, we conclude that male teachers in this state are paid a premium of $632 (or 6%) after accounting for variation in experience, education, and wealth of the school district in which the teacher is employed. Figure 8.14 illustrates the regression analysis of this example.

In Figure 8.13, the lower estimated line has an intercept of $4,269 if the gender variable is equal to 0 (for females). If the gender variable is equal to 1 (for males), the coefficient for the gender variable is added to the intercept, and the relationship between total years of experience and salary moves upward in a parallel line, as indicated on the graph. Also marked on the graph are various points for reference. A female schoolteacher with 10 years of experience receives a salary of $4,792 based on her experience only, but this is still $109 ($4901 – $4792) less than a male teacher with 0 years of experience.

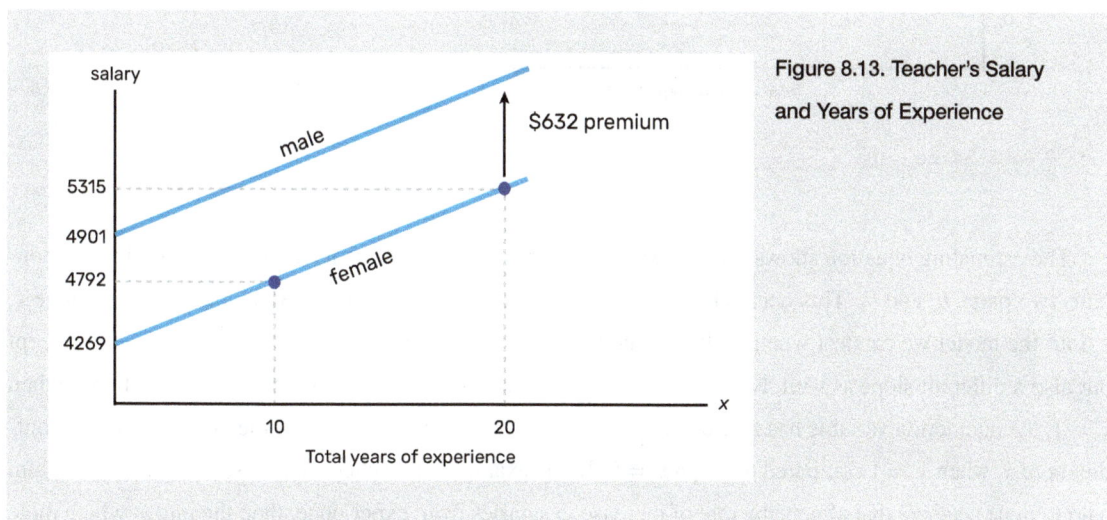

Figure 8.13. Teacher's Salary and Years of Experience

A more complex interaction between a categorical variable and the dependent variable can also be estimated. If the categorical variable interacts with one or more of the other continuous independent variables, it could have a larger effect on the dependent variable. While not tested in this example, what if the impact of gender on salary impacted the value of additional years of experience on salary, too? Hypothetically, what if female teachers' salaries were lower when they started teaching and grew slower than male teachers' salaries? If this were true, the regression line showing the relationship between total years of experience and salary would have a different slope for males than for females. The analysis would show that female teachers not only earn less than male teachers at the beginning of their careers but would fall further and further behind as time and experiences increased. Figure 8.14 shows how this hypothesis can be tested with the use of dummy variables and an interaction variable. The linear regrettion equation with the interaction variable is $\hat{y} = b_0 + b_2 x_2 + b_1 x_1 + b_3 x_2 x_1$

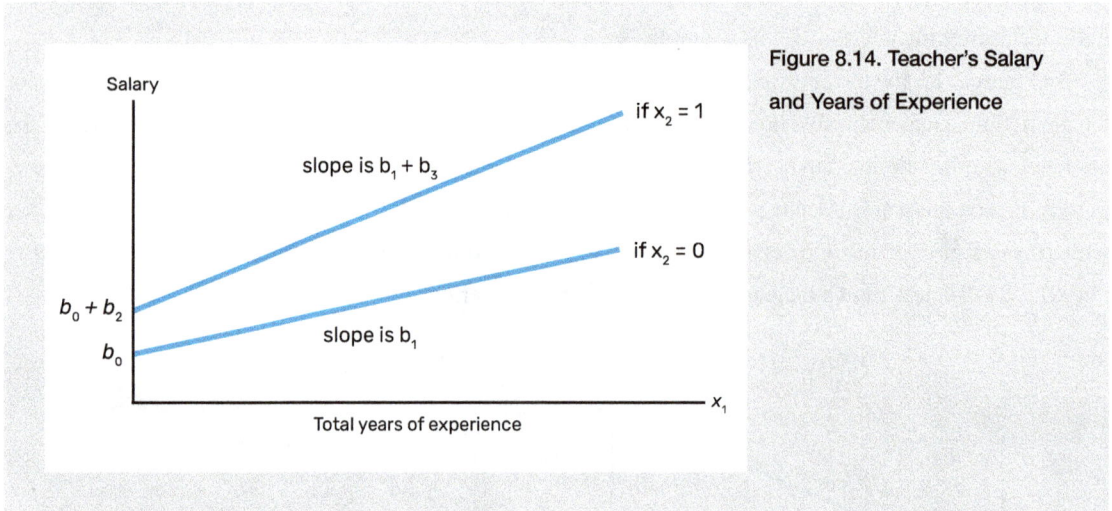

Figure 8.14. Teacher's Salary and Years of Experience

The estimating equation shows how the slope of x_1, the continuous random variable "experience," contains two parts, b_1 and b_3. This occurs because of the interaction variable $x_2 x_1$. By including the variable $x_2 x_1$ into the model we can test whether the female teacher's regression line has not only a different intercept but also a different slope as well. Note that when $x_2 = 0$, the interaction variable has a value of 0, but when $x_2 = 1$, the interaction variable has a value of x_1. The coefficient b_3 is an estimate of the difference in the coefficient of x_1 when $x_2 = 1$ compared to when $x_2 = 0$. In the example of teacher's salaries, if there is a premium paid to male teachers that affects the rate of increase in salaries from experience, then the rate at which male teachers' salaries rises would be $b_1 + b_3$ and the rate at which female teachers' salaries rise would be simply b_1. To test whether the slopes differ, perform a t-test using the test statistic for the parameter β_3.

$$H_0 : \beta_3 = 0$$
$$H_a : \beta_3 \neq 0$$

If we reject the null hypothesis that $\beta_3 = 0$, we conclude there is a difference between the rate of increase for the group with the binary variable of 1, or males in this example. The multiple regression equation can be combined with the linear one that tested only a parallel shift in the estimated line.

Exercises 8.5

1. Why is it better to use the adjusted R^2 value than the R^2 value in multiple linear regression model?

2. Consider the following estimated regression equation: $\hat{y} = 3536 + 1183x_1 - 1208x_2$. Suppose the model is changed to an estimated simple linear equation $\hat{y} = -10663 + 1386x_1$ after removing x_2 from the original equation.

   a. Interpret the meaning of the coefficient on x_1 in the estimated simple linear regression equation.

   b. Interpret the meaning of the coefficient on x_1 in the estimated multiple regression equation $\hat{y} = 3536 + 1183x_1 - 1208x_2$.

3. Suppose we regress the dependent variable y on 4 independent variables x_1, x_2, x_3, and x_4. After running the regression on $n = 16$ observations, we have the following information: SSR = 946.181 and SSE = 49.773. Please answer the following questions.

   a. What is the r^2?

   b. What is the adjusted R^2

   c. What is the F statistic?

   d. What is the p-value?

   e. Is the overall regression model significant? Test at $= 0.05$ level of significance.

4. We want to consider the correlation between the response variable, Intelligence (PIQ) and the three explanatory variables: height, weight, and brain size. Refer to the following statistical software output for the multiple regression model as follows:

```
Model Summary
S                R-sq         R=sq(adj)   R-sq(pred)
19.7944          24.49%       23.27%      12.76%


Coefficients
Term             Coef         SE Coef     T-Value     p-value
Constant         111.4        63          1.77        0.086
Brain            2.060        0.563       3.66        0.001
Height           -2.73        1.23        -2.22       0.033
Weight           0.001        0.197       0.00        0.998


Regression Equation
PIQ = 111.4 + 2.060 Brain -2.73 Height + 0.001 Weight
```

a. Conduct hypothesis tests at 0.05 significance level to test correlation each independent variable with the dependent variable.

b. Interpret the meaning of the r^2 and the adjusted R^2 value.

c. What is the effect of brain size on PIQ, after considering height and weight?

Try It Solutions

Try It 8.1

The outlier appears to be at (6, 58). The expected y value on the line for the point (6, 58) is approximately 82. Fifty-eight is 24 units from 82. Twenty-four is more than two standard deviations $(2s = (2)(8.6) = 17.2)$. So 82 is more than two standard deviations from 58, which makes (6, 58) a potential outlier.

Try It 8.2

1. $\hat{y} = -173.51 + 4.83(68) = 154.93$
2. The 95% prediction interval is $116.02 < y < 193.84$

Chapter 9
Hypothesis Testing with Two Samples

In the previous chapters, we studied inferential statistics about a population parameter, such as estimating the value of a population parameter using the confidence interval (Chapter 5) and hypothesis testing claims about a population parameter (Chapter 7). Hypothesis testing is important in inferential statistics where we use sample data to draw conclusions about a population. In this chapter, we continue with hypothesis testing that involves two populations. For hypothesis testing comparing two population parameters, the two groups are classified as independent or dependent (matched pairs). **Independent groups** have two samples where values selected from one population are unrelated to sample values selected from the other population. **Matched pairs** or **dependent groups** have two dependent samples, meaning that the values from one sample are related to values from another.

In Chapter 7, we conducted hypothesis tests on claims about a population mean and a population proportion. This chapter will expand that process and compare two means or two proportions using the same general procedure.

After reading this chapter, you will be able to do the following:

1. Determine whether a situation involves two independent samples or two dependent samples.
2. Conduct a hypothesis test of a claim about two population proportions.
3. Construct a confidence interval estimate of the difference between two population proportions.
4. Conduct a hypothesis test of a claim about two independent population means.
5. Construct a confidence interval estimate of the difference between two independent population means.
6. Conduct a hypothesis test of a claim about two dependent population means (matched pairs).
7. Construct a confidence interval estimate of the difference between the mean. difference between two dependent populations (matched pairs).
8. Conduct a hypothesis test of a claim about two population standard deviations or variances.

9.1 Comparing Population Proportions

Overview

The section will examine how to test differences in population proportions to understand percentage variations between two groups better. For example, do the students and teachers at your school prefer tea or coffee more? Or do particular age groups of online shoppers prefer standard or expedited shipping? This section will focus on hypothesis testing for comparing population proportions.

Hypothesis Testing for Two Population Proportions

The hypothesis test for population proportion checks if the difference in the proportions between two independent groups is statistically different. The hypothesis provides insights into meaningful distinctions between the groups. Usually, the null hypothesis states that the two proportions are the same or, $H_0: p_1 = p_2$. To conduct the test, we use a **pooled proportion**, $\bar{p}$, which is a weighted average of the proportions from the two samples.

Notation

$\hat{p}_1$ and $\hat{p}_2$ = the sample proportions

n_1 and n_2 = sample sizes

x_1 and x_2 = number of successes

$\hat{p}_1 = \dfrac{x_1}{n_1}$

$\hat{p}_2 = \dfrac{x_2}{n_2}$

$\hat{q}_1$ and $\hat{q}_2$ = the complement of $\hat{p}_1$ and $\hat{p}_2$

$\hat{q}_1 = 1 - \hat{p}_1$

$\hat{q}_1 = 1 - \hat{p}_1$

Pooled Sample Proportion ($\bar{p}$)

The pooled proportion combines the two sample proportions ($\hat{p}_1$ and $\hat{p}_2$) into one proportion.

$$\bar{p} = \frac{x_1 + x_2}{n_1 + n_2}$$

$$\bar{q} = 1 - \bar{p}$$

Test Statistic for Two Proportions

$$z = \frac{(\hat{p}_1 - \hat{p}_2) - (p_1 - p_2)}{\sqrt{\dfrac{\bar{p}\,\bar{q}}{n_1} + \dfrac{\bar{p}\,\bar{q}}{n_2}}}$$

where $p_1 - p_2 = 0$

Confidence Interval Estimate of $p_1 - p_2$

The $(1 - \alpha)\%$ confidence interval of $p_1 - p_2$ is

$$\hat{p}_1 - \hat{p}_2 \pm z_{\alpha/2} \frac{(\hat{p}_1 - \hat{p}_2) - (p_1 - p_2)}{\sqrt{\dfrac{\hat{p}_1\hat{p}_2}{n_1} + \dfrac{\hat{p}_2\hat{q}_2}{n_2}}}$$

Requirements

1. The sample proportions are from simple random samples.
2. The two samples are independent.
3. Each sample has at least 5 successes and at least 5 failures, (that is $n\hat{p} \geq 5$ *and* $n\hat{q} \geq 5$ for each of the samples).

Example 1

Two brands of allergy medication are tested in adults to determine if there *is a difference in the proportions* of patient reactions. Two independent, simple random samples of 200 adults are taken. Of the adults given medication A, 180 had no allergy symptoms an hour after taking the medication. Of the adults given medication B, 188 had no allergy symptoms an hour after taking the medication.

Use the *p*-value and the critical value methods and a 0.1 significance level to test the claim that medications A and B have the same proportions of allergy relief among patients.

Construct the 90% confidence interval estimate for the difference population proportion between medications A and B.

Solution 1

1. The problem asks for a *difference in proportions*. Let p_A and p_B be the population proportions for medications A and B.

Check the Requirements

The two independent samples are simple random samples, and the number of successes from samples:

$$n\hat{p}_A = 180 > 5 \text{ and } n\hat{p}_B = 188 > 5$$

The requirements are satisfied.

State the Hypotheses

Let $\hat{p}_A$ and $\hat{p}_B$ = the sample proportions of adult patients who responded to medication A and medication B an hour after application. The null and the alternative hypotheses are as follows:

$$H_0: p_A = p_B$$
$$H_a: p_A \neq p_B$$

Perform the Hypothesis Test Using *p*-value and Critical Value Methods

Like the one sample proportion test, where we used the normal distribution to test claims about a single population proportion, the difference of two proportions also follows an approximate normal distribution. To perform the test, we use a pooled proportion, $\bar{p}$.

$$\bar{p} = \frac{x_A + x_B}{n_A + n_B} = \frac{180 + 188}{200 + 200} = 0.08$$

$$\bar{q} = 1 - \bar{p} = 1 - 0.08 = 0.92$$

The sample proportions are as follows:

$$\hat{p}_A = \frac{x_A}{n_A} = \frac{180}{200} = 0.9, \text{ and } \hat{p}_B = \frac{x_B}{n_B} = \frac{188}{200} = 0.94$$

The value of $p_1 - p_2$ is 0 as stated in the null hypothesis. You can find this with the following calculation of the test statistic:

$$z = \frac{(\hat{p}_A - \hat{p}_B) - (p_A - p_B)}{\sqrt{\frac{\bar{p}\,\bar{q}}{n_A} + \frac{\bar{p}\,\bar{q}}{n_B}}}$$

$$= \frac{(0.9 - 0.94) - 0}{\sqrt{\frac{0.08 \cdot 0.92}{200} + \frac{0.08 \cdot 0.92}{200}}} = -1.474$$

The p-value is the probability of obtaining another test statistic z at least or more extreme than 1.474. That means z has a value less than -1.474 or greater than 1.474 because this is a two-tailed test.

$$p\text{-value} = P(z < -1.474) + P(z > 1.474) = 0.1404$$

Since the p-value is greater than the significance level, we fail to reject H_0.

$$p\text{-value} = 0.1404 > \alpha = 0.1$$

Because this is a two-tailed test, meaning that half of the α value will be in the right tail and half in the left tail. The critical value for the normal distribution at the 90% confidence level is ±1.645 found in the z-table. Since the test statistic z is well within the critical values of ±1.645, we reach the same conclusion: we fail to reject H_0.

$$-1.645 < z = 1.474 < 1.645$$

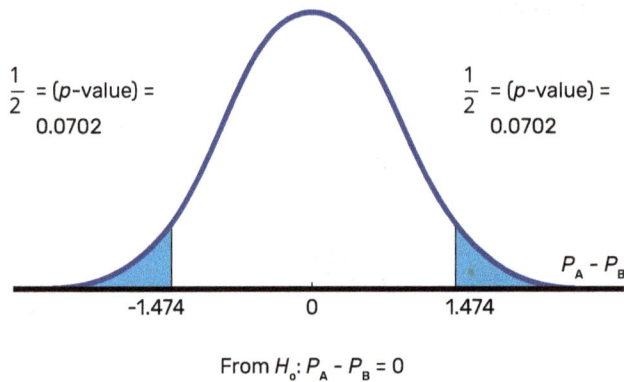

Figure 9.1. Distribution for Allergy Medication Hypothesis Tes

$$\frac{1}{2} = (p\text{-value}) = 0.0702$$

$$\frac{1}{2} = (p\text{-value}) = 0.0702$$

$P_A - P_B$

-1.474   0   1.474

From H_o: $P_A - P_B = 0$

Formal Conclusion

At a 0.1 level of significance, there is insufficient evidence to conclude that medications A and B do not have the same proportions of allergy relief among patients after one hour.

Using the TI-83, 83+, 84, 84+ Calculator

Press STAT.

Arrow over to TESTS and press 6: 2-PropZTest.

TI-84 Display

2-PropZTest	2-PropZTest
x_1 = 180	$P_1 \neq P_2$
n_1 = 200	z = -1.4744
x_2 = 188	P = .14037
n_2 = 200	$\hat{P}_1$ = .9
P_1: $\neq P_2$ < P_2 > P_2	$\hat{P}_2$ = .94
Color:	$\hat{P}$ = .92
Calculate Draw	n_1 = 2169
	n_2 = 2231

2. To construct the 90% confidence interval, we need to find $z_{\frac{\alpha}{2}}$. With $(1 - \alpha)\% = 90\%$, we find:

$$\alpha = 0.1 \text{ and } z_{\frac{\alpha}{2}} = z_{0.05} = 1.645$$

Now, using the confidence interval formula, we calculate the 90% confidence interval for the proportion difference between the two medications:

$$\hat{p}_1 - \hat{p}_2 \pm z_{\alpha/2}\sqrt{\frac{\hat{p}_1\hat{q}_1}{n_1} + \frac{\hat{p}_2\hat{q}_2}{n_2}}$$

$$= 0.9 - 0.94 \pm 1.645\sqrt{\frac{0.9 \cdot 0.1}{200} + \frac{0.94 \cdot 0.06}{200}}$$

$$= -0.04 \pm 0.0445 = (-0.0845, 0.0045)$$

Interpretation of Confidence Interval

We are 95% confident that the population proportion difference between the two medications is between -0.0845 and 0.0445. Since the confidence interval contains 0, there is no difference between the two medications in providing allergy relief among patients.

Try It 9.1

Two types of valves are being tested to determine if there is a difference in pressure tolerances. In a random sample of 100 Valve A, 15 cracked under 4,500 psi. In a random sample of 100 Valve B, 6 cracked under 4,500 psi. Test at a 5% level of significance.

Example 2

A research study was conducted to explore gender differences and dietary preferences. The researcher believed that the proportion of females who prefer fruit is less than the proportion of males who prefer fruit. The data collected among a random sample of adults aged between 30–50 is summarized in Table 9.1.

1. Is the proportion of females who prefer fruit less than the proportion of males who prefer fruit? Test at a 0.05 level of significance.
2. Construct the 95% confidence interval estimate for the difference population proportion between women and men who prefer fruit.

Table 9.1. Research Study About Dietary Preferences

	Males	Females
Prefer fruit	183	143
Total number surveyed	2231	2169

Solution 2

Check the Requirements

The two independent samples are simple random samples, and number of successes from samples:

$$n\hat{p}_M = 183 > 5 \text{ and } n\hat{p}_F = 143 > 5$$

The requirements are satisfied.

State the Hypotheses

This is a test of two population proportions where p_M and p_F are the population proportions for males and females. The phrase "less than" implies a left-tailed test. Let p_F and p_M be the population proportions of males and females who prefer fruit. The hypotheses are as follows:

$$H_0: p_F \geq p_M$$
$$H_a: p_F < p_M$$

Perform the Hypothesis Test Using p-value and Critical Value Methods

$$\bar{p} = \frac{x_F + x_M}{n_F + n_M} + \frac{143 + 183}{2169 + 2231} = 0.0741$$

$$1 - \bar{p} = 0.9259$$

The sample proportions are as follows:

$$\hat{p}_F = \frac{x_F}{n_F} = \frac{143}{2169} = 0.066$$

$$\hat{p}_M = \frac{x_M}{n_M} = \frac{183}{2231} = 0.082$$

The value of $p_F - p_M$ is 0 as stated in the null hypothesis and in the following calculation of the test statistic:

$$z = \frac{(\hat{p}_F - \hat{p}_M) - (p_F - p_M)}{\sqrt{\frac{\bar{p}\,\bar{q}}{n_F} + \frac{\bar{p}\,\bar{q}}{n_M}}}$$

$$= \frac{(0.066 - 0.082) - 0}{\sqrt{\frac{0.0741 \cdot 0.9259}{2169} + \frac{0.0741 \cdot 0.9259}{2231}}} = -2.038$$

The p-value is the probability of obtaining another test statistic z that is at least or more extreme than -2.038, such as a value less than -2.038 for left- tailed test.

Since the p-value is less than the significance level, p-value = 0.0207 < α = 0.05, we will reject H_0.

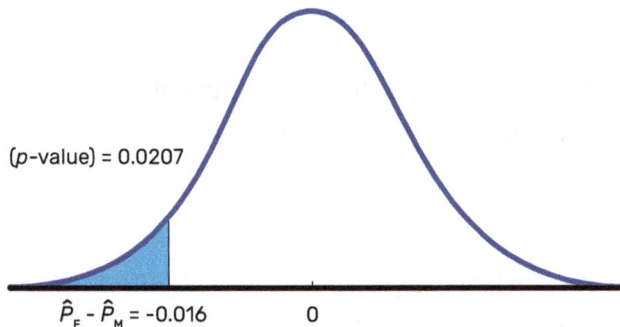

Figure 9.2. Distribution for Hypothesis Test on Fruit Preferences

(p-value) = 0.0207

$\hat{P}_F - \hat{P}_M = -0.016$ 0

Because this is a left-tailed test, the α value is the area in the left tail. The critical value is -1.645 can be found from the z-table. Since the test statistic z -2.038 is less than the critical values of -1.645 or $z = -2.038 < -1.645$, we reject H_0.

Formal Conclusion

At a 0.05 level of significance, there is sufficient evidence to conclude that the proportion of women who prefer fruit is less than the proportion of men who prefer fruit.

Using the TI-83, 83+, 84, 84+ Calculator

Press $STAT$, select $TEST$, then select $2-PropZTest$

TI Display

2-PropZTest	2-PropZTest
x_1 = 143	$P_1 < P_2$
n_1 = 2169	z = -2.0381
x_2 = 183	P = .020768
n_2 = 2231	$\hat{P}_1$ = .06593
$P_1: \ne P_2 \boxed{< P_2} > P_2$	$\hat{P}_2$ = .08202
Color:	n_1 = 2169
Calculate Draw	n_2 = 2231

2. To construct the 95% confidence interval, first find $z_{\alpha/2}$. With $(1 - \alpha)\% = 95\%$, we find:

$$\alpha = 0.05 \text{ and } z_{\alpha/2} = z_{0.025} = 1.96$$

Now, using the confidence interval formula, calculate the 95% confidence interval for the proportion difference between females and males who prefer fruit:

$$\hat{p}_1 - \hat{p}_2 \pm z_{\frac{\alpha}{2}} \sqrt{\frac{\hat{p}_1 \hat{q}_1}{n_1} + \frac{\hat{p}_2 \hat{q}_2}{n_2}}$$

$$= 0.066 - 0.082 \pm 1.96 \sqrt{\frac{0.066 \cdot 0.934}{2169} + \frac{0.082 \cdot 0.918}{2231}}$$

$$= -0.016 \pm 0.01545 = (-0.0315, -0.0006)$$

Interpretation of Confidence Interval

We are 95% confident that the population proportion difference between females and males who prefer fruit is between -0.0315 and -0.0006.

The confidence interval contains all negative values, indicating that the proportion of women who prefer fruit is less than that of men who prefer fruit.

Exercises 9.1

1. Two phone operating systems are tested to determine if there is a difference in the proportions of system failures (crashes). Fifteen out of a random sample of 200 phones with OS10 had system failures within the first eighty hours of operation, while 14 out of another random sample of 250 phones with OS12 had system failures. OS12 is believed to be more stable (have fewer crashes) than OS10.

   a. State the null and alternative hypotheses.

   b. At $\alpha = 0.05$, does it appear that OS12 is more stable (have fewer crashes) than OS10?

2. In the 2020 United States census, 10.2% of the population self-identified as multiracial. However, the percentage varied tremendously from state to state. Suppose that two random surveys are conducted. In the first random survey, out of 1,000 Oregonians, 42 people reported being of two or more races. In the second random survey, out of 500 Californians, 55 people reported being of two or more races. Conduct a hypothesis test to determine if the proportion for California is statistically higher than for Oregon.

   a. Define the random variable for this test.

   b. State the null and alternative hypotheses.

   c. Which distribution would you use for this hypothesis test?

   d. Calculate the test statistic.

   e. At $\alpha = 0.05$, does it appear that the proportion of multiracial Californians is higher than that of Oregonians?

3. A recent drug survey showed an increase in the use of drugs and alcohol among local high school seniors as compared to the national percentage. Suppose a survey of 100 local and 100 national seniors is conducted. Sixty-five local high school seniors reported using drugs or alcohol within the past month, while 60 national seniors reported using them. At the 0.1 level of significance, is there enough evidence that the proportion of drug and alcohol use is higher locally than nationally?

4. A year between 2015 and the present was randomly selected. That year, the proportion of Hispanic students at Portland Community College was 20%, with an enrollment of 48,000 students. At Chemeketa Community College, the proportion of Hispanic students was 28%, with an enrollment of 32,300 students. At the 0.05 significance level, do you think the percentage of Hispanic students at the two colleges is the same or different?

5. A group of friends debated whether more men or women use smartphones more, so they read a research study on smartphone use among adults. The survey results indicate that of the 950 men randomly sampled, 850 use smartphones. For women, 1265 of the 1,300 who were randomly sampled use smartphones. At the 0.01 significance level, test the claim that more women use smartphones than men.

6. Researchers conducted a study to determine if there is a difference in the use of e-readers by different age groups. Randomly selected participants were divided into two age groups. In the 16- to 29-year-old group, 17% of the 630 surveyed use e-readers, while 21% of the 2,300 participants 30 years old and older use e-readers. Use a 0.05 level of significance to perform an appropriate hypothesis test.

7. A high school principal claims that 20% of student athletes drive themselves to school, while 8% of non-athletes drive themselves to school. In a sample of 20 student athletes, 35% drive themselves to school. In a sample of 35 non-athlete students, 10% drive themselves to school. Is the percentage of student athletes who drive themselves to school more than the percentage of nonathletes? Use a 0.05 significance level to perform an appropriate hypothesis test.

9.2 Comparing Population Means: Independent Samples

Overview

Comparing two population means is very common. For example, comparing the effectiveness of two medications, two teaching methods, two manufacturing processes, or any scenario with two independent groups. In this section, we'll learn to conduct a hypothesis of a claim about two independent population means.

Hypothesis Testing with Unknown Standard Deviation

Most of the time, we don't know a population's standard deviation. If there is a lot of variation among the individual samples, vastly different means can occur by chance. To account for the variation, we take the difference of the sample means and divide it by the standard error to standardize the difference. The result is a t-score test statistic. For the hypothesis test, we calculate the estimated standard deviation, or **standard error**, as the difference in sample means. The hypothesis test procedure for population means is the same as for population proportions. You can use either the p-value method or the critical value method.

Notation

For population 1:

μ = population mean

σ_1 = unknown population standard deviation

n_1 = size of first sample

$\bar{x}_1$ = sample mean

s_1 = sample standard deviation

For population 2, $\sigma_2, n_2, \bar{x}_2$, and s_2 are the corresponding notations.

Test Statistic for Two Means of Independent Samples with Unknown σ_1 and σ_2

$$t = \frac{(\bar{x}_1 - \bar{x}_2) - (\mu_1 - \mu_2)}{\sqrt{\dfrac{(s_1)^2}{n_1} + \dfrac{(s_2)^2}{n_2}}}$$

where $\mu_1 - \mu_2 = 0$ as stated in H_0

Degrees of Freedom

$$df = \text{Smaller of } (n_1 - 1) \text{ and } (n_2 - 1)$$

More precise estimate:

$$df = \frac{(A + B)^2}{\dfrac{A^2}{n_{1-1}} + \dfrac{B^2}{n_{2-1}}}$$

where $A = \dfrac{(s_1)^2}{n_1}$ and $B = \dfrac{(s_2)^2}{n_2}$

Confidence Interval Estimate for Means of Independent Samples

$$(\bar{x}_1 - \bar{x}_2) - E < (\mu_1 - \mu_2) < (\bar{x}_1 - \bar{x}_2) + E$$

where

$$E = t_{\frac{\alpha}{2}} \sqrt{\frac{(s_1)^2}{n_1} + \frac{(s_2)^2}{n_2}}$$

The degrees of freedom are the same as for the hypothesis test: df Smaller of $(n_1 - 1)$ and $(n_2 - 1)$

Requirements

1. The samples are simple random samples.
2. The samples are independent.
3. The sample sizes are large ($n > 30$), or the samples come from normally distributed populations.

Example 3

The average amount of time boys and girls aged seven to 11 spend playing sports daily is believed to be the same. A study collects the data in Table 9.2. The samples come from populations that have normal distributions. At the 0.05 significance level, is there enough evidence to believe that boys and girls spend the same amount of time playing sports?

Table 9.2. Average Hours Spent Playing Sports per Day by Gender

	Sample Size	Average Number of Hours Playing Sports per Day	Sample Standard Deviation
Girls	9	2	0.866
Boys	16	3.2	1.00

Solution 3

Let μ_g be the population mean for girls and μ_b be the population mean for boys.

Check the Requirements

The two samples are simple random samples and independent. The samples come from normally distributed populations with unknown standard deviations.

State the Hypotheses

The random variable represents the difference in the mean amount of time girls and boys play sports daily. This is a two-tailed test.

$$H_0: \mu_g = \mu_b \quad (H_0: \mu_g - \mu_b = 0)$$
$$H_a: \mu_g \neq \mu_b \quad (H_a: \mu_g - \mu_b \neq 0)$$

From the data in Table 9.2, we know the following:

$$s_g = 0.866, \, s_b = 1, \, \bar{x}_g = 2, \, \text{and} \, \bar{x}_b = 3.2$$

Find the Test Statistic

$$t = \frac{(\bar{x}_g - \bar{x}_b) - (\mu_g - \mu_b)}{\sqrt{\frac{(s_g)^2}{n_g} + \frac{(s_b)^2}{n_b}}} = \frac{(2 - 3.2) - 0}{\sqrt{\frac{(0.866)^2}{9} + \frac{(1)^2}{16}}} = -3.142$$

The degrees of freedom is the smaller of $(9 - 1)$ *and* $(16 - 1)$, which is 8. From Table A-2, the critical value and the *p*-value are as follows:

$$t_c = -2.306$$

Since $t = -3.142 < t_c = -2.306$, reject H_0.

$$0.01 < p\text{-value} < 0.02$$

Since $\alpha > p$-value, reject H_0.

Note: since the critical value and *p*-value methods give equivalent results, it's not necessary to use both methods.

Find the Margin of Error

$$E = t_{\frac{\alpha}{2}} \sqrt{\frac{(s_1)^2}{n_1} + \frac{(s_2)^2}{n_2}} = 2.306 \sqrt{\frac{(0.866)^2}{9} + \frac{(1)^2}{16}} = 0.8806$$

Confidence Interval Estimate of $\mu_g - \mu_b$

$$(\bar{x}_1 - \bar{x}_2) - E < (\mu_g - \mu_b) < (\bar{x}_1 - \bar{x}_2) + E$$
$$(2 - 3.2) - 0.8806 < (\mu_g - \mu_b) < (2 - 3.2) + 0.8806$$
$$-2.081 < (\mu_g - \mu_b) < -0.3194$$

We are 95% confident that the population mean difference between the amount of time girls and boys spend playing sports daily is between -2.081 hours and -0.32 hours. The negative difference indicates that boys on average spend more time on sports than girls.

Formal Conclusion

At the 0.05 significance level, there is sufficient evidence to conclude that the mean number of hours that girls and boys aged seven to 11 play sports per day is different.

Using the Ti-83, 83+, 84, 84+ Calculator

Press STAT.

Arrow over to TESTS and press 4:2-SampleTTest.

TI Display

2-SampleTTest	2-SampleTTest
Inpt: Data Stats	μ_1: $\neq \mu_2$
$\bar{x}_1 = 2$	$t = -3.14239$
$s_{x1} = 0.866$	$p = 0.0054$
$n_1 = 9$	$df = 18.846$
$\bar{x}_2 = 3.2$	$\bar{x}_1 = 2$
$s_{x2} = 1$	$\bar{x}_2 = 3.2$
$n_2 = 16$	$s_{x1} = 0.866$
μ_1: $\boxed{\neq \mu_2}$ $< \mu_2$ $> \mu_2$	$s_{x2} = 1$
Pooled: $\boxed{No}$ Yes	$n_1 = 9$
	$n_2 = 16$

Try It 9.2

Two samples are shown in Table 9.3. Both have normal distributions. The means for the two populations are thought to be the same. Is there a difference in the means? Test at the 5% significance level.

Table 9.3. Samples with Normal Distributions

	Sample Size	Sample Mean	Sample Standard Deviation
Population A	25	5	1
Population B	16	4.7	1.2

Example 4

Many people claim that students who take online classes don't perform as well as those who take in-person classes. A professor at a large community college wanted to test this claim. The professor believed that the mean of the final exam scores for the online statistics class is lower than that of the face-to-face statistics class taught by the same professor. The randomly selected 20 final exam scores from each normally distributed group are listed in Table 9.4 and Table 9.5.

Table 9.4. Online Class Final Exam Scores

67	41	85	55	82	91	73
70	38	61	88	70	58	91
94	88	64	55	88	97	

Table 9.5. In-Person Class Final Exam Scores

77	95	81	74	98	88	85
69	57	69	67	97	85	88
98	61	92	90			

Use a 0.05 and a 0.01 significance level to test the claim that the mean of the final exam scores of the online class is lower than the mean of the final exam scores of the face-to-face class.

Solution 4

Let μ_1 be the population mean for online final exam scores and μ_2 be the population mean for the face-to-face class final exam scores.

Check the Requirements

The two samples are simple random samples and independent. The final exam scores come from normally distributed populations with unknown standard deviations. The requirements are satisfied.

State the Hypotheses

$$H_0: \mu_1 \geq \mu_2$$
$$H_a: \mu_1 < \mu_2$$

Perform the Hypothesis Test Using a Graphing Calculator and R Programming

Use TI-84

First, find the summary statistics

Press STAT then use EDIT to enter sample data.

TI Display

2-SampleTTest	2-SampleTTest
Inpt: Data Stats	μ_1: < μ_2
List1: L₁	t = -1.781
List2: L₂	P = 0.04185
Freq1: 1	df = 34.67
Freq2: 2	$\bar{x}_1$ = 72.8
μ_1: ≠ μ_2 < μ_2 > μ_2	$\bar{x}_2$ = 81.72
Pooled: No Yes	s_{x1} = 17.719
Calculate Draw	s_{x2} = 13.01
	n_1 = 20
	n_2 = 18

From the TI display, p-value = 0.0419. Since the p-value is less than the significance level 0.05, reject H_0.

There is sufficient evidence at α = 0.05 significance level that the mean of the final exam scores of the online class is lower than the mean of the final exam scores of the face-to-face class.

Use R Programming

In R, the two-sample *t*-test is called the Welch Two Sample *t*-test. We use the R function:

```
t.test (variable 1, variable 2, alternative = "  ")
```

where the alternative can be one of the following three options: "two.sided", "less", and "greater".

To proceed, enter the data for online and in-person exam scores in R as follows:

```
>online<-c(67,41,85,55,82,91,73,70,38,61,88,70,58,91,94,88,64,55,88,97)
>In_person<-c(77,95,81,74,98,88,85,69,57,69,67,97,85,88,98,61,92,90)
```

Next, type the following to run the Welch Two Sample t-test:

```
> t.test(online, In_person, alternative = "less")
```

The test result from R is displayed below:

```
                Welch Two Sample t-test
data:   online and In_person
t = -1.7808, df = 34.675, p-value = 0.04185
alternative hypothesis: true difference in means is less than 0
95 percent confidence interval:
        -Inf -0.4550112
sample estimates:
mean of x mean of y
 72.80000  81.72222n₂ = 18
```

Both TI-84 and R yield the same result, *p*-value = 0.04185. Since *p*-value < 0.05, we reject H_0

Formal Conclusion

There is sufficient evidence at 0.05 significance level that the mean of the final exam scores of the online class is lower than the mean of the final exam scores of the in-person class.

However, at the 0.01 significance level, we fail to reject H_0 because the p-value of 0.01485 > 0.01. Therefore, there is insufficient evidence at 0.01 significance level that the mean of the final exam scores of the online class is lower than the mean of the final exam scores of the in-person class. This conclusion is the opposite of the previous conclusion when $\alpha = 0.05$.

Note

The significance level for a hypothesis test can be set at a value depending on the consequences of a false positive. Increasing or decreasing the significance level can change the conclusion from a hypothesis test, as you see in Example 5. Remember that the conclusion we draw from a hypothesis test is related to the significance level of the test. Therefore, when reporting the results of a hypothesis test, you should always state the significance level α.

Hypothesis Testing with Known Population Standard Deviation

It's unlikely that we know the population standard deviation, but if we do, the procedure for hypothesis testing is the same as testing an unknown standard deviation. The main difference is that the test statistics are based on the normal distribution rather than the t-distribution, so we use slightly different formulas.

For population 1:

μ_1 = population mean

σ_1 = unknown population standard deviation

n_1 = size of first sample

$\bar{x}_1$ = sample mean

s_1 = sample standard deviation

For population 2:

$\mu_2, \sigma_2, n_2, \bar{x}_2$, and s_2 are the corresponding notations.

Standard Error

$$\sqrt{\frac{(\sigma_1)^2}{n_1} + \frac{(\sigma_2)^2}{n_2}}$$

Test Statistic for Two Means of Independent Samples with Known σ

$$z = \frac{(\bar{x}_1 - \bar{x}_2) - (\mu_1 - \mu_2)}{\sqrt{\frac{(\sigma_1)^2}{n_1} + \frac{(\sigma_2)^2}{n_2}}}$$

Margin of Error

$$E = z_{\frac{\alpha}{2}} \sqrt{\frac{(\sigma_1)^2}{n_1} + \frac{(\sigma_2)^2}{n_2}}$$

Example 5

A cleaning company wants to compare the durability of floor wax products to see which one lasts longer. Twenty floors are randomly assigned to test each wax. Both populations have normal distributions. The data are recorded in Table 9.6. Does the data indicate that wax 1 is more effective than wax 2? At $\alpha = 0.05$, state your conclusion.

Table 9.6. Study of Types of Floor Wax

Wax	Sample Mean Number of Months Floor Wax Lasts	Population Standard Deviation
1	3	0.33
2	2.9	0.36

Solution 5

This is a test of two independent groups' population means with known population standard deviations.

State the Hypotheses

$\bar{X}_1 - \bar{X}_2$ = difference in the mean number of months the competing floor waxes last

$H_0: \mu_1 \leq \mu_2$

$H_a: \mu_1 > \mu_2$

Find the Test Statistic

This is a right-tailed test. The words "is more effective" indicate that Wax 1 lasts longer than Wax 2.

The population standard deviations are given, so the distribution is normal. Using the formula, the distribution is:

$$z = \frac{(\bar{X}_1 - \bar{X}_2) - (\mu_1 - \mu_2)}{\sqrt{\frac{(\sigma_1)^2}{n_1} + \frac{(\sigma_2)^2}{n_2}}} = z = \frac{(3 - 2.9) - 0}{\sqrt{\frac{(.33)^2}{20} + \frac{(0.36)^2}{20}}} = 0.916$$

Since $\mu_1 \leq \mu_2$, then $\mu_1 - \mu_2 \leq 0$, and the mean for the normal distribution is zero.

p-value = 0.1799

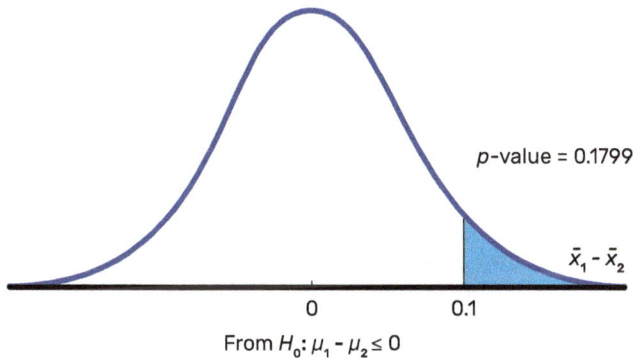

Figure 9.3. Distribution for Hypothesis Test of Floor Wax Efficacy

p-value = 0.1799

$\bar{x}_1 - \bar{x}_2$

0   0.1

From $H_0: \mu_1 - \mu_2 \leq 0$

Since $\alpha = 0.05 < p$-value, do not reject H_0.

Formal Conclusion

There is insufficient evidence to conclude that the mean time Wax 1 lasts is longer than the mean time Wax 2 lasts. At the 0.05 significance level, we conclude that Wax 1 isn't more effective than Wax 2.

Exercises 9.2

1. An instructor teaches two sections of a statistics course, one during the day and one at night. She believes that there is no significant difference between mean class scores on the final exam for day students and night students. She takes random samples from each of the populations. For 35 students taking the day class, the mean and standard deviation were 75.86 and 16.91, respectively. For 37 students taking the night class, the mean and standard deviation were 75.41 and 19.73. The "day" subscript refers to the statistics day students. The "night" subscript refers to the statistics night students. Which of the following is an appropriate alternative hypothesis for the hypothesis test?

    a. $\mu_{day} > \mu_{night}$

    b. $\mu_{day} < \mu_{night}$

    c. $\mu_{day} = \mu_{night}$

    d. $\mu_{day} \neq \mu_{night}$

2. Elijah wants to know whether textbook costs are different for different subjects. He selects a random sample of 33 sociology textbooks offered on a popular website. The mean price of his sample is $74.64 with a standard deviation of $49.36. He then selects a random sample of 33 math and science textbooks from the same site. The mean price of this sample is $111.56 with a standard deviation of $66.90. Is the mean price of a sociology textbook lower than the mean price of a math or science textbook? Use a 0.05 significance level to perform the hypothesis.

3. The means of the number of revolutions per minute (RPM) of two competing engine brands is compared. Thirty engines of each brand are randomly assigned to be tested. Both populations have normal distributions. Table 9.7 shows the result. Do the data indicate that Engine 2 has higher RPM than Engine 1? Test at a 5% level of significance.

Table 9.7. Comparison of Engines Brands' RPM

Engine	Sample Mean Number of RPM	Population Standard Deviation
1	1,500	50
2	1,600	60

4. Mean entry-level salaries for college graduates with mechanical engineering degrees and electrical engineering degrees are believed to be approximately the same. However, a recruiting office thinks that the mean mechanical engineering salary is lower than the mean electrical engineering salary. The recruiting office randomly surveys 50 entry-level mechanical engineers and 60 entry-level electrical engineers. Their mean salaries were $46,100 and $46,700, respectively. Their standard deviations were $3,450 and $4,210, respectively. Conduct a hypothesis test to determine if you agree that the mean entry-level mechanical engineering salary is lower than the mean entry-level electrical engineering salary at a 0.1 significance level.

5. Some manufacturers claim that non-hybrid sedan cars have a lower mean miles-per-gallon (mpg) than hybrid ones. Suppose that consumers test 21 hybrid sedans and get a mean of 31 mpg with a standard deviation of 7 mpg. Thirty-one non-hybrid sedans get a mean of 22 mpg with a standard deviation of 4 mpg. Suppose that the population standard deviations are known to be 6 and 3, respectively. Conduct a hypothesis test to evaluate the manufacturers claim.

9.3 Comparing Population Means with Matched Pairs

Overview

When comparing dependent samples, we explore the relationships within matched data pairs. As a refresher, dependent samples are those where values from one sample are related to values from the other. We could analyze the impact of a new training program on employee productivity by examining before-and-after performance scores. We might examine how changing a commuting route affects travel times by comparing the durations before and after the change.

Hypothesis Testing for Two Dependent Samples

Hypothesis testing for matched pairs examines the average differences within the values of each group. When comparing two dependent samples, the two measurements of pairs will be correlated. In analysis, the random variable is the mean of sample differences d_i, where $d_i = x_{1i} - x_{2i}$ and $x_{1i} - x_{2i}$ are individual values from each correlated sample. The method used in this section is the same as the method for testing a claim about a population mean.

Notation

d_i = differences between the two values of a matched pair

$\bar{d}$ = mean value of all *di* for the paired sample data

μ_d = mean value of all *di* for the population of all matched pairs

s_d = standard deviation of all *di* for the paired sample data

n = number of matched pairs in sample

Test Statistic for Matched Pairs (H_0: $\mu_d = 0$)

$$t = \frac{\bar{d} - \mu_d}{\frac{s_d}{\sqrt{n}}}$$

Confidence Interval Estimate for Mean of the Population Differences

$$\bar{d} - E < \mu_d < \bar{d} + E$$

Where the margin of error is:

$$E = t_{\frac{\alpha}{2}} \frac{s_d}{\sqrt{n}}$$

The degrees of freedom is $df = n - 1$.

Requirements
1. The samples are dependent.
2. The matched pairs are simple random samples.
3. The sample sizes are large ($n > 30$), or the samples come from approximately normally distributed populations.

Example 7

A study was conducted to investigate the effectiveness of hypnotism in reducing pain. Results for randomly selected subjects are shown in Table 9.8. The scores are sensory measurements, and a lower score indicates less pain. The "before" value is matched to an "after" value, and the differences are calculated. The differences have a normal distribution.

1. Use a 0.05 significance level to test the claim that the mean differences before and after hypnotism is less than zero, i.e. μ_d = after – before < 0.
2. Construct a 95% confidence interval for the mean difference of the matched pairs.

Table 9.8. Effects of Hypnotism in Pain Reduction

Subject	A	B	C	D	E	F	G	H
Before	6.6	6.5	9.0	10.3	11.3	8.1	6.3	11.6
After	6.8	2.4	7.4	8.5	8.1	6.1	3.4	2.0

Solution 7

1. The matched pairs of Before and After values are shown in Table 9.9. The data for the differences between the matched pairs is shown in the third row. Each matched pair difference must be in the same subtraction order. Here, the difference d_i is After – Before.

Table 9.9. Matched Pairs for Hypnotism Study

Subject	A	B	C	D	E	F	G	H
Before	6.6	6.5	9.0	10.3	11.3	8.1	6.3	11.6
After	6.8	2.4	7.4	8.5	8.1	6.1	3.4	2.0
d_i = After – Before	0.2	- 4.1	- 1.6	- 1.8	-3.2	- 2	- 2.9	- 9.6

The sample mean and the sample standard deviation can be found using TI-84 using "STAT", "CALC", and "1-Var Stats".

$\bar{d}$ = the mean difference of the sensory measurements, and s_d = the standard deviation of the sensory measurements

$$\bar{d} = \text{-}3.125, \text{ and } s_d = 2.911$$

Check for Requirements

The sample of 7 objects is a simple random sample and the differences have a normal distribution. The requirements for matched pair test are satisfied.

State the Hypotheses

Let μ_d be the population mean for the differences. The hypothesis is as follows:

H_0: $\mu_d \geq 0$ $(\mu_{after} \geq \mu_{before})$
H_a: $\mu_d < 0$ $(\mu_{after} < \mu_{before})$

The null hypothesis is zero or positive, meaning that there is the same or less pain felt after hypnotism. The alternative hypothesis is negative, meaning there is more pain felt after hypnotism, indicating no improvement.

Calculate the Test Statistic

$$t = \frac{\overline{d} - \mu_d}{\frac{s_d}{\sqrt{n}}} = \frac{-3.125 - 0}{\frac{2.911}{\sqrt{8}}} = -3.036$$

Perform the Hypothesis Test Using *p*-value and Critical Value Methods

First, let's use the critical value method. Since this is a left-tailed test, use Table A-3. In the column for 0.025 (area in one tail) with the *df* = 7, we find the critical value as follows:

$$t_c = -2.365$$

Comparing the test statistic with the critical value, it indicates that $t < t_c$.

$$t = -3.036 < t_c = -2.365$$

Reject H_0.

Next, let's use the TI-84 to find the *p*-value.

1. Press STAT, then use EDIT to enter sample data di in L1.
2. Press TEST, then choose2: T-Test and make sure to choose Data from Input.
3. Select for the alternative hypothesis.

TI-84 Display

T-Test	T-Test
Inpt: [Data] Stats	μ < 0
μ₀: 0	t = -3.036
List: L₁	P = 0.0095
Freq1: 1	x̄ = -3.125
Freq: 1	x̄₂ = 81.72
μ: ≠ μ₀ [< μ₀] > μ₀	sₓ = 2.9114
Calculate Draw	n = 8

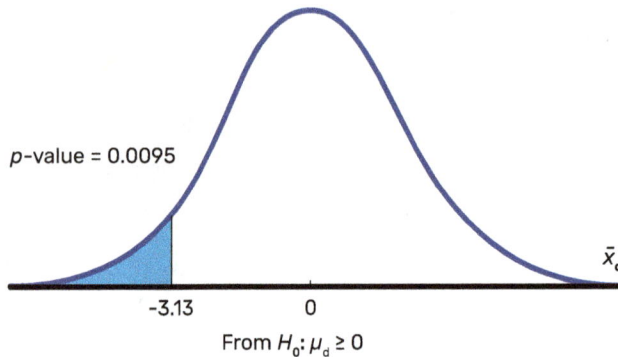

Figure 9.4. Distribution of Hypothesis Test of Hypnotism and Pain Relief

From the TI-84 display, we see that p-value $= 0.0095$. Since the p-value $< a$, we reach the same conclusion to reject H_0. This means that $\mu_d < 0$, and there is improvement.

Similarly, we can use R to analysis the data for the matched pairs. In R, type the sample data as follows with $x_1 = $ "after", and $x_2 = $ "before":

```
x1<-c(6.8,2.4,7.4,8.5,8.1,6.1,3.4,2)
x2<-c(6.6,6.5,9,10.3,11.3,8.1,6.3,11.6)
t.test(x1, x2, pair = T, alt = "less")
```

The paired test is performed in R and the test result is as follows:

```
                    Paired t-test
data: x₁ and x₂
t = -3.0359, df = 7, p-value = 0.009487
alternative hypothesis: true mean difference is less than 0
sample estimates:
mean difference
     -3.125
```

Formal Conclusion

At 0.05 significance level, there is sufficient evidence to conclude that the sensory measurements of pain, on average, are lower after hypnotism. Hypnotism appears to be effective in reducing pain.

2. To construct the 95% confidence interval, first find $t_{\alpha/2}$, and $df = 7$. With 95% confidence interval $(1 - \alpha)\% = 95\%$ we have $\alpha = 0.05$ and $t_{\alpha/2} = t_{0.025} = \pm\, 2.365$.

Now, using the confidence interval formula, calculate the 95% confidence interval for the population proportion mean difference.

Determine the Margin of Error

$$E = t_{\frac{\alpha}{2}} \frac{S_d}{\sqrt{n}} = 2.365 \frac{2.9114}{\sqrt{8}} = 2.4344$$

Find the Confidence Interval

$$\bar{d} - E < \mu_d < \bar{d} + E$$
$$-3.125 - 2.4344 < \mu_d < -3.12 + 2.4344$$
$$-5.5594 < \mu_d < -0.691$$

The confidence interval can also be found using a TI-84 by using "TInerval..." under TESTS.

Interpret the Confidence Interval

We are 95% confident that the population mean difference between the matched pairs of sensory measurements of pain before and after hypnotism is between -5.5594 and -0.691, indicating that hypnotism is effective in reducing pain.

Exercises 9.3

1. Why is it important to know whether samples are independent or dependent before conducting hypothesis tests?

2. How does the process of conducting a hypothesis test for two dependent population means differ from tests involving independent samples?

3. A new medicine is said to help improve sleep. Eight subjects are picked at random and given the medicine. The means hours slept for each person were recorded before starting the medication and after. They are listed in Table 9.10. Construct a 95% confidence interval for the mean of the before-and-after differences. Does the new medicine appear to be effective in improving sleep at $\alpha = 0.05$?

Table 9.10. Effects of Medications on Sleep

Subject	1	2	3	4	5	5	7	8
Before	4	8	5	8	6	7	4	5
After	6	7	9	7	5	8	6	6

4. A study investigated how effective a new diet was in lowering cholesterol. Results for the randomly selected subjects are shown in Table 9.11. The differences have a normal distribution. Are the subjects' cholesterol levels lower on average after the diet? At $\alpha = 0.05$, what is your conclusions?

Table 9.11. Study on Diet Effect on Lowing Cholesterol

Subject	A	B	C	D	E	F	G	H	I
Before	209	210	205	198	216	217	238	240	222
After	199	207	189	209	217	202	211	223	201

5. A study of marital satisfaction of dual-career couples examined the father's involvement in child development. Couples were asked to rate the statement, "I'm pleased with the way we divide the responsibilities for childcare." The rating scale was from one (strongly agree) to five (strongly

disagree). Table 9.12 contains ten of the paired responses for husbands and wives. Conduct a hypothesis test to see if the mean difference between the husband's satisfaction level and the wife's satisfaction level is negative. A negative difference would mean that the husband is happier than the wife about how childcare responsibilities are divided.

Table 9.12. Ratings of Parental Involvement in Childcare Responsibilities

Wife's Score	2	2	3	3	4	2	1	1	2	4
Husband's Score	2	2	1	3	2	1	1	1	2	4

9.4 Comparing Population Variances or Standard Deviations

Overview

Often it is more useful to compare two variances rather than two means. For instance, college administrators would like two college professors grading exams to have the same variation in their grading. For a lid to fit a container, the variation in the lid and the container should be the same. A supermarket might be interested in the variability of check-out times for two lines. In this section, we will use the F-test to test claims about two population variances or standard deviations.

F-Test for Two Variances or Standard Deviations

The distribution for the hypothesis test for two variances or standard deviations is called the F-distribution, named after the English statistician, Ronald Fisher. The F-statistic is a fraction with two types of degrees of freedom: the numerator degrees of freedom df_N and the denominator degrees of freedom; df_D. The F-test may be left, right, or two-tailed. However, it is easier to make sample 1 variance s_1^2 to be the larger variance and sample 2 variance s_2^2 to be the smaller variance so that we would only need to work with a right tail test and find the right critical value.

Unlike most other tests in this book, the F test for equality of two variances is very sensitive to deviations from normality. If the two distributions are not normal, the test can give higher p-values than it should, or lower ones, in ways that are unpredictable.

Notation

s_1^2 = The larger sample variance

s_2^2 = The smaller sample variance

σ_1^2 = The population variance from which the sample with the larger sample variance was drawn

σ_2^2 = The population variance from which the sample with the smaller sample variance was drawn

n_1 = sample size of the larger sample

n_2 = sample size of the smaller sample

Numerator degrees of freedom $df_N = n_1 - 1$

Denominator degrees of freedom $df_D = n_2 - 1$

Test Statistic

$$F = \frac{s_1^2}{s_2^2}$$

where s_1^2 is the larger sample variance of the two sample variances

Requirements

1. The populations from which the two samples come from are normally distributed.
2. The two samples are simple random samples.
3. The two populations are independent of each other.

If the two populations have the same variances, then $\frac{s_1^2}{s_2^2}$ is close to one. But if the two population variances are very different, then $\frac{s_1^2}{s_2^2}$ tend to be very different too. Therefore, if F is close to one, the evidence favors the null hypothesis, meaning the two population variances are equal. But if F is much larger than one, then the evidence is against the null hypothesis.

Example 8

Two college instructors are interested in whether there is any variation in the way they grade math exams. They each grade the same set of 30 exams selected from normally distributed population. The first instructor's grades have a variance of 52.3. The second instructor's grades have a variance of 89.9. Test the claim that the first instructor's variance is smaller. The level of significance is 0.1.

Solution 8

Check the Requirements

The two populations are independent, the samples from the two instructors are simple random samples, and the population has a normal distribution. The requirements for the F-test are satisfied.

State the Hypotheses

Let $s_1^2 = 89.9$ the variance from second instructor's grades.

Let $s_2^2 = 52.3$ be the variance from the first instructor's grades. This is a right-tailed test with $n_1 = n_2 = 30$

$$H_0: \sigma_1^2 \le \sigma_2^2$$
$$H_1: \sigma_1^2 > \sigma_2^2$$

Calculate the Test Statistic

$$F = \frac{s_1^2}{s_2^2} = \frac{89.9}{52.3} = 1.7189$$

Perform the Hypothesis Test Using the Critical Value and *p*-value Methods

First, let's find the critical value. The numerator and denominator degrees of freedom are $df_N = df_D = 29$.

Using the *F*-Distribution table from the Appendix (Table A.4), we find the critical value:

$$F_c \approx 1.8543$$

Since $F < F_c \approx 1.8543$, we fail to reject H_0.

Next, let's find the *p*-value on the TI-84. Press TESTS, then go to E:2-SampleFTest.

TI-84 Display

```
2-SampleFTest            2-SampleFTest
Inpt: Data Stats         σ₁ > σ₂
 s    = √89.9            F = 1.7189
  x1
 n  = 30                 P = 0.0753
  1
 s    = √52.3            s  = 9.482
  x1                      1
 n  = 30                 s   = 7.232
  2                       x2
 σ: ≠ σ₂ < σ₂ > σ₂       n  = 30
                          1
 Calculate Draw          n  = 30
                          2
```

The *p*-value = 0.0753. Since the *p*-value > 0.05, we reach the same conclusion of failing to reject H_0.

Formal Conclusion

At the 0.05 significance level, there is insufficient evidence to conclude that the variance for the first instructor is smaller.

Note: The *F*-test can be performed in R with the function:

```
var.test (var 1, var 2, alternative = " ")
```

where the alternative can be one of these three: "two.sided", "greater", and "less".

Example 9

Is the variance of money spent ($) by mall shoppers spend on Saturdays the same as the variance of money spent by mall shoppers on Sundays? Table 9.13 shows the results of a randomized study. Assume the populations are normally distributed. Perform the *F*-test for variances. Use a 0.1 significance level.

Table 9.13. Amount of Money Mall Shoppers Spent on Weekends

Saturday	75	18	150	94	62	73	62	0	124	50	31	118	
Sunday	44	58	61	19	99	60	89	137	82	39	127	141	73

Solution 9

Check the Requirements

Since Saturday and Sunday populations are normally distributed, samples are independent of each other, and are simple random samples, the requirements are satisfied.

State the Hypotheses

H_0: The population variance for Saturday shoppers is the same as for Sunday shoppers

H_a: The population variance for Saturday shoppers is not the same as that for Sunday shoppers

Calculate the *F*-test Statistic and *p*-value

This is a two-tailed test. We will use R to conduct the F-test by entering the data values for Saturday and Sunday as follows:

```
> Saturday<-c(75,18,150,94,62,73,62,0,124,50,31,118)
> Sunday<-c(44,58,61,19,99,60,89,137,82,39,127,141,73)
> var.test(Saturday, Sunday, alternative = "two.sided")
```

```
F test to compare two variances
data: Saturday and Sunday
F = 1.3543, num df = 11, denom df = 12, p-value = 0.609
alternative hypothesis: true ratio of variances is not equal to 1
95 percent confidence interval:
0.4077396 4.6447236
sample estimates:
ratio of variances
         1.3543
```

The *p*-value = 0.609 > 0.1, so we fail to reject the null hypothesis.

Formal Conclusion

We conclude that at the 0.1 significance level, there is no difference between the population variances for Saturday shoppers and Sunday shoppers.

Exercises 9.4

1. Two cyclists are comparing the variances of their overall uphill speeds. Each cyclist records their speeds going up 35 hills. The first cyclist has a variance of 23.8 and the second cyclist has a variance of 32.1. The cyclists want to see if their variances are the same or different. Assume that populations are normally distributed.

    a. State the null and alternative hypotheses.

    b. What is the *F*-statistic?

c. At the 5% significance level, what can we say about the cyclists' variances?

2. Three students, Linda, Tuan, and Javier, are given five laboratory rats each for a nutritional experiment. At the beginning of the experiment, the weight of each rat is recorded in grams. Linda feeds her rats Formula A, Tuan feeds his rats Formula B, and Javier feeds his rats Formula C. At the end of a specified time, each rat is weighed again and the net gain in grams is recorded in Table 9.14. Assume all three rat populations are normal. At the 0.1 significance level, determine whether the variance in weight gain is the same among Javier's and Linda's rats.

Table 9.14. Net Weight Gain (grams) of Rats

	Rat 1	Rat 2	Rat 3	Rat 4	Rat 5
Linda's rats	43.5	39.4	41.3	46	38.2
Tuan's rats	47	40.5	38.9	36.3	44.2
Javier's rats	51.2	40.9	37.9	45	48.6

3. Table 9.15 shows the results of a study of incomes on the East Coast and West Coast. Income is shown in thousands of dollars. Assume that both distributions are normal. Are the variances for incomes on the East Coast and the West Coast the same? Use a level of significance of 0.05.

Table 9.15. Income (in thousands of dollars) of East and West Coasts

East Coast	38	47	30	82	75	52	115	67
West Coast	71	126	42	51	44	90	88	

4. A packing company wants to know if the variation in packing time is the same between the new machine and the old machine. Independent samples were selected from each machine, and packaging times of the two machines are recorded in Table 9.16. At the 0.05 significance level, does the new packing machine have a lower population variance than the old packing machine?

Table 9.16. Packing Time from New and Old Machines

New Machine	42.1	43.3	41.4	42.2	41.1	41	41.6	42.8	42.3	41.7
Old Machine	42.7	43.8	43.5	43.1	42.8	44.6	43.3	43.5	42.7	44

5. Two types of car engines are compared. Engines of each type are randomly assigned to be tested, and the outcomes are the speed of engine measured in revolutions per minute (RPM). Twenty data values from Engine 1 and 25 from Engine 2 are recorded. Both samples are from populations with normal distributions. Table 9.17 gives the mean and the standard deviation of the two samples. Does the data suggest that Engine 2 has a higher population standard deviation in RPMs than Engine 1? Test at a 0.05 level of significance.

Table 9.17. Engines of Car Factory

Engine	Sample Mean Number of RPM	Sample Standard Deviation of RPM
1	1,500	50
2	1,600	60

Chapter 10
Hypothesis Testing for Categorical Data

This chapter will explore the chi-square distribution, a statistical tool that allows us to analyze categorical data associations and patterns. Imagine you're investigating whether there's a connection between smartphone preference and age groups in a community. You may want to know about the distribution of favorite ice cream flavors among students from two schools. The chi-square distribution becomes your guide to determine whether these variables are truly independent or have a significant link.

You've already been introduced to the chi-square distribution through confidence intervals of population variance (Chapter 6) and hypothesis testing for variance (Chapter 7). The chi-square distribution is a statistical probability distribution that describes the variation in categorical data. It is commonly used to assess the association or dependence between two categorical variables. The distribution's shape depends on the degrees of freedom, which influences the likelihood of different outcomes. Chi-square tests employing this distribution help analyze observed and expected frequencies in categorical data sets, providing insights into patterns and associations within diverse sets of non-numerical information. The chi-square test is also called Pearson's chi-squared test, named after the mathematician Karl Pearson. The chi-square test is applied to categorical data in three types of comparison: goodness-of-fit, independence, and homogeneity. We'll also look at Fisher's exact and McNemar's tests.

After reading this chapter, you will be able to do the following:

1. Determine whether a categorical variable fits a distribution using the chi-square goodness-of-fit test.

2. Determine whether there is a significant association (dependency) between two categorical variables using the chi-square test or Fisher's exact test for independence.

3. Test whether different populations have the same proportions of specific characteristics using the chi-square test for homogeneity or McNemar's test.

10.1 Goodness-of-Fit Test

Overview

The **goodness-of-fit test** determines whether an observed single row or column of frequency counts from a categorical variable has a specific distribution, such as a uniform or normal distribution. In the goodness-of-fit test, data in one row or column of a frequency table is compared with an expected distribution. In this case, only one qualitative variable for a single population is measured. Goodness-of-fit is typically used to see if the population is uniform (all outcomes occur with equal frequency), if the population is normal, or if the population is the same as another population with a known distribution. The goodness-of-fit test is usually right-tailed test.

Observed and Expected Frequencies

To perform the goodness-of-fit test, first determine an outcome's expected frequency (E). The **expected frequency** is the value of an outcome that we'd expect to get if the claimed distribution is accurate. You know the observed frequency (O) because you can count and record it in a frequency table. There are two paths to finding the expected frequency. A goodness-of-fit test can determine if the observed frequencies match the expected frequencies and if the distribution claims are true.

Expected Frequency (E) with Equally Likely Outcomes

$$E = \frac{n}{k}$$

Expected Frequency (E) with Unequally Likely Outcomes

$$E = np$$

If all outcomes are equally likely, the expected frequency is the total number of trials divided by the number of categories. For example, a vending machine claims to have an equal distribution of three types of energy bars: A, B, and C. To test this claim, suppose that 60 energy bars are purchased. Assuming an equal likelihood for each energy bar, the expected frequencies should be 20 bars of each type. Table 10.1 lists the expected frequency as well as the observed frequency.

Table 10.1. Observed and Expected Frequencies of Energy Bars in Vending Machine

Type	Observed Frequency (O)	Expected Frequency (E)
A	18	$\frac{1}{3} \times 60 = 20$
B	22	$\frac{1}{3} \times 60 = 20$
C	20	$\frac{1}{3} \times 60 = 20$

If all outcomes are not equally likely, then the expected frequency is the total number of trials multiplied by the probability of one specific outcome. For example, suppose a snack mix bag contains an unequal distribution of pretzels, peanuts, and chocolate chips, and the manufacturer says they include twice as many pretzels as peanuts or chocolate chips. To find out if this is true, we count the number of each snack in 75 snack mix bags. Table 10.2 shows the expected frequency and the actual observed frequency.

Table 10.2. Observed and Expected Frequencies of Snack Mix

Snack	Observed Frequency (O)	Expected Frequency (E)
Pretzels	30	$\frac{1}{3} \times 75 = 25$
Peanuts	20	$\frac{1}{6} \times 75 = 12.5$
Chocolate Chips	25	$\frac{1}{6} \times 75 = 12.5$

Notation

O = observed frequency of an outcome found in the sample data

E = expected frequency of an outcome if the claimed distribution is accurate

k = number of categories

n = total number of trials

p = probability that a sample value is within a specific category

Null and Alternative Hypotheses

H_0: The frequency count follows the claimed distribution.

H_a: The frequency count does not follow the claimed distribution.

Test Statistic for Goodness-of-Fit Test

$$\chi^2 = \sum \frac{(O - E)^2}{E}$$

where the degrees of freedom are $df = k - 1$

Requirements

1. The data are randomly selected.
2. The samples have frequency counts for each category.
3. The expected frequency in each category is at least 5. The observed outcome does not need to be 5.

Example 1

Corporate executives want to know if employees are equally absent on different days of a five-day workweek. Most executives believe that employees are absent equally during the week. A random sample of managers is surveyed, and the number of employee absences each day is recorded in Table 10.3. Use a 0.05 significance level to test the claim that the number of employee absences is uniformly distributed among the five working days of the week.

Table 10.3. Day of the Week Employees Were Most Absent

	Mon	Tues	Wed	Thurs	Fri	Total
Number of Absences	15	12	9	9	15	60

Solution 1

If the absent days occur with a uniform distribution, then it is expected that there will be 12 absences on each day of a five-day workweek. The values in Table 10.4 show the observed (O) and expected (E) values. We also add a row for expected proportions.

Table 10.4. Observed and Expected Values for Employee Absences by Day of the Week

	Mon	Tues	Wed	Thurs	Fri	Total
Observed Absences (O)	15	12	9	9	15	60
Expected Absences (E)	12	12	12	12	12	60
Expected Proportions (p)	0.2	0.2	0.2	0.2	0.2	1

Check the Requirements

The data is randomly selected. The data are frequency counts for each category (day). The expected frequency is 12 and greater than 5. The requirements are satisfied.

State the Hypotheses

H_0: The absent days occur with equal frequencies. They are uniformly distributed.

H_a: The absent days occur with unequal frequencies. They are not uniformly distributed.

Calculate the Test Statistic

$$\chi^2 = \sum \frac{(O-E)^2}{E} = \frac{(15-12)^2}{12} + \frac{(12-12)^2}{12} + \frac{(9-12)^2}{12} + \frac{(9-12)^2}{12} + \frac{(15-12)^2}{12} = 3$$

We can use the p-value or the critical value method to draw a conclusion.

As stated in the overview, the goodness-of-fit test is right-tailed. Using a computer or calculator, we find the p-value = 0.5578. Since the p-value > 0.05, we fail to reject H_0.

Use the chi-square table (Table A-X) with degrees of freedom = 4 (df = number of cells − 1). We find the critical value $\chi_c^2 = 9.488$. Since the test statistic $\chi^2 < \chi_c^2$, we would reach the same conclusion.

Formal Conclusion

At the 0.05 significance level, there is insufficient evidence to conclude that the absent days do not occur with equal frequencies.

Using Technology

To use the TI-84 for the goodness-of-fit text, press "STAT", "EDIT", then enter the observed and expected frequencies in L1 and L2. Press "quit". Go to "Tests", choose "χ^2 GOF-Test".

TI-84 Display

χ^2 GOF − Test	χ^2 GOF − Test
Observed: L_1	χ^2 = 3
Expected: L_2	p = 0.5578
df: 4	df = 4
Color:	Color:
Calculate Draw	Calculate Draw

The chi-square goodness-of-fit test in R uses the **chisq.test( )** function. We will enter the function as **chisq.test(x, p)**, where x = the observed frequencies, and p = expected proportions. Perform the chi-square goodness-of-fit test by typing the following in R:

```
> observed <- c(15, 12, 9, 9, 15)
> expected <- c(0.2, 0.2, 0.2, 0.2, 0.2)
> chisq.test(x = observed, p = expected)
```

You will see this chi-square goodness-of-fit test result in R:

```
              Chi-squared test for given probabilities
data: observed
X-squared = 3, df = 4, p-value = 0.5578
```

Try It 10.1

Teachers want to know which night of the week students do most of their homework. Most teachers think students do homework equally throughout the week. Suppose a random sample of 56 students were asked on which night of the week they did the most homework. The results were distributed as in Table 10.5.

From the population of students, do the nights for the highest number of students doing most of their homework occur with equal frequencies during a week? What type of hypothesis test should you use?

Table 10.5. Night of the Week Students Do Most Homework

	Sun	Mon	Tues	Wed	Thurs	Fri	Sat
Number of Students	11	8	10	7	10	5	5

Example 2

A study found that 54% of homes in the United States have 3 or more smart TVs, 28% have 2 smart TVs, and only 18% have 1 smart TV, as seen in Table 10.6. A random sample of 600 Oregon homes was selected. Of the 600 Oregon homes, 76 have 1 smart TV, 229 homes have 2 smart TVs, and 295 homes have 3 or more smart TVs.

At the 0.05 significance level, does the distribution "number of smart TVs" for Oregon households appear to differ from the distribution for the American population?

Table 10.6. Distribution of Televisions in American Homes

Number of Smart TVs	Percent	Observed Frequency
1	18	96
2	28	209
3 or more	54	295

Solution 2

We need to test whether the Oregon household distribution fits the distribution of the American households.

Table 10.6 contains expected percentages. To get the expected (E) frequencies, multiply the percentage by 600, which is the number of households surveyed in Oregon. The expected frequencies are shown in Table 10.7.

Table 10.7. Expected Frequency of Number of Smart TVs in Oregon Homes

Number of Smart TVs	Percent	Expected Frequency
1	18	(0.18)(600) = 108
2	28	(0.28)(600) = 168
3 or more	54	(0.54)(600) = 324

Therefore, the expected frequencies are 108, 168, and 324.

State the Hypotheses

H_0: The "number of smart TVs" distribution is the same for Oregon homes and the American population.

H_a: The "number of smart TVs" distribution differs for Oregon homes and the American population.

Use the chi-square distribution, χ_4^2, where the degrees of freedom $df = 3 - 1 = 2$.

Note: $df \neq 600 - 1$.

Calculate the Test Statistic

$$\chi^2 = \sum \frac{(O - E)^2}{E} = \frac{(96 - 108)^2}{108} + \frac{(209 - 168)^2}{168} + \frac{(295 - 324)^2}{324} = 13.94$$

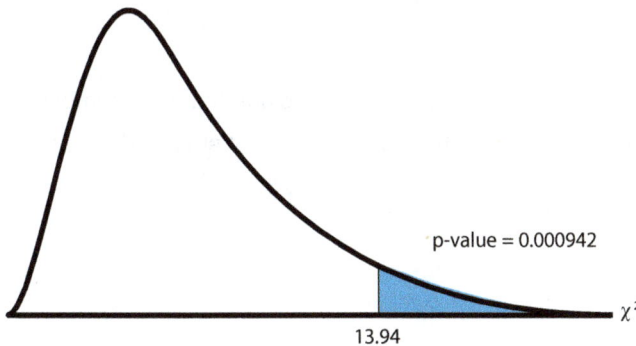

Figure 10.1. Number of Smart TVs Distribution

p-value = 0.000942

13.94

χ^2

We Use the Critical Value and p-value Methods to Draw a Conclusion

We can find the critical value from the chi-square table with $df = 2$. $\chi_c^2 = 5.991$.
Because $5.991 < \chi^2 = 13.94$, we reject H_0.

Using Technology

From TI -84:
"χ^2 GOF-Test", p-value = 0.000942. Because $0.00942 < a = 0.05$, we reach the same conclusion of rejecting H_0

Perform the χ^2 GOF-Test in R:

```
> observed <-c(96, 209, 295)
> expected <-c(0.18, 0.28, 0.54)
> chisq.test (x = observed, p = expected)
```

```
Chi-squared test for given probabilities

data: observed

X-squared = 13.935, df = 2, p-value = 0.000942
```

Formal Conclusion

At the 0.05 significance level, there is sufficient evidence to conclude that the "number of smart TVs" distribution for Oregon homes differs from the "number of smart TVs" distribution for the American population.

Exercises 10.1

For exercises 1–5, determine if each statement is true or false. If false, rewrite it as a true statement.

1. In a goodness-of-fit test, the expected values are the values we would expect if the null hypothesis were true.

2. In general, if the observed values and expected values of a goodness-of-fit test are not close to-gether, then the test statistic can get very large and will be way out in the right tail.

3. Use a goodness-of-fit test to determine if high school principals believe that students are absent equally during the week or not.

4. The test to use to determine if a six-sided die is fair is a goodness-of-fit test.

5. In a goodness-of fit test, if the p-value is 0.0113, in general, do not reject the null hypothesis.

6. The expected percentage of the number of pets U.S. students have in their homes is distributed as in Table 10.8. A random sample of 1,000 students from Nebraska resulted in the data in Table 10.9. At the 0.01 significance level, does it appear that the distribution "number of pets" of stu-dents in Nebraska differs from the distribution for the United States student population? What is the p-value?

Table 10.8. Number of Pets Expected in a Student's Home

Number of Pets	Percent
0	18
1	25
2	30
3	18
4+	9

Table 10.9. Number of Pets Observed in Students' Homes

Number of Pets	Frequency
0	210
1	240
2	320
3	140
4+	90

7. A sample of 212 commercial businesses was surveyed to determine the number of businesses that recycled plastic or aluminum. Table 10.10 shows the business categories in the survey, the sample size of each category, and the number of businesses in each category that recycle one of the materials. Based on the study, on average, half of the businesses were expected to recycle either plastic or aluminum. As a result, the last column shows the expected number of businesses in each category that recycle one of the two. At the 0.05 significance level, perform a hypothesis test to determine if the observed number of businesses that recycle one of the two follows the distribution of the expected values.

Table 10.10. Recycling by Business Type

Business Type	Number in class	Observed Number of recycling	Expected number of recycling
Office	35	19	17.5
Retail/Wholesale	48	27	24
Food/Restaurants	53	35	26.5
Manufacturing/ Medical	52	21	26
Hotel/Mixed	24	9	12

8. Table 10.11 shows results from a local survey of 499 participants that measured obesity rates by age group. The second column shows the percentage of obese people per age class among participants. The third column shows the percentages of obese people in the same age classes in the United States. Conduct a hypothesis test at the 0.05 significance level to determine whether the survey participants are a representative sample of the United States obese population.

Table 10.11. Local and National Obesity Rates

Age class (years)	Local obesity rate (percentage)	Expected U.S. obesity rate (percentage)
20–30	15.0	32.6
31–40	26.5	32.6
41–50	13.6	36.6
51–60	21.9	36.6
61–70	21.0	39.7

9. A local survey of 1,000 public school students measures how many use mass transit by race/ethnicity, with the data collected shown in Table 10.12. The table also shows the percentage of the overall U.S. student population who use mass transit and the local frequency counts.

Table 10.12. Mass Transit Use by Public School Students, by Race/Ethnicity

Race/Ethnicity	Percentage using Mass Transit	Percentage of Overall Student Population	Survey Frequency
Asian, Asian American, or Pacific Islander	10.2	5.4	113
Black or African-American	8.2	14.5	94
Hispanic or Latino	15.5	15.9	136
American Indian or Alaska Native	0.6	1.2	10
White	59.4	61.6	604
Not reported/other	6.1	1.4	43

a. Perform a goodness-of-fit test to determine whether the local results follow the distribution of the U.S. overall student population based on race/ethnicity.

b. Perform a goodness-of-fit test to determine whether the local results follow the distribution of those who use mass transit based on race/ethnicity.

10.2 Tests of Independence

Overview

Tests of independence are used to determine if there is any association between two categorical variables. The frequency of each category of one variable is compared with the frequency of a second variable. The data can be displayed in a contingency table where each row represents a category for one variable, and each column represents a category for the other variable. The two categorical variables are also called the row variable and the column variable. These tests determine if a contingency table's row and column variables are independent. This section will show you how to conduct a hypothesis test with the chi-square test of independence and Fisher's exact test to see if there is a statistical relationship between row and column variables. This helps determine if changes in one variable are independent of changes in the other, providing insights into patterns or connections within the data.

Chi-Square Test of Independence

In the **chi-square test of independence**, you will determine whether row and column variables in a contingency table are independent by using the chi-square distribution.

Notation

O = observed frequency of an outcome in a cell of a contingency table

E = expected frequency of an outcome in a cell, based on the assumption that the row and column variables are independent

r = number of rows in a contingency table

c = number of columns in a contingency table

Null and Alternative Hypothesis

The chi-square test of independence is right tailed.

H_0: There is no association between the row and column variables.

H_a: There is no association between the row and column variables.

Expected Frequency (E)

$$E = \frac{(\text{row total})(\text{column total})}{(\text{grand total})}$$

Test Statistic for Chi-Square Test of Independence

$$\chi^2 = \sum \frac{(O - E)^2}{E}$$

where the degrees of freedom $df = (r - 1)(c - 1)$

Requirements

1. The samples are randomly selected.
2. The samples have frequency counts in a two-way contingency table in which the row and column variable are to be checked for independence.
3. The expected frequency in each cell in the contingency table is at least 5. The observed outcome does not need to be 5.

Example 3

Pets4Friends is a volunteer group where adults spend time with shelter pets for 1 to 9 hours a week. Volunteers are classified into 4 types: community college students, four-year college students, and nonstudents. Table 10.13 shows a sample of the adult volunteers and the number of hours they volunteer per week. The table shows the observed values (O). Is the number of hours volunteered independent of the type of volunteer? Use a 0.05 significance level.

Table 10.13. Number of Volunteer Hours per Week, By Volunteer Type (Observed Values)

Type of Volunteer	1–3 Hours	4–6 Hours	7–9 Hours	Row Total
Community College Students	111	96	48	255
Four-Year College Students	96	133	61	290
Nonstudents	91	150	53	294
Column Total	298	379	162	839

Solution 3

Table 10.14 and the question at the end of the problem, "Is the number of hours volunteered independent of the type of volunteer?" tell you this is a test of independence. The two factors are the number of hours volunteered and the type of volunteer. This test is always right tailed.

State the Hypotheses

H_0: The number of hours volunteered is independent (no association) of the type of volunteer.
H_a: The number of hours volunteered is dependent (association) on the type of volunteer.

Calculate Expected Frequencies

For example, the calculation for the expected frequency for the top left cell of Table 10.14 is as follows:

$$E = \frac{\text{(row total)(column total)}}{\text{(grand total)}} = \frac{(255)(298)}{839} = 90.57$$

Table 10.14 shows the expected (E) values of the contingency table

Table 10.14. Expected Volunteer Hours per Week, by Volunteer Type

Type of Volunteer	1-3 Hours	4-6 Hours	7-9 Hours
Community College Students	90.57	115.19	49.24
Four-Year College Students	103.00	131.00	56.00
Nonstudents	104.42	132.81	56.77

Calculate the Test Statistic

$$\chi^2 = \Sigma \frac{(0-90.57)^2}{90.57} = \frac{(111-115.19)^2}{115.19} + \frac{(96-49.24)^2}{49.24} + \dots + \frac{(0-56.77)^2}{56.77} = 12.99$$

$df = (3-1)(3-1) = (2)(2) = 4$

Using Technology

TI-84

First, enter the observed frequencies in matrix of domination of 3 x 3. It is not necessary to enter expected values as the TI calculator computes the expected frequencies.

Next select "TEST", then "χ^2 - Test." To view the expected frequencies, select "Matrix" and enter on "[B] 3 x 3."

MATRIX[A] 3x3	χ^2 - Test	χ^2 - Test
[111 96 48]	Observed: [A]	χ^2 = 12.991
[96 133 61]	Expected: [B]	P = 0.01132
[91 150 53]	Color:	df = 4
	Calculate Draw	

The test statistic 12.9909 and the p-value = 0.0113.

Using R

It is easier to input data in the matrix in R. We will create a contingency table for the data from the matrix. Then, convert it to the table using the as.table( ) command. Here we call it "Volunteer". Type the following in R:

```
> A=matrix(c(111, 96, 91,96,133,150,48,61,53), nrow=3, ncol=3)
> Volunteer = as.table(A)
> chisq.test(Volunteer)
```

```
                Pearson's Chi-squared test

 data: Volunteer
 X-squared = 12.991, df = 4, p-value = 0.01132
```

We can view the contingency table using the code "print"

```
> print(Volunteer)
```

```
      A   B   C

A   111  96  48
B    96 133  61
C    91 150  53
```

Find the Critical Value

Using the χ^2 table and $df = 4$, you find $\chi_4^2 = 9.488$

p-value $= P(\chi^2 > 12.99) = 0.0113$

Since $a = 0.05 > p$-value $= 0.0113$, we reject H_0. This means that the variables are dependent.

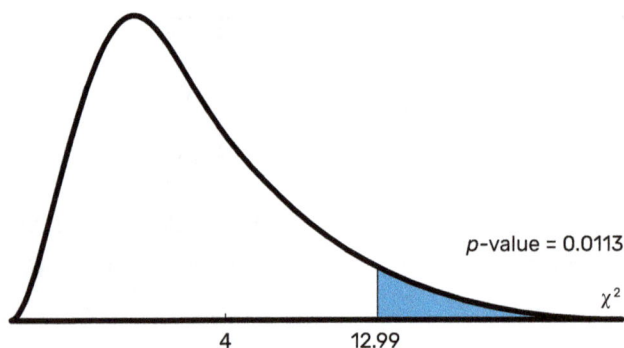

Figure 10.2. Distribution of Volunteer Hours

Formal Conclusion

At the 0.05 significance level, there is sufficient evidence to conclude that the number of hours volunteered and the type of volunteer are dependent on one another.

Fisher's Exact Test

The chi-square test of independence requires that all expected frequencies are greater than five for the chi-squared distribution to approximate the exact distribution of the test statistic and the p-value. If this requirement is not satisfied, the test result is inaccurate. If any expected frequencies are 5 or less, use **Fisher's exact test**, which provides precise p-values rather than approximations. Fisher's exact test is valid for all sample sizes. However, it is often applied in analyzing small samples in 2×2 contingency tables (2 rows and 2 columns) when researchers have small data sets to work with. Because it provides reliable conclusions, Fisher's exact test is important in many fields, including medicine, biology, agriculture, and social sciences. Use Fisher's exact test if any of these are true:

1. The row totals and column totals are fixed by the study or survey design.
2. Contingency table data comes from small samples.
3. More than 20% of expected frequencies are less than 5, but all expected frequencies are at least 1.

Manual calculations of Fisher's exact test aren't feasible because they involve more difficult probability distribution formulas. However, specialized software and programming like R offer efficient and accurate results.

Example 4

A simple random sample of 30 voters is selected. The voters answer questions on their political party preference. Table 10.15 shows the results of the survey. We want to know whether gender is associated with political party preference. Use Fisher's exact test and a 0.05 significance level.

Table 10.15. Political Party Preference by Gender

	Democrat	Republican	Total
Male	4	11	15
Female	9	6	15
Total	13	17	30

Solution 4

State the Hypotheses

H_0: Voter gender is independent (no association) of the political party preference.

H_a: Voter gender is not independent (no association) of the political party's preference.

Using Technology

We use the 'fisher.test( )' function in R programming. We store voters' data in a 2 x 2 contingency table called "data", then perform the test and obtain the p-value. Type the following in R:

```
> data = matrix(c(4,9,11,6), nrow = 2)
> fisher.test(data)
```

```
                Fisher's Exact Test for Count Data

data: data
p-value = 0.1394
alternative hypothesis: true odds ratio is not equal to 1
95 percent confidence interval:
0.03860422 1.40824396
sample estimates:
odds ratio
0.2551243
```

Since the p-value = 0.1394 > 0.05, we fail to reject H_0 at a 0.05 significance level.

Formal Conclusion

We conclude that there is not sufficient evidence to conclude voter gender is dependent on political party preference.

Exercises 10.2

1. A group of skiers argued about where the best skiing is in the United States. Their debate prompted a survey that asked randomly selected skiers where skiing is the best and what level of skier they are. Table 10.16 shows the results. At a 0.01 significance level, conduct a hypothesis test to see if the best ski area is independent of the level of the skier.

    Table 10.16. Ski Area Preference by Skier Level

U.S. Ski Area	Beginner	Intermediate	Advanced
Tahoe	20	30	40
Alta	10	30	60
Aspen/Snowmass	10	40	50

2. Car manufacturers are interested in whether there is a relationship between the size of the car an individual drives and the number of people in the driver's family—that is, whether car size and family size are independent. To test this, suppose that 800 car owners were randomly surveyed with the results in Table 10.17. At the 0.05 significance level, conduct a test of independence.

    Table 10.17. Car Size and Family Size

Family Size	Sub & Compact	Mid-size	Full-size	Van & Truck
1	20	35	40	35
2	20	50	70	80
3–4	20	50	100	90
5+	20	30	70	70

3. College students want to know if their major affects starting salaries after graduation. Suppose that 300 recent graduates were surveyed about their college majors and starting salaries after graduation. Table 10.18 shows the survey results. At the 0.1 significance level, perform a test of independence.

Table 10.18. College Major and Starting Salaries

Major	< $50,000	$50,000 – $68,999	$69,000 +
English	5	20	5
Engineering	10	30	60
Nursing	10	15	15
Business	10	20	30
Psychology	20	30	20

4. We ask 6 men and 6 women whether they prefer cycling or running. Only two men chose running, and only three women chose cycling. The remaining people chose the other option. Table 10.19 shows the data in a 2 × 2 contingency table. At the 0.05 significance level test if the gender is independent of the type of activity.

Table 10.19. Exercise Preference

	Running	Cycling
Men	2	4
Women	3	3

10.3 Tests of Homogeneity

Overview

The tests of independence and homogeneity are two sides of the same chi-square test. Tests of homogeneity determine whether two or more populations have the same proportions of a characteristic. If the row variable and column variable are independent, then the row variable is distributed for the column variable. For example, if two animal shelters wanted to compare their volunteer populations to see if they had the same proportion of community college students and four-year college students volunteering, they could use the test of homogeneity to find the answer. The chi-square test and McNemar's test for matched pairs are described in this section. These tests apply to two or more populations and test multiple categories.

Chi-Square Test for Homogeneity

The **chi-square test of homogeneity** determines if two populations with unknown distributions have the same distribution. The procedure is almost identical to the chi-square test of independence. It uses the same notation, test statistic, critical value, and data requirements. The differences are listed below in the formula box.

Degrees of Freedom for Chi-Square Test of Homogeneity

The degrees of freedom $df = (r - 1)(c - 1)$

Null and Alternative Hypothesis

The chi-square test of homogeneity is right-tailed.

H_0: There is no association between the row and column variables.

H_a: There is no association between the row and column variables.

Requirements

1. The samples are randomly selected.
2. The samples have frequency counts in a two-way contingency table in which the row and column variable are to be checked for independence.
3. The expected frequency in each cell in the contingency table is at least 5. The observed outcome does not need to be 5.

Example 5

Do male and female college students have the same distribution of living arrangements? Use a 0.01 significance level. Suppose that 250 randomly selected male college students and 300 randomly selected female college students were asked about their living arrangements: dormitory, apartment, with parents, or other. The results are shown in Table 10.20. Do male and female college students have the same distribution of living arrangements?

Table 10.20. Distribution of Living Arrangements for College Students by Sex

	Dormitory	Apartment	With Parents	Other
Males	72	84	49	45
Females	91	86	88	35

Solution 5

We want to test at the 0.01 significance level that males and females are distributed equally among the four living arrangements.

State the Hypotheses

H_0: The proportion of living arrangements for male college students is the same as for female college students.

H_a: The proportion of living arrangements for male college students is different for female college students.

$$df = \text{number of categories} - 1 = 4 - 1 = 3$$

Calculate the Test Statistic

We use "χ^2-Test" on the TI -84 calculator to find the test statistic.

$\chi^2 = 10.1287$

$p\text{-value} = P(\chi^2 > 10.1287) = 0.0175$

Since $\alpha = 0.01$ and $p\text{-value} = 0.0175 > \alpha$, we fail to reject H_0.

Formal Conclusion

At the 0.01 significance level, there is insufficient evidence to conclude that the distributions of living arrangements for male and female college students differ.

Notice that the conclusion is only that the distributions are not the same. We can't use the test for homogeneity to draw any conclusions about how they differ because this is an observational study, not an experiment.

McNemar's Test for Matched Pairs

McNemar's test for matched pairs tests for consistency in responses from two variables, the row and column variables. It's generally used in repeated measures or match pair data. It is a right-tailed chi-square test. In McNemar's test, we can compare counts directly because the comparison is not based on row totals like the chi-square test of homogeneity.

The null hypothesis for McNemar's test is that the probability for each outcome is the same. The data in a 2 × 2 contingency table, like Table 10.21, has the null hypothesis that the proportion of b is the same as the proportion of c. The cell frequencies equal the number of matched data pairs. The table is set up with the variable counts before and after applying the treatment.

Table 10.21. Sample Contingency Table for McNemar's Test for Matched Pairs

		Treatment 2	
		Improvement	No Improvement
Treatment 1	Improvement	a	b
	No Improvement	c	d

Test Statistic for McNemar's Test for Matched Pairs

$$\chi^2 = \frac{(|b - c| - 1)^2}{b + c}$$

where

$$b + c \geq 10$$

Example 6

Table 10.22 records the frequency of patients who experience knee joint pain before and after knee surgery. Researchers are interested in whether the surgery treatment was effective in changing patient answers from "Yes" before treatment to "No" after treatment. Use the significance level of 0.05 to test if the proportion of knee pain after the surgery is the same as before.

Table 10.22. Frequency Count of Knee Pain Experienced Before and After Surgery

		Knee Pain After Surgery		
		No	Yes	Total
Knee Pain Before Surgery	No	40	28	68
	Yes	60	20	80
	Total	100	48	148

Solution 6

State the Hypotheses

H_0: The proportion of knee pain after the surgery is the same as the proportion of knee pain before the surgery.

H_a: The proportion of knee pain after the surgery is not the same as the proportion of knee pain before the surgery.

Calculate the Test Statistic

$$\chi^2 = \frac{(|b-c|-1)^2}{b+c} = \frac{(|13-105|-1)^2}{13+105} = 70.18$$

In R programming, Mcnemar's test can be performed using the function mcnemar.test( ). We enter the data in matrix form.

```
> knee_data <- matrix(c(40, 60, 28, 20), nrow=2)
> mcnemar.test(knee_data)
```

```
        McNemar's Chi-squared test with continuity correction

data: knee_data
McNemar's chi-squared = 10.92, df = 1, p-value < 0.0009511
```

Since the p-value <0.05, we reject H_0. There is sufficient evidence at a 0.05 significance level that the proportion of patients experiencing knee pain after the surgery differs from the proportion experiencing knee pain before the surgery.

In general, if some counts in the contingency table are small, a continuity correction needs to be used, as seen in Example 5. If continuity correction is not necessary, Mcnemar's test can be conducted using the R function mcnemar.test(x,y= NULL, correct = False). For Example 6, the R function mcnemar.test(knee_data, correct = False) can be used.

Exercises 10.3

1. Do multi-person and single-person households have the same distribution of types of cars? Use a 0.05 significance level. Suppose that 100 randomly selected multi-person households and 200 randomly selected single-person households were asked what type of car they drove: sport, sedan, hatchback, truck, van/SUV. The results are shown in Table 10.23.

    Table 10.23. Distribution of Car Type by Household Size

	Sport	Sedan	Hatchback	Truck	Van/SUV
Multi-person households	5	15	35	17	28
Single-person households	45	65	37	46	7

2. A psychologist is interested in testing whether there is a difference in the distribution of personality descriptions for business majors and social science majors. The results of the study are shown in Table 10.24. Conduct a chi-square test of homogeneity. Test at a 0.01 level of significance.

    Table 10.24. Personality Descriptions and College Major

	Open	Conscientious	Extrovert	Agreeable	Neurotic
Business	41	52	46	61	58
Social Science	72	75	63	80	65

3. Do men and women select different breakfasts? The breakfasts ordered by randomly selected men and women at a popular breakfast restaurant is shown in Table 10.25. Conduct a chi-square test for homogeneity at a 0.05 level of significance.

    Table 10.25. Breakfast Orders by Gender

	French Toast	Pancakes	Waffles	Omelets
Men	47	35	28	53
Women	65	59	55	60

4. A fish and wildlife technician is interested in whether the distribution of fish caught in Green Valley Lake is the same as those caught in Echo Lake. Of the 191 randomly selected fish caught in Green Valley Lake, 105 were rainbow trout, 27 were other trout, 35 were bass, and 24 were catfish. Of the 293 randomly selected fish caught in Echo Lake, 115 were rainbow trout, 58 were other trout, 67 were bass, and 53 were catfish. Perform a chi-square test for homogeneity at a 0.1 level of significance.

5. When looking at energy consumption, we are often interested in detecting trends over time and how they correlate among different countries. Table 10.26 shows the average energy use (in units of kg of oil equivalent per capita) in the United States and European Union countries for a specific six-year period. Do the energy use values in these two areas come from the same distribution? Perform the analysis at the 0.05 significance level.

Table 10.26. Average Energy Use in the United States and European Union

Year	European Union	United States	Row Total
6	3,413	7,164	10,557
5	3,302	7,057	10,359
4	3,505	7,488	10,993
3	3,537	7,758	11,295
2	3,595	7,697	11,292
1	3,613	7,847	11,460
Column Total	20,965	45,011	65,976

6. A researcher tests a new medication and records if the drug worked on patients from a clinical trial, as seen in Table 10.27. A "yes" means the medication worked, and "no" means it didn't. At the 0.05 significance level, perform Mcnemar's test and determine if the medication improved patients' condition.

Table 10.27. Medication Efficacy

Before Medication		
No	Yes	
80	100	No
90	110	Yes

After Medication

Chapter 11
Analysis of Variance (ANOVA)

This chapter will explore a powerful statistical technique that helps us understand how different factors influence data. Imagine you're baking cookies and want to know if changing the type of flour affects their taste. You bake four batches of cookies with different types of flour and taste one cookie from each batch. You are analyzing the variance between batches. Analysis of variance, or ANOVA, is like the chef's secret recipe for comparing multiple batches of cookies at once. It allows us to determine if variations in data are due to genuine differences or random fluctuations. By the end of this chapter, you'll be able to use ANOVA to analyze data in various real-world situations, from cookie recipes to scientific experiments. After reading this chapter, you will be able to do the following:

1. Conduct one-way ANOVA for three or more population means.
2. Analyze data from populations with different categories of factors or characteristics.
3. Conduct two-way ANOVA for interaction between two factors.
4. Conduct two-way ANOVA for row factor and column factor.
5. Describe the F-distribution.
6. Apply the F-distribution to ANOVA.

11.1 Basic Concepts of Analysis of Variance (ANOVA)

Overview

Many statistical applications in science or business involve comparing several groups to find differences. For example, an environmentalist is interested in knowing if the average amount of pollution varies among the Great Lakes. A sociologist is interested in knowing if a child's use of libraries impacts their literacy test scores in high school. When buying a new car, you will probably compare the average gas mileage of several models.

In Chapter 9, we used the F-distribution to compare two population variances. Using the data from the samples taken from two independent populations, the F-test statistic was calculated and then used in the hypothesis testing about claims of equality of two population variances. In this chapter, we'll use the F-distribution to compare three or more population means in a hypothesis test. The one-way ANOVA analyzes variances between three or more population means with data categorized into only one factor. For example, if you want to determine if the average heights of college basketball players (factor) vary between NCAA conferences (populations), use one-way ANOVA. The two-way ANOVA compares three or more populations with data categorized into two factors. With our college basketball example, you'd use two-way ANOVA if you wanted to look at the variance of average height (factor 1) by player gender (factor 2) in NCAA conferences (populations). As a reminder, the F-distribution has three key properties:

1. It is not symmetric and is skewed to the right.
2. All values are non-negative.
3. It's exact shape depends on two different degrees of freedom, one for the numerator and one for the denominator.

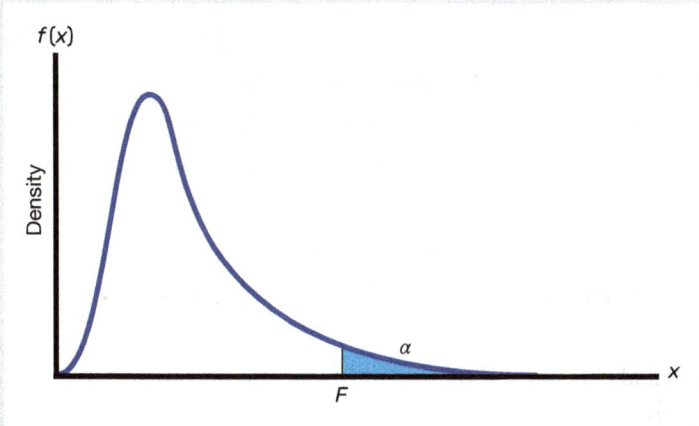

Figure 11.1. The *F*-distribution is Non-negative and Skewed to the Right

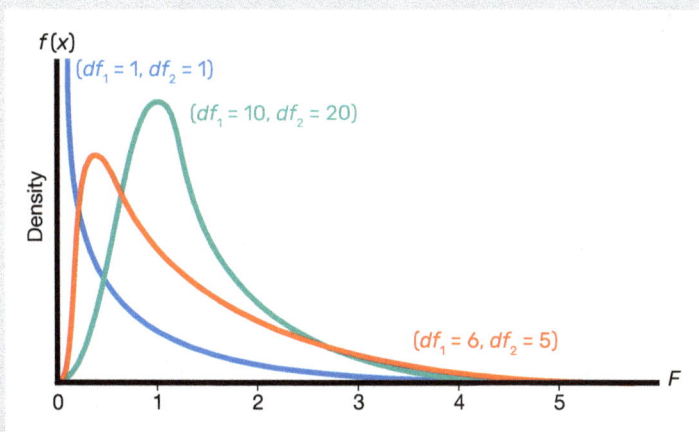

Figure 11.2. The *F*-distribution with Different Numerator and Denominator Degrees of Freedom

---

NOTE

The *F*-distribution is derived from the Student's *t*-distribution. While the *t*-statistic indicates if one variable is statistically significant, one-way ANOVA expands the *t*-test to see if a group of variables are jointly significant. It's better to use ANOVA when there are more than two groups instead of performing pairwise *t*-tests because performing multiple tests introduces the likelihood of making a Type 1 error.

Exercises 11.1

1. Suppose data from three types of soil treatments are sampled. If a soil expert wants to know if different soil impact plant growth differently, what statistical test should she use for her analysis?

2. If the objective is to test the claim that the students in three sections of statistics class have the same mean exam scores. How would ANOVA help you analyze if there are significant differences among the classes?

3. Describe a scenario where you will need to use the one-way ANOVA to analyze the data.

4. Use the F distribution table to find the critical F value for the one-sided hypothesis test if $\alpha = 0.05$, $df_1 = 24$, and $df_2 = 30$

5. Use the F distribution table to find the critical F value for the two-sided hypothesis test if $\alpha = 0.05$, $df_1 = 24$, and $df_2 = 30$

11.2 One-Way ANOVA

Overview

A **one-way ANOVA** test aims to determine the existence of a statistically significant difference among several population means. The test uses variances to help determine if the means are equal or not.

One-Way ANOVA Procedure

Because one-way ANOVA relies on so many calculations, data is typically put into a table for easy viewing, like in Table 11.1. One-way ANOVA results are often displayed like this in statistical software such as R. The display result on a TI calculator looks different.

Test Statistic

$$F = \frac{\text{Variance between groups}}{\text{Variance within groups}} = \frac{MSB}{MSW}$$

where MSB = Mean Squares Between, and MSW = Mean Squares Within

Requirements

1. Each population has an approximately normal distribution. A nonparametric test (chapter 12) will be required if a population is far from normal.
2. Samples are simple random samples and are independent from each other.
3. The populations are assumed to have approximately same standard deviations (or variances).

First, we'll learn how to set up the table, and then we'll test our hypothesis using one-way ANOVA. Our goal is to calculate the F test statistic: $F = \frac{MSB}{MSW}$. The MSB and MSW are defined as mean squares between and within as follows:

Mean Squares (MS)

$$\text{Mean Squares Between } (MSB) = \frac{SSB}{df_B}$$

$$\text{Mean Squares Within } (MSW) = \frac{SSW}{df_W}$$

---

where

$$df_B = \text{degree of freedom between groups} = k - 1$$
$$df_W = \text{degree of freedom within groups} = n - k$$
$$k = \text{the number of groups}$$
$$n = \text{total sample size} = n_1 + n_2 + \dots + n_k$$

The *SSB* and *SSW* are defined as sum of squares between and within as follows:

$$SSB = \sum \left[\frac{(s_j)^2}{n_j} \right] - \frac{(\sum s_j)^2}{n}$$

where s_j = the sum of the values in the *j*th group, j = 1, 2, ..., *k*

The total sum of squares (SST) is the sum of *SSB* and *SSW* such as:

$$SST = SSB + SSW$$

where

$$SST = \sum x^2 - \frac{(\sum x)^2}{n}$$

Therefore, the sum of square within (*SSW*) is:

$$SSW = SST - SSB$$

Note

It is best to find SSB and SST, and then use the formula to find SSW.

Table 11.1. Summary Table of One-Way ANOVA

Source of Variation	Sum of Squares (SS)	Degrees of Freedom (df)	Mean Square (MS)	F
Between (Treatment)	SSB	$k - 1$	$MSB = \frac{SSB}{df_B}$	$F = \frac{MSB}{MSW}$
Within (Residual)	SSW	$n - k$	$MSW = \frac{SSW}{df_W}$	
Total	SST	$n - 1$		

Analysis of Variance (ANOVA)

In a one-way ANOVA, the null hypothesis is that all the group population means are the same. The alternative hypothesis is that at least one pair of means is different. Let's look at an example to illustrate how ANOVA works. Figures 11.3 and 11.4 show two sets of graphs. They are a set of box plots representing the distribution of values within three independent populations. A horizontal dotted line indicates the group means to help understand the hypothesis test.

Figure 11.3 shows the first set of populations, $H_0: \mu_1 = \mu_2 = \mu_3$, and we can see that the three populations have the same distribution if the null hypothesis can be rejected. The variance of the combined data is approximately the same as the variance of each of the populations. We fail to reject H_0 because all the means are about the same. Figure 11.4 shows what the group means would look like if H_0 is rejected. The variance of the combined data is larger. We reject H_0 because all the means are not the same.

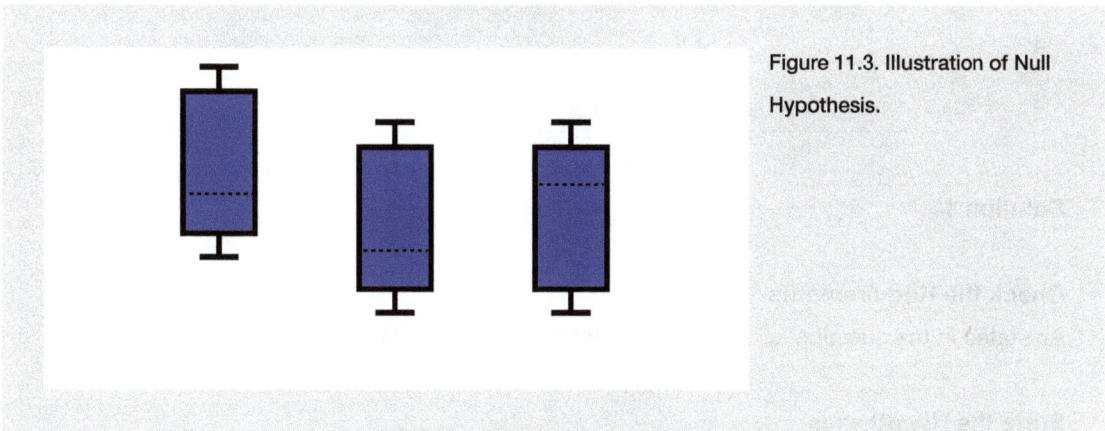

Figure 11.3. Illustration of Null Hypothesis.

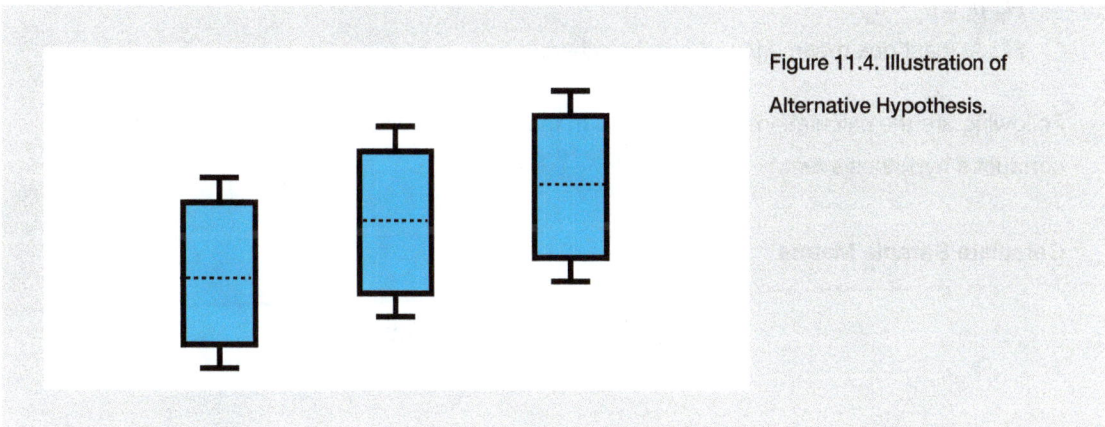

Figure 11.4. Illustration of Alternative Hypothesis.

Example 1

The data values in the following table are the weight losses (in pounds) for three different diet plans. The three samples are randomly selected from populations that have approximately normal distributions whose variances can be assumed equal. Use and one-way ANOVA to test if the population mean weight losses for the three diet plans are the same.

Table 11.2. Diet Plan Mean Weight Loss

Plan 1: $n_1 = 4$; $s_1 = 16.5$	Plan 2: $n_2 = 3$; $s_2 = 15$	Plan 3: $n_3 = 3$; $s_3 = 15.5$
5	3.5	8
4.5	7	4
4	4.5	3.5
3		

Solution 1

Check the Requirements

As stated in the question, all requirements for one-way ANOVA are satisfied.

State the Hypotheses

$H_0: \mu_1 = \mu_2 = \mu_3$
H_a: At least one mean differs from the others

Following are the calculations needed to fill in the one-way ANOVA table. The table is used to conduct a hypothesis test.

Calculate Sample Means

$$\overline{X}_1 = 4.125, \quad \overline{X}_2 = 5, \quad \overline{X}_3 = 5.167$$

Calculate Total Sum of Squares (SST)

$$n = n_1 + n_2 + n_3 = 10$$

$$SST = \sum x^2 - \frac{(\sum x)^2}{n} = 5^2 + 4.5^2 + 4^2 + 3^2 + 3.5^2 + 7^2 + 4.5^2 + 8^2 + 4^2 + 3.5^2$$

$$- \frac{(5 + 4.5 + 4 + 3 + 3.5 + 7 + 4.5 + 8 + 4 + 3.5)^2}{10}$$

$$= 244 - \frac{47^2}{10} = 244 - 220.9 = 23.1$$

Calculate Sum of Squares and Mean Square

$$SSB = \sum \left[\frac{(s_j)^2}{n_j}\right] - \frac{(\sum s_j)^2}{n} = \frac{s_1^2}{4} + \frac{s_2^2}{3} + \frac{s_1^2}{3} - \frac{(s_1 + s_2 + s_3)^2}{10}$$

$$= \frac{16.5^2}{4} + \frac{15^2}{3} + \frac{15.5^2}{3} - \frac{(16.5 + 15 + 15.5)^2}{10} = 2.2458$$

$$SSW = SST - SSB = 23.1 - 2.2458 = 20.8542$$

$$MSB = \frac{SSB}{k-1} = \frac{2.2458}{2} = 1.1229$$

$$MSW = \frac{SSW}{n-k} = \frac{20.8542}{7} = 2.9792$$

Calculate F test statistic

$$F = \frac{MSB}{MSW} = \frac{1.1229}{2.9792} = 0.3769$$

Find the F Critical Value

$$F_c = 4.7374$$

Since $F_c = 4.7374 > 0.3769$, we fail to reject the null hypothesis.

Formal Conclusion

At a 0.05 significance level, there is not sufficient evidence that at least one of the three population mean weight losses is different from the means of the other two diet plans.

Table 11.3. One-way ANOVA Summary Table for Diet Plan Study

Source of Variation	Sum of Squares (SS)	Degrees of Freedom (df)	Mean Square (MS)	F
Between (Treatment)	SSB =2.2458	df_B = 2	MSB = 1.1229	F = 0.3769
Within (Residual)	SSW = 20.8542	df_w = 7	MSW = 2.9792	
Total	SST = 23.1	df = 9		

Using Technology

From TI-84

To find these results on the TI-83, 83+, 84, 84+ calculator:

Press STAT. Press 1:EDIT. Enter the data into the lists L1, L2, and L3.

Press STAT, and arrow over to TESTS, and arrow down to ANOVA.

Press ENTER, and then enter L1, L2, L3).

Press ENTER. You will see that the values in the ANOVA table are easily produced by the calculator, including the test statistic and the p-value of the test.

TI-84 Display

```
    Enter data under stat, EDIT          One-way ANOVA
  L₁    L₂    L₃                      F = 0.376923
  5     3.5   8                       P = 0.6991
  4.5   7     4                       Factor
  4     4.5   3.5                       df = 2
  3                                      SS = 2.24583
  ANOVA(L₁, L₂, L₃)                      MS = 1.12292
                                       Error
                                         df = 7
                                         SS = 20.8542
                                         MS = 2.9792
```

From R

Type the following steps in R:

1. Define vectors to enter data

```
> Plan1 <- c(5, 4.5, 4, 3)
> Plan2 <- c(3.5, 7, 4.5)
> Plan3 <- c(8, 4, 3.5)
```

2. Pad shortest vector with NA's to have same length as the longest vector. This step is needed only if all vectors don't have the same lengths.

```
> length(Plan2) <- length(Plan1)
> length(Plan3) <- length(Plan1)
```

3. Create data frame using vectors as columns

```
> df <- data.frame(Plan1, Plan2, Plan3)
```

4. View resulting data frame to make sure all columns have same lengths

```
> df
```

R Display

```
  Plan1 Plan2 Plan3
1   5.0   3.5   8.0
2   4.5   7.0   4.0
3   4.0   4.5   3.5
4   3.0    NA    NA
```

5. It's mandatory to stack the individual groups to a single column:

```
> df_stacked <- stack(df[1:3])
```

6. Use the aov() function to calculate ANOVA in R

```
> anova <- aov(values ~ ind, data = df_stacked)
```

7. ANOVA in R summary

```
> summary(anova)
```

R Display for ANOVA

	Df	Sum Sq	Mean Sq	F Value	Pr (>F)
ind	2	2.246	1.123	0.377	0.699
Residuals	7	20.854	2.979		
2 observations deleted due to missingness					

Note: In R it is called "ind" as referring to "Between" or "Treatment." Both TI-84 and R yield the same p-value of 0.699 which leads to the same conclusion that population mean weight losses from the three diet plans appear to be the same.

Example 2

As part of an experiment to see how different types of soil cover would affect tomato production, students grew tomato plants under different soil cover conditions. Groups of three plants each had one of the following treatments: bare soil, a commercial ground cover, black plastic, straw, and compost.

All plants grew under the same conditions and were of the same variety. Students recorded the weight (in grams) of tomatoes produced by each of the $n = 15$ plants and recorded the data in Table 11.4.

Table 11.4. Weight in grams of tomatoes

Bare: $n_1 = 3$	Ground Cover: $n_2 = 3$	Plastic: $n_3 = 3$	Straw: $n_4 = 3$	Compost: $n_5 = 3$
2,625	5,348	6,583	7,285	6,277
2,997	5,682	8,560	6,897	7,818
4,915	5,482	3,830	9,230	8,677

Let the population means of the tomato yields under the five mulching conditions be μ_1, μ_2, μ_3, μ_4, and μ_5. Use a significance level of 0.05 to test the claim that there is no difference in population mean yields among the five groups. Assume all requirements for one-way ANOVA are satisfied.

Solution 2
Define the Hypotheses

H_0: $\mu_1 = \mu_2 = \mu_3 = \mu_4 = \mu_5$
H_a: At least one of the five population means *is different from the others*

Since all requirements for one way ANOVA are satisfied, we can proceed with the test. The degrees of freedom for the F-distribution are:

$$df_B = 5 - 1 = 4$$
$$df_W = 15 - 5 = 10$$

Calculate the Test Statistic

$$F = \frac{9,162,140}{2,044,672.6} = 4.481$$

Calculations for others are shown in Table 11.5.

Table 11.5. One-Way ANOVA Table for Tomato Weights

Source of Variation	Sum of Squares (SS)	Degrees of Freedom (df)	Mean Square (MS)	F
Between	36,648,561	5 − 1 = 4	$\frac{36,648,561}{4} = 9,162,140$	4.4810
Within	20,446,726	15 − 5 = 10	$\frac{20,446,726}{10} = 2,044,672.6$	
Total	57,095,287	15 − 1 = 14		

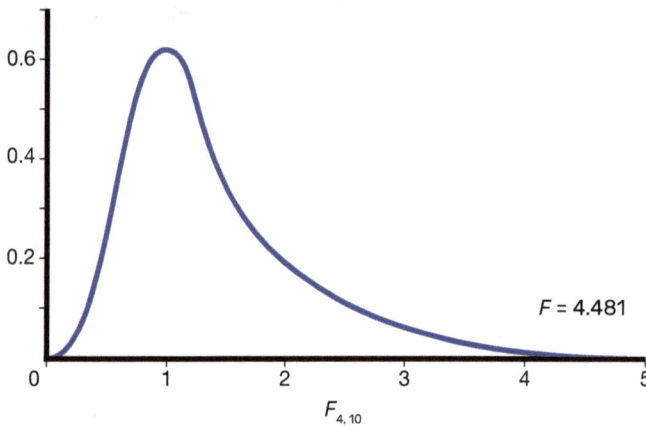

Figure 11.5. Distribution for Tomato Weights

Using Technology

From TI-84

TI -84 Display

Enter data under stat, EDIT	F = 4.48098
	P = 0.024784
L_1 L_2 L_3 L_4 L_5	Factor
2625 5348 6583 7285 6277	df = 4
2997 5682 8560 6897 7818	SS = 36648560.9
4915 5482 3830 9230 8677	MS = 9162140.23
ANOVA(L_1, L_2, L_3, L_4, L_5)	Error
	df = 10
	SS = 20446726
	MS = 2044672.6

From R

We can perform a full ANOVA in R, like in Example 1. But here, we use the R function "pf(f-stat, df_B, df_W, lower.tail = FALSE)" to find the p-value = $P(F > 4.481)$ because the F statistic is already calculated. Type the following in R:

```
> pf(4.481, 4, 10, lower.tail = FALSE)
```

R will display the p-value as follows:

```
[1] 0.02479332
```

Since the p-value $= 0.02479332 < a = 0.05$, we reject H_0.

Formal Conclusion

At the 0.05 significance level, we conclude that at least one type of soil cover resulted in different mean yield of the tomato plant.

Identify the Different Means

The one-way ANOVA is used to determine if any group mean is different from the rest of the means of three or more independent groups. If the overall p-value from the ANOVA table is less than a chosen significance level, then we have sufficient evidence that at least one of the means of the groups is different from the others. However, this doesn't tell us which group or groups are different from the others because a one-way ANOVA can only tell us that not all the group means are equal.

To find out exactly which groups are different from each other, we must conduct a post-hoc test to identify which means are different. A post-hoc test is done after a study is conducted and data is collected. Three of the most used post-hoc tests are the Tukey test, the Scheffe test, and the Bonferroni test. Here, we will use the Tukey test to demonstrate how a post-hoc test can detect the groups whose means differ from others.

Example 3

In Example 2, the one-way ANOVA test has concluded that at least one type of soil cover resulted in different mean yield of the tomato plant. Conduct the Tukey test to identify which types of soil cover would affect the growth of tomato plants. Use a 0.05 significance level.

Solution 3

First, we conduct the one-way ANOVA in R.

```
> Bare <- c(2625, 2997, 4915)
```

```
> Ground <- c(5348, 5682, 5482)
> Plastic <- c(6583, 8560, 3830)
> Straw <- c(7285, 6897, 9230)
> Compost <- c(6277, 7818, 8677)
> df <- data.frame (Bare, Ground, Plastic, Straw, Compost)
> df
```

R Display of data

	Bare	Ground	Plastic	Straw	Compost
1	2625	5348	6583	7285	6277
2	2997	5682	8560	6897	7818
3	4915	5482	3830	9230	8677

```
> df_stacked <-stack(df[1:5])
> anova <- aov(values ~ ind, data = df_stacked)
> summary(anova)
```

R Display for ANOVA

	Df	Sum Sq	Mean Sq	F Value	Pr (>F)
ind	4	36648561	9162140	4.81	0.0248
Residuals	10	20446726	2044673		

Since the *p*-value of 0.0248 is less than 0.05, we conclude that at least one type of soil cover resulted in different mean yields of the tomato plant.

Next, we will use R code "TukeyHSD( )" to perform the Tukey test. The Tukey test performs pairwise comparisons among each of the groups and find the groups which are statistically different from one another.

```
> TukeyHSD(anova)
```

The R output of the TukeyHSD is follows:

```
              Tukey multiple comparisons of means

95% family-wise confidence level
Fit: aov(formula = values ~ ind, data = df_stacked)

$ind
                    diff        lwr         upr        p adj
Ground-Bare       1991.6667   -1850.7547   5834.088   0.4719445
Plastic-Bare      2812.0000   -1030.4214   6654.421   0.1899778
Straw-Bare        4291.6667     449.2453   8134.088   0.0275202
Compost-Bare      4078.3333     235.9120   7920.755   0.0365334
Plastic-Ground     820.3333   -3022.0880   4662.755   0.9512434
Straw-Ground      2300.0000   -1542.4214   6142.421   0.3445369
Compost-Ground    2086.6667   -1755.7547   5929.088   0.4302174
Straw-Plastic     1479.6667   -2362.7547   5322.088   0.7153263
Compost-Plastic   1266.3333   -2576.0880   5108.755   0.8104685
Compost-Straw     -213.3333   -4055.7547   3629.088   0.9997035
```

The table lists the pairwise differences among groups for the independent variable. Under the "$ind: section, we see the adjusted p-value for multiple pairwise comparisons. The pairwise comparisons show that both straw and compost have a significantly higher mean tomato weight than Bare. The p-values are 0.0275202 and 0.0365334, and both are less than 0.05. No other pairwise differences are statistically significant.

Exercises 11.2

1. Four professors teach sections of Composition 100 at a local college. They want to compare the mean grades of students in their classes. They took a random sample of students' mean grades for the past term. The results are shown in Table 11.6. Using a significance level of 1%, is there a difference in mean grades among the sections?

    Table 11.6. Mean Grades for Four Sections of Composition 100

Section 1	Section 2	Section 3	Section 4
2.17	2.63	2.63	3.79
1.85	1.77	3.78	3.45
2.83	3.25	4.00	3.08
1.69	1.86	2.55	2.26
3.33	2.21	2.45	3.18

2. Girls from four different soccer teams are to be tested for mean goals scored per game. The entries in Table 11.7 are the goals per game for the different teams. The one-way ANOVA results are shown in the following table.

    Table 11.7. Mean Goals per Game

Team 1	1	2	0	3	2
Team 2	2	3	2	4	4
Team 3	0	1	1	0	0
Team 4	3	4	4	3	2

    a. Find the degrees of freedom for the numerator and denominator.
    b. Find SSB and MSB.
    c. Find SSW and MSW.
    d. Calculate the F statistic.
    e. Perform a one-way ANOVA test at the 0.05 significance level.

3. Suppose a group is interested in determining whether teenagers obtain their driver's licenses at approximately the same average age across the country. The data in Table 11.8 is randomly collected from five teenagers in each region of the country. The numbers represent the age at which teenagers obtained their driver's licenses.

Table 11.8. Mean Age of Getting a Driver's License by Region

Northeast	South	West	Central	East
16.3	16.9	16.4	16.2	17.1
16.1	16.5	16.5	16.6	17.2
16.4	16.4	16.6	16.5	16.6
16.5	16.2	16.1	16.4	16.8

a. State the null and alternative hypotheses.
b. Use the 0.05 significance level to test the claim that there is no difference in the average age teenagers obtain their driver's licenses across the country.

4. A program director wants to compare three different exam prep programs in terms of the mean scores on an exam. To test this, she recruits 30 students to participate in a study and splits them into three groups. The students in each group are randomly assigned to use one of the three exam prep programs for the next four weeks to prepare for an exam. At the end of the four weeks, all students take the same exam. The exam scores for each group are shown in Table 11.9. State the hypotheses and perform a one-way ANOVA using a 0.05 significance level.

Table 11.9. Exam Scores by Prep Program

Group 1	85	86	88	75	78	94	97	78	75	82
Group 2	90	91	93	86	85	84	81	88	94	95
Group 3	80	78	88	94	91	85	83	84	80	79

5. A researcher wants to know if the mean times (in minutes) that people watch their favorite news station are the same. The results of the study are shown in Table 11.10. Assume that all distributions are normal, the three population standard deviations are approximately the same, and the data were collected independently and randomly. Use a level of significance of 0.05 to conduct a one-way ANOVA test.

Table 11.10. Mean Minutes Spent Watching Favorite News Station

ABC	CBS	Local
45	15	72
12	43	37
18	68	56
38	50	60
23	31	51
35	22	

6. Thirty male college students were randomly divided into three groups of 10, and the groups received different doses of caffeine. Two hours after consuming the caffeine, each participant tapped a finger as rapidly as possible, and the number of taps per minute was recorded. Table 11.11 is a partially completed ANOVA table for the experiment.

Table 11.11. Partially Completed ANOVA Table for Caffeine Experiment

Source of Variation	DF	SS	MS	F	p-value
Caffeine	----	61.4	----	----	0.006
Error	----	-----	4.937		
Total	----	195.5			

a. Fill in the missing values to complete Table 11.11.

b. If you are examining the effect of caffeine on the number of taps per minute in the population for the three caffeine amounts, define the random variable and state the hypotheses.

c. Based on the p-value given in the ANOVA table, what conclusion can be made about the mean taps per minute in the population for the three caffeine amounts with a level of significance of 0.05?

7. Are the means for the final exams the same for all class delivery types? Table 11.12 shows the scores on final exams from several randomly selected classes that used the different delivery types. Assume that all distributions are normal, the three population standard deviations are approximately the same, and the data were collected independently and randomly. Use a level of significance of 0.05.

Table 11.12. Final Exam Scores by Class Delivery Type

Online	Hybrid	Face-to-Face
72	83	80
84	73	78
77	84	84
80	81	81
81		86
		79
		82

11.3 Two-Way ANOVA

Overview

Two-way ANOVA is like one-way ANOVA, but **two-way ANOVA** examines how two different categorical variables can influence the outcome of the dependent or response variable. In a two-way ANOVA, we focus on the main effects and the interaction effect of the two categorical variables on the dependent variable. For example, a study on how three levels of fertilizer *and* two watering frequencies affect plant growth (dependent variable) would examine the following three effects: the main effect of three levels of fertilizer on plant growth, the main effect of two watering frequencies on plant growth, and the interaction effect between the fertilizer and watering frequency on plant growth.

Two-Way ANOVA Procedure

Main effect deals with each factor separately. In the case of plant growth, the main effect of the independent variable, or factor A (fertilizer), is the difference in the mean plant growth for three levels of fertilizer application between two watering frequencies. The main effect of the independent variable, or factor B (watering frequency), is the difference between the mean plant growth for two types of watering frequency among the three levels of fertilizer.

An **interaction** effect occurs when the independent variables' combined effect on the dependent variable is more than just adding up their individual effects. In the plant growth example, the interaction effect is the difference in the mean plant growth due to both factors. After determining that there was an interaction effect, a test for the two main effects should be carried out. We use row factor and column factor for the two main effects.

Assumptions

1. There is one continuous dependent variable and two categorical independent variables with two or more levels.
2. The dependent variable is approximately normally distributed for each level of the independent variables.
3. The observations from each treatment are independently selected from a normal distribution with equal variance for each treatment.
4. Samples from different treatments are independent of each other.

Table 11.13. Summary Table for Two-Way ANOVA

Source of Variation	df	Sum of Squares	Mean square	F
Factor (A)	$a-1$	SS(A)	$MS(A) = \dfrac{SS(A)}{a-1}$	$F_A = \dfrac{MS(A)}{MSE}$
Factor (B)	$b-1$	SS(B)	$MS(B) = \dfrac{SS(B)}{b-1}$	$F_B = \dfrac{MS(B)}{MSE}$
Interaction (AB)	$(a-1)(b-1)$	SS(AB)	$MS(AB) = \dfrac{SS(AB)}{(a-1)(b-1)}$	$F_{AB} = \dfrac{MS(AB)}{MSE}$
Error	$ab(c-1)$	SSE	$MSE = \dfrac{SSE}{ab(c-1)}$	
Total	$abc-1$			

Notation

a = number of levels of factor A

b = number of levels of factor B

c = number of observations of each treatment

The Null and Alternative Hypothesis

H_0: There is no interaction between factor (A) and factor (B)

H_a: There is a significant interaction between factor (A) and factor (B)

H_0: There is no effect of factor (A) on the response variable

H_a: There is an effect of factor (A) on the response variable

H_0: There is no effect of factor (B) on the response variable

H_a: There is an effect of factor (B) on the response variable

There are three sets of p-values because we perform three sets of hypothesis tests. If a p-value is less than a chosen significance level, the null hypothesis of the test will be rejected to reach the conclusion that the main or interaction or both effects are significant.

We use Excel to perform a two-way ANOVA. If you have not used Excel data analysis, then first select the Data Analysis Toolpak. This is a free add-in program of Excel that provides statistical, financial, or engineering analysis.

1. Click the "File" tab from the top left corner of your Excel window, then click "Options."

2. Under Add-Ins, click the "Analysis ToolPak," then click "Go." Once you have added the "Analysis ToolPak," it will always be there.

3. Click the tab "Data," select "Data Analysis."

4. Select "Anova: Two-Factor With Replication" then click "OK" (Note: Replication means the repeated observations in each group).

5. Select Input Range by highlighting the cell range where the data values are located on the Excel worksheet including the headings. Enter the number of observations in each group under "Rows per sample."

Example 4

An optical engineer wants to examine the effect of temperature level and glass type on light output. She randomly selected a sample of 36 light output measurements under different glass types and temperature levels, shown in Table 11.14. At a 0.05 significance level, perform a two-way ANOVA using R.

Table 11.14. Light Output Measurements by Glass Type and Temperature

Glass Type	1000°C			1250°C			1500°C		
A	580	568	570	1090	1087	1085	1392	1380	1386
B	550	530	579	1070	1035	1000	1328	1312	1299
C	546	575	599	1045	1053	1066	867	904	889

Solution 4

State the Null and Alternative Hypotheses

H_0: There is no effect of the glass type on the light output.

H_a: There is an effect of the glass type on the light output.

H_0: There is no effect of the temperature type on the light output.

H_a: There is an effect of the temperature type on the light output.

H_0: There is no interaction effect of the glass and temperature on the light output.

H_a: There is an interaction effect of the glass and temperature on the light output.

Data Analysis in Excel

Enter the data in Table 11.15 into an Excel worksheet exactly as it appears.

Table 11.15. Glass Type and Temperature Data for Excel

	Temperature (F)		
Glass Type	Low	Medium	High
A	580	1090	1392
	568	1090	1392
	570	1090	1392
B	550	1090	1392
	530	1090	1392
	579	1090	1392
C	546	1090	1392
	575	1090	1392
	599	1090	1392

Use Excel's data analysis tool to run a two-way ANOVA.

Go to Data menu, click Data Analysis.

In the dialog box, Click on "Anova: Two-Factor With Replication."

In the dialog box, select Input Range by highlighting the data values from the entire data table. Make sure to enter "3" for "Rows per sample."

Table 11.16 shows how Excel displays the two-way ANOVA results for this data set.

Table 11.16. Anova: Two-Factor With Replication

ANOVA

Source of Variation	SS	df	MS	F	p-value	F crit
Sample	266.8889	2	133.4444	0.887876	0.428801	3.554557
Columns	3141469	2	1570734	10450.92	2.585E-28	3.554557
Interaction	533.7778	4	133.4444	0.887876	0.49108	2.927744
Within	2705.333	18	150.2963			
Total	3144975	26				

The ANOVA table from Excel gives three p-values: one for the main effect of Glass type, one for the Temp level, and one for the interaction between Glass and Temp. The p-value is < 0.05 for the temperature effect, so we reject the null hypothesis that temperature has no effect on light output. The p-value is > 0.05 for the glass effect and interaction between glass and temperature, so we fail to reject the other two null hypotheses.

Formal conclusion

At the 0.05 significance level, Glass type has no significant effect on light output, temperature level has a significant effect on light output, and the interaction between glass and temperature has no significant effect on the light output.

Exercises 11.3

1. Imagine you're studying how both room temperature and mattress type affect sleep duration. How would a two-way ANOVA help you explore if there's an interaction between room temperature and mattress type?

2. In a factory with two shifts and two machine types, how would you use a two-way ANOVA to see if both shifts and machines impact product quality?

3. The Portland Marathon takes place annually on the first Sunday of October in Portland, Oregon. Table 11.17 lists running times (in hours) from randomly selected runners who completed the marathon. Use a 0.05 significance level to test if running time is affected by age groups, gender, or the interaction between age groups and gender.

Table 11.17. Running Times by Age and Gender

	Age		
	20-29	30-39	40 and over
Male	4.11	4.24	4.15
	4.42	5.42	5.03
	3.97	2.98	6.13
	5.62	3.16	4.01
	3.02	4.64	3.51
Female	5.84	5.12	5.07
	3.82	5.48	5.82
	4.24	4.02	6.33
	4.55	4.73	4.5
	4.28	3.51	3.54

4. A botanist wants to know whether plant growth is influenced by sunlight exposure and watering frequency. She plants 40 seeds and lets them grow for two months under different conditions for sunlight exposure and watering frequency. After two months, she records the height of each plant. There were five plants grown under each combination of conditions. She performs a two-way ANOVA in Excel and ends up with the output in Table 11.18.

   a. State the hypothesis for testing the main effects and the interaction effects.
   b. State the p-value for each set of hypotheses.
   c. Use 0.05 significance level to write a formal conclusion.

Table 11.18. Excel Output for Plant Growth by Sunlight Exposure and Watering Frequency

	G		H	I	J	K	L	M
SUMMARY			None	Low	Medium	High	Total	
		Daily						
Count			5	5	5	5	20	
Sum			20.7	24.9	28.6	28.9	103.1	
Average			4.14	4.98	5.72	5.78	5.155	
Variance			0.378	0.232	0.447	0.412	0.775237	
		Weekly						
Count			5	5	5	5	20	
Sum			20	26.1	30.3	26.6	103	
Average			4	5.22	6.06	5.32	5.15	
Variance			0.085	0.137	0.163	0.317	0.722632	
		Total						
Count			10	10	10	10		
Sum			40.7	51	58.9	55.5		
Average			4.07	5.1	5.89	5.55		
Variance			0.211222	0.18	0.303222	0.382778		
ANOVA								
Source of Variation			SS	df	MS	F	P-value	F crit
Sample (Watering)			0.00025	1	0.00025	0.000921	0.975975	4.149097
Columns (Sunlight)			18.76475	3	6.254917	23.04898	3.9E-08	2.90112
Interaction			1.01075	3	0.336917	1.241517	0.310898	2.90112
Within			8.684	32	0.271375			
Total			28.45975	39				

Chapter 12
Nonparametric Statistics

Sometimes, we need to analyze information that doesn't follow a specific pattern or shape. Until now, we've learned about statistical methods that require populations to behave in certain ways, which we call parametric statistics. For example, one-way ANOVA (and most of the other tests we've discussed so far) requires the population distribution to be relatively normal. **Nonparametric statistics** is a method for understanding and analyzing data without relying on specific assumptions about how the information is distributed, like whether it follows a particular pattern or shape. Let's say you have information that doesn't follow the typical patterns—maybe it's about people's preferences or rankings. Nonparametric methods make sense of data without the requirements of parametric statistics.

After reading this chapter, you will be able to do the following:

1. Describe the difference between parametric and nonparametric tests
2. Identify when to use a nonparametric test and the limitations of nonparametric tests
3. Convert data into ranks
4. Conduct sign tests for matched pairs of samples, nominal data, or population median
5. Conduct Wilcoxon signed-rank test for matched pairs
6. Conduct Wilcoxon rank-sum test for samples from independent populations
7. Conduct the Kruskal-Wallis test for samples from three or more independent populations

12.1 Basic Concepts of Nonparametric Tests

Overview

Nonparametric statistics are not based on assumptions about the population distribution from which the sample was drawn. Because nonparametric statistics procedures use no or few assumptions about the shape or parameters of the population distribution, they have less statistical power than parametric tests. So, a nonparametric test will need a larger sample to have the same power as its corresponding parametric test. Unlike parametric tests, which use the actual values in the sample, most nonparametric tests use rankings of the values in the sample. Interpreting a nonparametric test result is not as easy as the corresponding parametric test. The advantage of nonparametric tests is their flexibility. When a parametric test can't be used because the data doesn't meet test requirements, we can turn to the corresponding nonparametric test to analyze the data. Because of this, nonparametric tests are used in more situations. They can also be used with more data types than parametric tests, like categorical and ranked data. However, the TI calculators and Excel do not have built-in nonparametric tests.

Nonparametric Test Basics

When we sort data, we arrange them in a particular order based on criteria like smallest to largest or worst to best. A **rank** is the number given to data based on its position in the organized list. The first item is number 1, the second is 2, and so on. This helps compare and analyze information without needing specific rules about how the data should look.

One essential test is the Wilcoxon signed-rank test, which is useful when comparing pairs of things. For example, when examining teaching methods to see if they help students improve their test scores, with the Wilcoxon signed-rank test, there is no need to assume that the scores are distributed in a specific way. This nonparametric test corresponds to the t-test for matched pairs.

When dealing with more than two groups, the Kruskal-Wallis test is handy. Imagine that a researcher wants to determine if there are any significant differences in job satisfaction among employees working in different departments, such as HR, Marketing, and IT. Not knowing the specific distribution of the population from which data is drawn, the Kruskal-Wallis test is an appropriate nonparametric test to use. The corresponding parametric test is the one-way ANOVA.

A rank correlation method, such as Spearman's rank correlation coefficient, is the nonparametric test for nonlinear correlations between an explanatory variable and a response variable. If we were investigating any correlation, linear or nonlinear, between study hours and exam scores of students, this would be the test to use. The corresponding parametric test is Pearson's correlation coefficient.

Lastly, randomization tests are useful when other methods don't work. These tests involve mixing up and rearranging data to ensure our conclusions are solid, even when we don't know how the data is spread out.

Limitations of Nonparametric Tests

Nonparametric statistical tests have some drawbacks. They aren't good at finding subtle differences in small datasets and can be sensitive to sample size. These tests often simplify the data by ranking it, which means losing some details from the original information. They also assume that each piece of data is separate, which isn't the best for truly continuous data.

Nonparametric tests have fewer choices for complex analyses compared to parametric tests and may struggle when some data is missing. Despite these issues, nonparametric tests are still useful, especially when data doesn't follow normal patterns or the necessary assumptions of parametric tests can't be made. Table 12.1 shows a comparison of the parametric and nonparametric tests.

Table 12.1. Commonly Used Parametric and Nonparametric Tests

Parametric Test (Normal Distributions)	Nonparametric Test (Non-normal Distributions)	Purpose of Test
t-test for independent samples	Rank sum test	Compares means of two independent samples
Paired t-test	Sign test	Examines a set of differences of means
Pearson correlation coefficient	Rank correlation test	Assesses the linear association between two variables
One-way analysis of variance (F-test)	Kruskal-Wallis test	Compares three or more groups
Two-way analysis of variance	Runs test	Compares groups classified by two different factors

Exercises 12.1

1. Explain the fundamental difference between parametric and nonparametric tests.

2. When would a nonparametric test be used? What are the key limitations of nonparametric tests, and when might they be preferred over parametric tests?

3. If you have a dataset with numerical values, explain how these values are converted into ranks. Why is ranking important in nonparametric statistics?

12.2 Sign Test

Overview

The **sign test** is a nonparametric test that converts data values to positive or negative signs to determine if the positive sign occurs more significantly than the negative sign, or vice versa. The sign test has two applications. First, testing a claim about a single population median and second, testing a claim about the difference in the medians of matched pairs.

Sign Test Procedure

The sign test does not measure the magnitude of the data values. It simply tests whether the differences between the observations are equally likely to be positive or negative. The only requirement is that the data is from a simple random sample.

Notation

x = the number of times the less frequent sign occurs

n = the total number of positive and negative signs

M = median

Null and Alternative Hypotheses

Two-Tailed test:
$$H_0: M = M_0$$
$$H_a: M \neq M_0$$

One-tailed test:
$$H_0: M \leq M_0 \text{ or } H_0: M \geq M_0$$
$$H_0: M > M_0 \text{ or } H_0: M < M_0$$

Test Statistic for Sign Test

If $n \leq 25$, the test statistic is x = the number of times the less frequent sign occurs. Use Table A.6 in the Appendix to find the critical x values.

If $n > 25$, use this test statistic formula. Use Table A.6 in the Appendix to find the critical z values.

$$z = \frac{(x + 0.5) - \frac{n}{2}}{\frac{\sqrt{n}}{2}}$$

Example 1

A college student claims the median rental rate for a one-bedroom apartment near her campus is $950. Her parents took a random sample of 28 one-bedroom apartments and recorded the rent in Table 12.2. Test the student's claim using $a = 0.10$.

Table 12.2. One-bedroom Apartment Rents

850	750	900	975	855	985	859	940	950	915
925	980	995	1025	850	915	740	985	970	885
900	925	1000	875	950	895	900	925		

Solution 1

State the Hypothesis

$$H_0\text{: Median} = 950 \ (M = 950)$$
$$H_a\text{: Median} \neq 950 \ (M \neq 950)$$

Calculate the Test Statistic

Compare each value to the median. If the value is below the median, give it a negative sign. If the value is above the mean, give it a positive sign. If the value is tied with the median, give it a zero. Table 12.3 shows the given sign below the rent value.

Table 12.3. Signs of Rents

850	750	900	975	855	985	859	940	950	915
-	-	-	+	-	+	-	-	0	-
925	980	995	1025	850	915	740	985	970	885
-	+	+	+	-	-	-	+	+	-
900	925	1000	875	950	895	900	925		
-	-	+	-	0	-	-	-		

Count the number of positive and negative signs. Positive signs = 8, negative signs = 18, and 2 data values yield a difference of 0, so $n = 26$, and $x = 8$ (the number of less frequent signs). Since n is greater than 25, we convert the test statistic x to a z-score:

$$z = \frac{(x + 0.5) - \frac{n}{2}}{\frac{\sqrt{n}}{2}}$$

$$= \frac{(8 + 0.5) - \frac{26}{2}}{\frac{\sqrt{26}}{2}} = -1.765$$

We find the critical value using the normal distribution table from the Appendix (Table A.1). $z_c = -1.645 > -1.765$. We will reject the null hypothesis because the z-score is less than the critical value.

Formal Conclusion

At the 0.1 significance level, there is sufficient evidence that the median rate for a one-bedroom apartment near the campus is not $950.

Example 2

We want to test a claim that students who receive tutoring get better exam scores than those who don't. Table 12.4 shows a sample set of exam scores from an accounting class where Group A receives tutoring and Group B doesn't. The exam scores are paired by students. At a 0.05 significance level, use the sign test to test the claim that there is no difference between the exam scores of Group A and Group B.

Table 12.4. Exam Scores for Groups A and B

Group A	87	95	85	92	88	96	78	87	86	93
Group B	88	75	85	89	90	90	83	79	81	90
Sign of Difference	-	+	0	+	-	+	-	+	+	+

Solution 2

Subtract each value of Group B from the corresponding value of Group A. Record the signs of difference in the last row of Table 12.4. Note: The "0" sign of difference means no differ-ence between Groups A and B on the pair with an exam score of 85.

State the Hypothesis

H_0: There is no difference in the median exam scores between Group A and Group B.

H_a: There is a difference in the median exam scores between Group A and Group B.

Calculate the Test Statistic

From the 10 pairs of data values, there are 6 positive signs and 3 negative signs. Since the sample size $n = 10$ is less than 25, the test statistic which is the less frequent sign that oc-curs is the negative signs, so $x = 3$. Using Table A.6 in the Appendix, we find the critical value $x = 1$. We fail to reject the null hypothesis because the test statistic value of 2 is greater than the critical value of 1.

Formal Conclusion

At the 0.05 significance level, there is no sufficient evidence that the median exam scores between Group A and Group B differ.

Exercises 12.2

1. A light bulb manufacturer claims that a certain type of light bulb's median lifetime (in hours) is at least 730 hours. A random sample of 12 light bulbs was selected from a large population, and the lifetime of these light bulbs was recorded in Table 12.5. State the hypothesis and use the sign test at $\alpha = 0.05$ significance level to test the manufacturer's claim.

Table 12.5. Light Bulb Lifetime Hours

735	687	789	735	700	695	705	795	740	730	701	720

2. A bank manager claims that the median number of customers walking into the bank is 160 per day. A random sample of 15 days was selected from last year, and the number of customers on those days is listed in Table 12.6. At $\alpha = 0.05$, test the hypothesis that the medium number of customers is greater than 160.

Table 12.6. Customers Walking Into Bank

155	180	170	165	159	185	148	178	200	140	138	157	166	150	177

3. Your friend claims that a fuel additive improved gas mileage in their car. To check your friend's claim, you randomly select 16 new cars of the same car model from a car manufacturer. Eight cars were given fuel additives, and the other eight were not. Table 12.7 shows the gas mileage (in miles per gallon) of 16 new cars. At $\alpha = 0.05$, use the sign test to test your friend's claim.

Table 12.7. Gas Mileage and Fuel Additive

Cars without fuel additive	35.8	37.7	39.4	36.8	36.6	33.7	38.4	37.3
Cars with fuel additive	36.2	39.8	40.1	39.3	36.9	34.5	38.8	39.1

4. To measure the effectiveness of a pain-relief drug, 10 patients were randomly selected for a study. Each patient was given the pain medicine at the onset of pain and the placebo at another onset of pain. Thirty minutes after administering the medication, the pain index value was obtained from each patient, and the results are recorded in Table 12.8. Without knowing the population distribution where the sample was drawn, test the claim that the pain-relief drug significantly reduces the pain index among pain sufferers with a 0.05 significance level.

Table 12.8. Pain Index of Patients

Patient	1	2	3	4	5	6	7	8	9	10
Pain medicine	25	30	28	31	18	17	33	21	19	23
Placebo	28	32	29	33	15	18	30	24	17	25

12.3 Wilcoxon Tests

Overview

Wilcoxon tests are nonparametric tests that compare two sets of related data when a population does not follow a normal distribution. An alternative to the parametric paired t-test, the Wilcoxon signed-rank test looks at the changes between pairs, like before-and-after scores, without focusing on the actual numbers. It investigates whether there's a clear improvement or difference in the medians rather than the means. The Wilcoxon rank-sum test does a similar thing but for two separate groups. For example, comparing the movie theater box office earnings of two superhero movies to see if there's a significant difference in their popularity. The data in these tests are assumed to be continuous.

Wilcoxon Signed-Rank Test for Matched Pairs Procedure

The **Wilcoxon signed-rank test** for matched pairs looks at samples from within the same population to test the claim that the difference in medians is zero. Do the same workers perform better after a training session? Do customers who used one of the products prefer the new-and-improved version? We have matched pairs when comparing the same group in two different situations.

Notation

T = the smaller of the two sums below, excluding nonzero differences:

1. The sum of the positive ranks of the differences d, or
2. The absolute value of the sum of the negative ranks of the differences d

n = the total number of positive and negative signs

Null and Alternative Hypotheses

H_0: The samples come from populations with same medians.

H_a: The samples come from populations with different medians.

Test Statistic for Wilcoxon Signed-Rank Test

If $n \leq 30$, the test statistic is T, and the critical T-value can be found in Table A.7 in the Appendix.

If $n > 30$, the test statistic value is

$$z = \frac{T - \frac{n(n + 1)}{4}}{\sqrt{\frac{n(n + 1)(2n + 1)}{24}}}$$

and the critical value can be found from the normal table in the Appendix (Table A.1).

As the name suggests, the Wilcoxon signed-rank test uses signs and ranks. Here, the samples are ranked according to the difference in their values, and each rank is given a sign. For example, if the difference is positive, the rank is positive. If the difference is negative, the rank is negative. But if the difference is 0, then there is no signed-rank. The data must come from a simple random sample and consist of matched pairs of observations. In addition, the distribution of the differences between paired observations should be roughly symmetric so that extreme values on one side are balanced by extreme values on the other. But that doesn't mean the data needs to be normally distributed, just that it is roughly symmetrical. Let's look at an example to walk through the procedure.

Example 3

To compare the delivery times of two food delivery services, your classmates collect data on the time it takes for each service to deliver from their favorite restaurant. Table 12.9 lists the delivery times (in minutes) of each order for Service A and Service B, Use a 0.05 significance level and the signed-rank test to determine if there is a difference between the two delivery services.

Table 12.9. Signs and Ranks of Delivery Times by Service

Order	1	2	3	4	5	6	7	8
Service A	25	28	24	31	36	34	27	33
Service B	31	36	29	27	36	36	33	30

Solution 3

The Wilcoxon signed-rank test is a nonparametric test that doesn't rely on assumptions about the specific time distribution.

State the Hypotheses

H_0: There is no difference in the median delivery times between the services

H_a: There is a difference in the median delivery times between services

Calculate Differences, Rank, and Assign Signs

These steps are carried out and shown in the last three rows of Table 12.10. First, calculate the differences (d) between Service A and B. Second, rank and assign signs (+ or -) for the absolute difference value ($|d|$). The differences are the same for order numbers 1 and 7, so they receive the same rank, in this case 4.5, the average of ranks 4 and 5. For order number 5, the difference between the two services is 0, so no rank is assigned to that order. Finally, note that the positive or negative signs are consistent for the Difference (d) and the signed-rank.

Table 12.10. Signed-Rank for Delivery Orders

Order	1	2	3	4	5	6	7	8
Service A	25	28	24	31	36	34	27	33
Service B	31	36	29	27	36	36	33	30
Difference (d = Service A − Service B)	-6	-8	-5	+4	0	-2	-6	+3
Rank of |d|	5.5	7	4	3	na	1	5.5	2
Signed-Rank	-5.5	-7	-4	+3	na	-1	-5.5	+2

Calculate the Test Statistic (T)

Since n is less than 30, the test statistic T = the smaller of the two sums below:

1. The sum of the positive ranks of the differences is

$$2 + 3 = 5$$

2. The absolute value of the sum of the negative ranks of the differences is

$$|(-1) + (-4) + (-5.5) + (-5.5) + (-7)| = |-23| = 23$$

Therefore, $T = 5$

Determine Critical Value

There are 7 non-zero differences, so $n = 7$. Using Table A.7 in the Appendix, we find the critical value for T is 2. Comparing the test statistic T with the critical T, we conclude that we will reject the null hypothesis since the test statistic T is greater than the critical T.

Formal Conclusion

At a 0.05 significance level, we conclude that there is a difference in delivery times between Service A and Service B.

Using Technology: R Programming

We can use "wilcox.test( )" function to perform the Wilcoxon signed rank test on matched pairs. The test gives approximated p-values if there is tie in the ranks or 0 differences in matched pairs.

```
> service_A< -c (25, 28, 24, 31, 36, 34, 27, 33)
> service_B<-c (31, 36, 29,27, 36, 36, 33, 30)
> wilcox.test (service_A, service_B, paired=TRUE)
```

Using R

```
        Wilcoxon signed rank test with continuity correction

data: service_A and service_B
V = 5, p-value = 0.1501
alternative hypothesis: true location shift is not equal to 0
```

Wilcoxon Rank-Sum Test for Two Independent Populations Procedure

The **Wilcoxon rank-sum test** ranks values from two independent samples to test whether the samples come from populations with equal medians. It assesses if one population tends to have larger values than the other, considering the order rather than the precise values. This nonparametric method is valuable when facing skewed data or unequal variances. This test is equivalent to the Mann-Whitney U test described in other textbooks. The Wilcoxon rank-sum test requires two independent simple random samples with more than 10 values. The samples do not need to have the sample number of values. As the non-parametric equivalent of the t-tests, the Wilcoxon rank-sum test doesn't require the population to have a particular distribution.

Notation

n_1 = the size of sample 1

n_2 = the size of sample 2

$R_1 = R$ = sum of ranks for sample 1

R_2 = sum of ranks for sample 2

μ_R = expected mean of sample R values when populations have equal medians

σ_R = expected standard deviation of sample R values when populations have equal medians

---

Null and Alternative Hypotheses

H_0: The samples come from populations with equal medians.

H_a: The samples come from populations with medians that aren't equal.

Test Statistic for Wilcoxon Rank-Sum Test

$$z = \frac{R - \mu_R}{\sigma_R}$$

where $\mu_R = \dfrac{n_1(n_1 + n_2 + 1)}{2}$ and $\sigma_R = \sqrt{\dfrac{n_1 n_2(n_1 + n_2 + 1)}{12}}$

Example 4

The game scores of two basketball teams, the Comets and the Supernovas, over a two-month period are listed in Table 12.11. The managers of the teams want to know if there's a significant difference in the team's scores. At a 0.05 significance level, perform the Wilcoxon rank-sum test.

Table 12.11. Scores from the Comets and Supernovas Last Two Months of Games

Comets' Scores	Supernovas' Scores
85	78
92	75
88	80
90	79
86	82
78	87
84	76
89	81
91	87
88	76
91	86
83	
$n_1 = 12$	$n_2 = 11$

Solution 4
State the Hypotheses

H_0: The population median scores from two teams are equal

H_a: The population median scores from two teams are unequal

Combine and Rank

Combine the two teams' scores into one large sample and rank them from the smallest to the largest. The smallest team score, 75, is ranked 1, and the largest team score, 92, is ranked 21. Tied scores get the average rank. The sum of ranks by each team is shown in Table 12.12.

Table 12.12. Combined and Ranked Scores for Comets and Supernovas

Comets' Scores	Combined Rank	Supernovas' Scores	Combined Rank
85	12	78	4.5
92	23	75	1
88	17.5	80	7
90	20	79	6
86	13.5	82	9
78	4.5	87	15.5
84	11	76	2.5
89	19	81	8
91	21.5	87	15.5
88	17.5	76	2.5
91	21.5	86	13.5
83	10		
$n_1 = 12$	$R_1 = 191$	$n_2 = 11$	$R_2 = 85$

Calculate the Test Statistic

We use the Comet's score as Sample 1, then $R = R_1 = 191$

$$\mu_R = \frac{n_1(n_1 + n_2 + 1)}{2} = \frac{12(12 + 11 + 1)}{2} = 144$$

$$\sigma_R = \sqrt{\frac{n_1 n_2(n_1 + n_2 + 1)}{12}} = \sqrt{\frac{(12 \cdot 11)(12 + 11 + 1)}{12}} = 16.248$$

$$z = \frac{R - \mu_R}{\sigma_R} = \frac{191 - 144}{16.248} = 2.893$$

Find the *p*-value

This is a two-tailed test, so:

$$p\text{-value} = P(z > 2.89) + P(z < -2.89)$$
$$= 0.002 + 0.002 = 0.004$$

Since the *p*-value < 0.05, we reject the null hypothesis.

Formal Conclusion

At the 0.05 significance level, we conclude that population distributions of the two team scores are different, and therefore, the medians of the team's score are not the same.

Using Technology: R Programming

We can run the Wilcoxon Rank Sum Test in R using the "wilcox.test( )" function for Example 4. First of all we enter the team scores in R:

> Comets < – c (85, 92, 88, 90, 86, 78,84, 89, 91, 88, 91, 83)
> Supernovas < – c (78, 75, 80, 79, 82, 87, 76, 81, 87, 76, 86)
> wilcox.test (Comets, Supernovas, alternative = "two.sided")

Using R

```
        Wilcoxon rank sum test with continuity correction

data:   Comets and Supernovas
W = 113, p-value = 0.004155
alternative hypothesis: true location shift is not equal to 0

Warning message:
In wilcox.test.default(Comets, Supernovas, alternative = "two.
sided"): cannot compute exact p-value with ties
```

The *p*-value from R programming is the same as the *p*-value from the calculated test statistic (z). Using R saves time in obtaining results. A boxplot (Figure 12.1) created in R shows the distribution of the scores from the two teams, indicating the median scores of the teams are quite different.

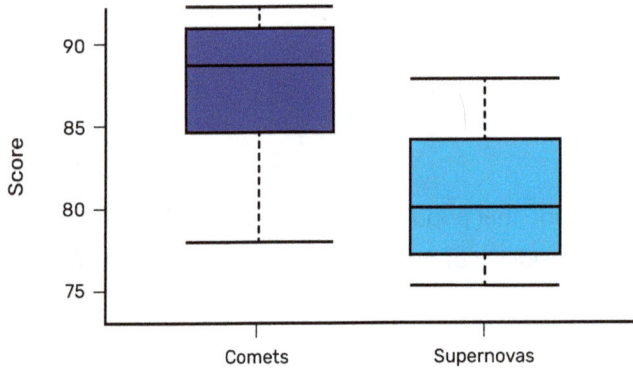

Figure 12.1. Boxplot for Basketball Team Game Scores

Exercises 12.3

1. A manager plays soft music in the office to improve the working environment. Ten employees are randomly selected within the company, and each person's stress level (1–10) on a specific day is recorded. After one week of music, the same group is surveyed again. The data are shown in Table 12.13. At $\alpha = 0.05$, can the manager conclude that listening to music has improved the working environment?

Table 12.13. Employee Stress Levels

Employee	1	2	3	4	5	6	7	8	9	10
Before	3	7	5	9	6	10	1	2	6	8
After	2	5	6	7	8	7	1	3	4	6

2. Consider a Phase II clinical trial designed to investigate the effectiveness of a new drug to reduce asthma symptoms in children. A total of 16 participants are randomized to receive either the new

drug or a placebo. Participants are asked to record the number of episodes of shortness of breath over a one-week period following the assigned treatment. The data are shown in Table 12.14. At $\alpha = 0.05$, is there a difference in the number of episodes of shortness of breath over a one-week period in participants receiving the new drug compared to those receiving the placebo?

Table 12.14. Shortness of Breath Episodes and Treatment

Placebo	7	5	6	4	12	8	10	9
New Drug	3	6	4	2	4	1	6	5

3. A pharmaceutical company is testing to see if there's a significant difference in patient pain relief between two new pain medications. Researchers randomly assigned two different pain medications to 35 patients with chronic pain and recorded the pain rating for each patient one hour after each dose. The pain ratings are on a sliding scale of 1–10. The results are listed in Table 12.15. Use a Wilcoxon test to see if there is a significant difference at $\alpha = 0.05$.

Table 12.15. Pain Medications and Pain Relief Measurements

Patient	Drug 1	Drug 2	Patient	Drug 1	Drug 2	Patient	Drug 1	Drug 2
1	2.3	2.5	9	5	3.4	17	5	5
2	4.5	3.2	10	5.7	5.4	18	3.1	5.1
3	1.2	5.1	11	1.9	5.2	19	3.2	3.2
4	58	5.5	12	3.1	4.3	20	3.3	3.3
5	4.3	4.9	13	4	6	21	3	5.9
6	4.1	5.2	14	2.3	2.9	22	5.4	3.2
7	2.6	4.7	15	2.7	4.2	23	3.6	5.9
8	3	2.4	16	3	3.3	24	2.2	5.6

4.	At Rachel's eleventh birthday party, eight kids were timed to see how long (in seconds) they could hold their breath in a relaxed position. After a two-minute rest, they timed themselves while jumping. The results are shown in Table 12.16. The kids thought the median difference between their jumping and relaxed times would be zero. Test their claim at $\alpha = 0.01$.

Table 12.16. Time Holding Breath Relaxed and After Jumping

Relaxed time (seconds)	Jumping time (seconds)
26	21
47	40
30	28
22	21
23	25
45	43
37	35
29	32

12.4 Kruskal-Wallis Test

Overview

The **Kruskal-Wallis test** uses ranks to compare three or more samples from independent populations to determine if the populations have equal medians. The Kruskal-Wallis test checks to see if there's a change in the response variable based on the various levels of treatment. It's a useful test when looking at big-picture questions that span many populations when we don't know how the data is distributed.

Kruskal-Wallis H Test for Three or More Populations Procedure

The Kruskal-Wallis H test makes no assumptions about the distribution of the populations, so it is easy to use, especially when a data set is small. The variable for the test has three or more independent groups, and data are randomly selected independent samples with at least five observations in each group sample. The response variable of the test is an ordinal or continuous variable. An example of an ordinal variable is a survey response measured on a 5-point scale from "strongly disagree" to "strongly agree." An example of a continuous variable is height in feet.

The H test is the nonparametric equivalent of one-way ANOVA. The test checks to see if there is a significant difference between groups. However, it can't tell which group or groups are different from others. The H test is used when the assumptions for ANOVA aren't met, like the assumption of normality. Therefore, if the data follows a normal distribution, the one-way ANOVA, not the Kruskal-Wallis test, should be used to analyze the data.

Notation

N = the total number of observations in all samples

k = the number of samples

$R_1, R_2, \ldots R_k$ = sum of ranks for sample 1, sample 2, ...sample k

$n_1, n_2, \ldots n_k$ = number of observations in sample 1, sample 2, ...sample k

Null and Alternative Hypotheses

H_0: medians of each group are the same

H_a: At least one of the group medians is different from the others

Test Statistic for Kruskal-Wallis Test

$$H = \frac{12}{N(N + 1)} \left(\frac{R_1^2}{n_1} + \frac{R_2^2}{n_2} + ... + \frac{R_k^2}{n_k} \right) - 3(N + 1)$$

where

$$N = n_1 + n_2 + n_3$$

Example 5

Table 12.17 shows the growth rate (measured in millimeters per day) of lettuce plants in three different soil types. At a 0.05 significance level, test the claim that there is no difference in the median growth rate of lettuce among the three types of soil.

Table 12.17. Growth of Lettuce Sprouts (mm/day) in Different Soils

Soil A	Soil B	Soil C
10	15	12
8	14	11
8	17	13
9	14	10
10	16	13
11	13	9
7		12
		11
		11
		10

Solution 5

State the Hypotheses

H_0: The median lettuce growth rate is equal across soil groups ($M_1 = M_2 = M_3$)

H_a: At least one median lettuce growth rate is different from the others

The boxplot of lettuce growth rates (Figure 12.2) from the three soil types shows clearly the different distributions of growth rates.

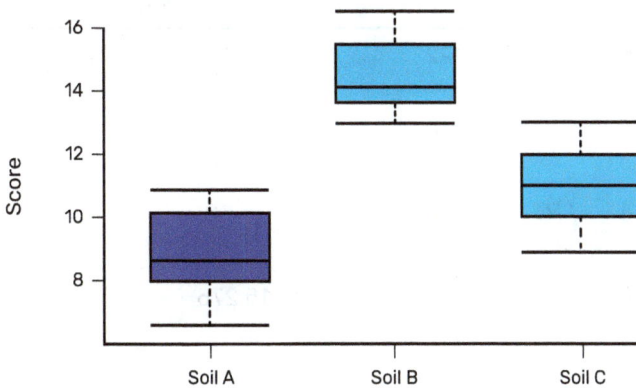

Figure 12.2. Boxplot of Lettuce Growth Rates

Combine and Rank Data

Combine three growth rates into one large sample and rank them from the smallest to the largest. The smallest growth rate, 7, is ranked 1, and the largest growth rate, 17, is ranked 23. Tied scores get the average rank. The sum of ranks by each team is shown in Table 12.18.

Table 12.18. Combined Ranks of Soils

Soil A	Combined Rank	Soil B	Combined Rank	Soil C	Combined Rank
10	7.5	15	21	12	14.5
8	2.5	14	19.5	11	11.5
8	2.5	17	23	13	17
9	4.5	14	19.5	10	7.5

Soil A	Combined Rank	Soil B	Combined Rank	Soil C	Combined Rank
10	4.5	16	22	13	17
11	11.5	13	17	9	4.5
7	1			12	14.5
				11	11.5
				11	11.5
				10	7.5
$n_1 = 7$	$R_1 = 34$	$n_2 = 6$	$R_2 = 122$	$n_3 = 10$	$R_3 = 117$

Calculate the Test Statistic

$$H = \frac{12}{N(N+1)} \left(\frac{R_1^2}{n_1} + \frac{R_2^2}{n_2} + \dots + \frac{R_k^2}{n_k} \right) - 3(N+1)$$

$$= \frac{12}{23(23+1)} \left(\frac{34^2}{7} + \frac{122^2}{6} + \frac{117^2}{10} \right) - 3(23+1) = 15.276$$

Find the Critical Value

Use the chi-square table with $df = k-1 = 2$, we find the critical value $\chi^2 = 5.991$. Since $H >$ 5.991, we reject the null hypothesis.

Formal Conclusion

At a 0.05 significance level, there is enough evidence that the at least one median growth rate (likely Soil type B) is different from the other two medians.

Note: We can also perform the Kruskal-Wallis Test in R using "Kruskal.test( )," `Kruskal.test (values ~ ind, data = df_stacked)`

Exercises 12.4

1. Which nonparametric test is equivalent to one-way ANOVA? Discuss the advantages of the Kruskal-Wallis test.

2. When comparing data from three or more independent groups, how does the Kruskal-Wallis test work, and when is it a suitable alternative to parametric analyses?

3. What are the differences between the H test and the one-way ANOVA?

For exercises 4 and 5, determine if the Kruskal-Wallis test is appropriate in analyzing the following data.

4. We want to find out how test anxiety affects test scores. The independent variable "test anxiety" has three levels: no anxiety, low–medium anxiety, and high anxiety. The dependent variable is the exam score, rated from 0 to 100%.

5. You want to determine how socioeconomic status affects attitude toward sales tax increases. The independent variable is "socioeconomic status," which has three levels: working class, middle class, and wealthy. The dependent variable is measured on a 5-point scale from strongly agree to strongly disagree.

6. A researcher wants to know if three drugs affect knee pain differently. A sample of 30 knee pain sufferers are randomly assigned into three groups to receive either Drug 1, Drug 2, or Drug 3. After one month of taking the drug, the researcher asks each individual to rate their knee pain on a scale of 1 to 100, with 100 indicating the most severe pain. The ratings for all 30 individuals are shown in Table 12.19. The researcher conducts a Kruskal-Wallis Test using a .05 significance level to determine if there is a statistically significant difference between the median knee pain ratings across these three groups.

Table 12.19. Knee Pain Ratings

Drug 1	78	65	63	44	50	78	70	61	50	44
Drug 2	71	66	40	55	31	45	66	47	42	56
Drug 3	57	88	58	78	65	61	62	44	48	77

7. A researcher wants to know if the mean times (in minutes) people watch their favorite news station are the same. The results of the study are shown in Table 12.20. Suppose the assumptions for the one-way ANOVA are not met. Use a 0.05 significance level to conduct the H test, interpret the test result, and draw a conclusion.

Table 12.20. Mean Time Watching News Stations

ABC	45	12	18	38	23	35
CBS	15	43	68	50	31	22
Local	72	37	56	60	51	

12.5 Rank Correlation

Overview

The **rank correlation test** (also called Spearman's rank correlation test) uses ranks of paired data to determine if there's an association between two variables. For example, investigating whether the time spent exercising correlates with overall fitness in a group of gym members would use a rank correlation test. It focuses on the overall order of exercise time and fitness levels. It's like figuring out if people who exercise more tend to be more fit without knowing how the population is distributed.

Rank Correlation Test Procedure

The rank correlation test is a nonparametric test with the test statistic r_s. Unlike linear correlation from Chapter 8, which specifies one variable as the explanatory variable and another as the response variable and tests claims about a linear relationship between the two variables, the rank correlation test looks for any relationship (not just a linear relationship) between the two variables. The subscript s is named after the English psychologist Charles Spearman, who developed rank correlation. The requirements for the rank correlation test are that the data come from a simple random sample, are matched pairs, and are ranks or can be converted to ranks.

Notation

r_s = rank correlation coefficient for sample paired data; the sample statistic

ρ_s = rank correlation coefficient for all population data; the population parameter

n = number of pairs of sample data

d = difference between ranks of paired data values

Null and Alternative Hypotheses

$$H_0: \rho_s = 0 \text{ (There is no rank correlation)}$$
$$H_a: \rho_s \neq 0 \text{ (There is a rank correlation)}$$

Test Statistic for Rank Correlation Test

When paired data is converted to ranks, use formula 1 to find r_s:

$$r_s = \frac{n(\sum xy) - (\sum x)(\sum y)}{\sqrt{n(\sum x^2) - (\sum x)^2} \; \sqrt{n(\sum y^2) - (\sum y)^2}} \qquad (1)$$

If there aren't any ties among ranks of the first variable or ties among ranks of the second variable, use formula 2:

$$r_s = 1 - \frac{6\sum d^2}{n(n^2 - 1)} \qquad (2)$$

Critical Value Formula if $n > 30$

$$r_s = \pm \frac{z}{\sqrt{n - 1}}$$

where z corresponds to the significance level

Example 6

A fitness trainer observed 6 people in a gym by the hours they exercised per week and their fitness level, as shown in Table 12.21. The population where the sample of 6 come from all go to the same gym. The people sampled had an overall fitness assessment and were assigned a fitness level. Is there a connection between the amount of exercise and fitness levels at the significance level of 0.05?

Table 12.21. Exercise Time in Hours (x) and Fitness Level (y)

Person	Exercise Time (hours)	Fitness Level
A	2	3
B	6	2
C	4	5
D	10	3
E	12	8

Solution 6

State the Hypotheses

H_0: $\rho_s = 0$ (There is no correlation between the ranks of exercise time and fitness level)

H_a: $\rho_s \neq 0$ (There is a correlation between the ranks of exercise time and fitness level)

Convert Data to Ranks

Rank both sets of data separately, from the smallest to the largest. Use average ranks for tied values. We do have ties (2.5) in ranks in the data shown in Table 12.22.

Table 12.22. Ranked Data of Exercise and Fitness Level

Person	Exercise Time (hours)	Exercise Time Rank (x)	Fitness Level	Fitness Level Rank (y)
A	2	1	3	2.5
B	6	3	2	1
C	4	2	5	4
D	10	4	3	2.5
E	12	5	8	5

Calculate the Test Statistic

Since there are ties, we use formula (1) to calculate . First, find the values of each component in the formula:

$$\Sigma xy = 2 \cdot 3 + 6 \cdot 2 + 4 \cdot 5 + 10 \cdot 3 + 12 \cdot 8 = 164$$
$$\Sigma x = 2 + 6 + \ldots + 12 = 34$$
$$\Sigma x^2 = 2^2 + 6^2 + \ldots + 12^2 = 300$$
$$(\Sigma x)^2 = 34^2 = 1156$$
$$\Sigma y = 3 + 2 + \ldots + 8 = 21$$
$$\Sigma y^2 = 3^2 + 2^2 + \ldots + 8^2 = 111$$
$$(\Sigma y)^2 = 21^2 = 441$$

Now find r_s:

$$r_s = \frac{n(\sum xy) - (\sum x)(\sum y)}{\sqrt{n(\sum x^2) - (\sum x)^2}\sqrt{n(\sum y^2) - (\sum y)^2}}$$

$$= \frac{6(164) - (34)(21)}{\sqrt{6(300) - 1156}\sqrt{6(111) - 441}}$$

$$= \frac{984 - 714}{\sqrt{644}\sqrt{225}} = \frac{270}{380.657} = 0.7093$$

Find the Critical Value

Since $n = 6 < 30$, the critical value for r_s from Table A.8 in the Appendix is 0.886 and -0.886 for the two-tailed test. Since the test statistic 0.7093 is between -0.886 and 0.886, we fail to reject the null hypothesis.

Formal Conclusion

At the 0.05 significance level, there's no correlation between exercise time and fitness level.

Example 7

A professor collected a random sample of 11 students' midterm and final exam scores. She produces the data in Table 12.23, where x is the midterm exam score (out of 100 possible points), and y is the final exam score (out of 200 possible points). Use the rank correlation test to check the professor's claim that there is connection between the ranks of the midterm and the final exams.

Table 12.23. Midterm and Final Exam Scores

x (midterm exam score)	88	67	71	78	66	75	68	90	71	89	79
y (final exam score)	175	133	185	163	126	198	153	166	159	151	169

Solution 7
State the Hypotheses

H_0: $\rho_s = 0$ (There is no correlation between the ranks of midterm and final exams)

H_a: $\rho_s \neq 0$ (There is a correlation between the ranks of midterm and final exams)

Convert Data to Ranks

Rank both sets of data separately, from the smallest to the largest. The data in Table 12.24 shows no ties in ranks.

Table 12.24. Ranked Test Scores

Midterm score	88	67	71	78	66	75	68	90	73	89	79
Rank of midterm score	9	2	4	7	1	6	3	11	5	10	8
Final exam score	175	133	185	163	126	198	153	166	159	151	169
Rank of Final score	9	2	10	6	1	11	4	7	5	3	8
d = Final exam score – Midterm score	0	0	6	-1	0	5	1	-4	0	-7	0

Calculate the Test Statistic

Since there are no ties in either set of ranks, we use formula (2) to calculate r_s. First find values of $\sum d^2$:

$$\sum d^2 = 0^2 + 0^2 + 6^2 + (-1)^2 + 0^2 + 5^2 + 1^2 + (-4)^2 + 0^2 + (-7)^2 + 0^2 = 128$$

Now find r_s:

$$r_s = 1 - \frac{6\sum d^2}{n(n^2-1)} = 1 - \frac{6(128)}{11(11^2-1)} = -0.4182$$

Find the Critical Value

Since $n = 11 < 30$, the critical value for r_s from Table A.8 in the Appendix is 0.618 and -0.618 for the two-tailed test. Since the test statistic -0.4182 is between -0.886 and 0.886, we fail to reject the null hypothesis.

Formal Conclusion

At the 0.05 significance level, there's no correlation between scores from the midterm and final exams.

Using Technology: R Programming

Use "cor.test( )" function in R, we can perform the rank correlation test to check for a relationship between the two variables.

> Midterm < -c (88, 67, 71, 78, 66, 75, 68, 90, 71, 89, 79)
> Final < -c (175, 133, 185, 163, 126, 198, 153, 166, 159, 151, 169)
> cor.test (Final, Midterm, method = 'spearman')

R-Display

```
                Spearman's rank correlation rho

  data:  Final and Midterm
  S = 128, p-value = 0.2031
  alternative hypothesis: true rho is not equal to 0

  sample estimates:
        rho
  0.4181818
```

Since the p-value = 0.2031 > 0.05, we reach the same conclusion of not rejecting the null hypothesis.

Exercises 12.5

1. Suppose the rank correlation test concludes that there is a correlation between the two variables. Can we use the methods from Chapter 8.3 to find the regression equation to model the linear correlation between the two variables and make predictions about the response variable?

2. It's commonly believed that the more you study for an exam, the higher you'll score. To test this belief, we evaluate the relationship between the number of hours studied and students' test scores. The data in Table 12.25 shows study hours and exam scores obtained from a sample of 10 students. Test for correlation between hours studied and students' test scores at $\alpha = 0.1$.

Table 12.25. Study Hours and Test Scores for 10 Students

Study Hours	Test Scores
3	70
6	80
2	78
10	85
4	80
5	88
9	92
12	90
8	79
10	82

3. Can crickets feel temperature? The relation between air temperature and the number of times a cricket chips in 15 seconds was studied. The study data is listed in Table 12.26. Is there evidence of correlation between temperature and the cricket chips? Use $\alpha = 0.05$.

Table 12.26. Air Temperature and Cricket Chirps

Current temperature(F)	78	75	73	76	82	72	71	79	83	70	73	75	77
Cricket chirps in 15 seconds	44	40	38	43	48	37	34	45	51	32	35	42	43

12.6 Runs Test for Randomness

Overview

The **runs test for randomness** is used to analyze the distribution of a dataset to ensure that it can be described as random without any trends. It's good for spotting patterns in data. Suppose we want to know if a number sequence is random. We can apply the runs test for randomness to decide if the sequence is random or has a subtle pattern.

Runs Test Procedure

The runs test categorizes each result into two groups and looks at the **runs**, or sequences of data from the same category, to see if they are random or have a noticeable order. The runs test requires that the data have some order or arrangement. This could be the order in which the data was collected or occurred. For instance, in the case of a coin-toss experiment, did heads or tails come up first? In a weather experiment where you're recording whether it was sunny or rainy over the course of a month, the order of days matters.

Notation

n_1 = the number of items in a sequence with characteristic A

n_2 = the number of items in a sequence with characteristic B

R = the number of runs

Null and Alternative Hypotheses

H_0: The data are in a random order.

H_a: The data are in an order that's not random.

Test Statistic for Runs Test

For Small Samples: If n_1 and n_2 are both ≤ 20, then the test statistic is the number of runs (R).

For Large Samples: If either n_1 or $n_2 > 20$, or both n_1 and $n_2 > 10$, then the distribution of R can be approximated by a normal distribution, then the test statistic is:

$$z = \frac{R - \mu_R}{\sigma_R}$$

where

$$\mu_R = \frac{2n_1 n_2}{n_1 + n_2} + 1$$

and

$$\sigma_R = \sqrt{\frac{(2n_1 n_2)(2n_1 n_2 - n_1 - n_2)}{(n_1 + n_2)^2 (n_1 + n_2 + 1)}}$$

Example 8

The results of tossing a fair coin 20 times are recorded in Table 12.27. At a 0.05 significance level, determine if the results of heads or tails are in a random order.

Table 12.27. Results of 20 Coin Tosses

Toss	1	2	3	4	5	6	7	8	9	10	11	12	13	14	15	16	17	18	19	20
Result	H	T	H	T	T	T	H	H	T	H	T	H	H	T	T	H	H	H	T	H

Solution 8

State the Hypotheses

H_0: The sequence of heads and tails is random

H_a: The sequence of heads and tails is not random

Determine Values

There are 11 heads and 9 tails. So, $n_1 = 11$ and $n_2 = 9$. To find R, count the number of consecutive tosses with the same outcome.

H	T	H	TTT	HH	T	H	T	HH	TT	HHH	T	H
1st run	2nd run	3rd run	4th run	5th run	6th run	7th run	8th run	9th run	10th run	11th run	12th run	13th run

There are 13 runs. $R = 13$

Calculate the Test Statistic

R is the test statistic since both n_1 and $n_2 < 20$. Therefore, the test statistic is $R = 13$.

Find the Critical Value

Use Table A.9 in the Appendix and $n_1 = 11$ and $n_2 = 9$, we find the critical values are 6 and 16. Since $6 < R = 13 < 16$, according to the instructions from the table, we fail to reject the null hypothesis that the sequence is random.

Formal Conclusion

At a 0.05 significance level, we find there is not enough evidence that the sequence of heads and tails is not random.

Example 9

Table 12.28 shows a sequence of political party affiliations of recent presidents of the United States, where R = Republican and D = Democrat.

Table 12.28. Presidential Party Affiliation

Party Affiliations	R	R	R	R	D	R	D	R	R	R	D	R	R	R	D	D	R	D	D	R	R	D	R	R	D	R	D	D

Use a 0.1 significance level to test the claim that the sequence is random.

Solution 9

State the Hypotheses

H_0: The sequence is random

H_a: The sequence is not random

Determine Values

There are 17 Rs and 11 Ds. So, $n_1 = 17$ and $n_2 = 11$. To find R, count the series of consecutive tosses with the same outcome.

RRRR	D	R	D	RRR	D	RRR	DD	R	DD	RR	D	RR	D	R	DD
1st run	2nd run	3rd run	4th run	5th run	6th run	7th run	8th run	9th run	10th run	11th run	12th run	13th run	14th run	15th run	16th run

There are 16 runs. $R = 16$

Calculate the Test Statistic

Since both n_1 and $n_2 > 10$, we use a normal distribution to approximate the test statistic

$$\mu_R = \frac{2n_1 n_2}{n_1 + n_2} + 1 = \frac{2(17)(11)}{17 + 11} + 1 = 14.357$$

and

$$\sigma_R = \sqrt{\frac{(2n_1 n_2)(2n_1 n_2 - n_1 - n_2)}{(n_1 + n_2)^2(n_1 + n_2 + 1)}} = \sqrt{\frac{2(17)(11)(2(17)(11) - 17 - 11)}{(17 + 11)^2 (17 + 11 + 1)}}$$

$$= \sqrt{\frac{(374)(346)}{(784)(29)}} = 2.386$$

$$z = \frac{R - \mu_R}{\sigma_R} = \frac{16 - 14.357}{2.386} = 0.689$$

We can use $z = 0.6886$ to find the p-value. For the two-tailed test,

$$\text{p-value} = P(z > 0.6889) + P(z < -0.689) = 0.2451 + 0.2451 = 0.4902$$

Since the p-value > 0.1, we fail to reject the null hypothesis.

Formal Conclusion

At the 0.1 significance level, there is not enough evidence that the political party affiliation is not random.

Example 10

A quality control engineer believes that the weights of paint cans (in ounces) are not random-ly distributed. The data in Table 12.29 records the weights of 16 paint cans in order of time taken off the production line at a fixed point.

Table 12.29. Table 12.29. Paint Can Weight in Ounces

68.2	71.6	69.3	71.6	70.4	65.0	63.6	64.7
65.3	64.2	67.6	68.6	66.8	68.9	66.8	70.1

Use the run test and a 0.1 significance level to determine whether the weights of the paint cans on the production line are random.

Solution 10

State the Hypotheses

H_0: The weights of paint cans are random

H_a: The weights of paint cans are not random

Using Technology: R Programming

We can use "runs.test" function in R to perform the test for randomness. We enter the following code in R:

> weights < -c (68.2, 71.6, 69.3, 71.6, 70.4, 65, 63.6, 64.7, 65.3, 64.2, 67.6, 68.6, 66.8, 68.9, 66.8, 70.1)

> runs.test(weights)

R-Display

```
                        Runs Test

data: weights
statistic = -1.0351, runs = 7, n1 = 8, n2 = 8, n = 16,
p-value = 0.3006
alternative hypothesis: nonrandomness
```

The display indicates that the p-value of 0.3006 is greater than the significance level of 0.1. We fail to reject the null hypothesis.

Formal Conclusion

At a 0.1 significance level, there is no evidence that the weights of the paint cans are not random.

Exercises 12.6

1. Twenty numbers are randomly selected from 0 through 9, and their values are recorded in Table 12.30. Use a 0.1 significance level to determine whether the occurrence of the numbers is random.

Table 12.30. Twenty Numbers 0–9

2	0	5	8	3	0	1	1	2	6	9	6	4	6	4	9	7	5	2	8

A train conductor wants to see if passengers get on the train randomly by gender. He observes the first 25 people, noting the sequence of males (M) and females (F) recorded in Table 12.31. Test for randomness at $\alpha = 0.05$.

Table 12.31. Order of People Entering Train

F	F	F	M	M	F	F	F	F	M	M	F	M	M	F	F	F	F	M	M	F	F	F	M

2. Table 12.32 shows the departures from normal (0^0 C) of daily temperatures recorded at a weather station in Illinois in November 2018. We would like to know if the pattern of departures above and below normal results from a nonrandom process. Test for randomness at $\alpha = 0.1$.

Table 12.32. Departure by Degrees from Normal Daily Temperatures

Day	Departure from Normal
1	12
2	13
3	12
4	11
5	5
6	2
7	-1
8	2
9	-1
10	3
11	2
12	-6
13	-7
14	-7
15	-12
16	-9
17	6
18	7
19	10
20	10
21	1
22	1
23	3
24	7
25	-2
26	-6
27	-6
28	-5
29	-2
30	-2

Table A.1. Standard Normal Distribution z Scores

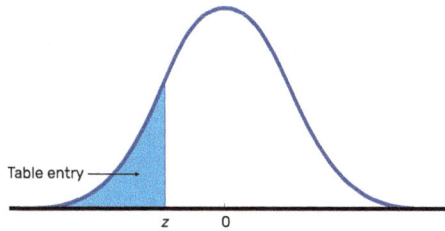

z Score	Area
-1.645	0.050
-2.575	0.005

Z	.00	.01	.02	.03	.04	.05	.06	.07	.08	.09
-3.9	.00005	.00005	.00004	.00004	.00004	.00004	.00004	.00004	.00003	.00003
-3.8	.00007	.00007	.00007	.00006	.00006	.00006	.00006	.00005	.00005	.00005
-3.7	.00011	.00010	.00010	.00010	.00009	.00009	.00008	.00008	.00008	.00008
-3.6	.00016	.00015	.00015	.00014	.00014	.00013	.00013	.00012	.00012	.00011
-3.5	.00023	.00022	.00022	.00021	.00020	.00019	.00019	.00018	.00017	.00017
-3.4	.00034	.00032	.00031	.00030	.00029	.00028	.00027	.00026	.00025	.00024
-3.3	.00048	.00047	.00045	.00043	.00042	.00040	.00039	.00038	.00036	.00035
-3.2	.00069	.00066	.00064	.00062	.00060	.00058	.00056	.00054	.00052	.00050
-3.1	.00097	.00094	.00090	.00087	.00084	.00082	.00079	.00076	.00074	.00071
-3.0	.00135	.00131	.00126	.00122	.00118	.00114	.00111	.00107	.00104	.00100
-2.9	.00187	.00181	.00175	.00169	.00164	.00159	.00154	.00149	.00144	.00139
-2.8	.00256	.00248	.00240	.00233	.00226	.00219	.00212	.00205	.00199	.00193
-2.7	.00347	.00336	.00326	.00317	.00307	.00298	.00289	.00280	.00272	.00264
-2.6	.00466	.00453	.00440	.00427	.00415	.00402	.00391	.00379	.00368	.00357
-2.5	.00621	.00604	.00587	.00570	.00554	.00539	.00523	.00508	.00494	.00480
-2.4	.00820	.00798	.00776	.00755	.00734	.00714	.00695	.00676	.00657	.00639
-2.3	.01072	.01044	.01017	.00990	.00964	.00939	.00914	.00889	.00866	.00842
-2.2	.01390	.01355	.01321	.01287	.01255	.01222	.01191	.01160	.01130	.01101
-2.1	.01786	.01743	.01700	.01659	.01618	.01578	.01539	.01500	.01463	.01426
-2.0	.02275	.02222	.02169	.02118	.02068	.02018	.01970	.01923	.01876	.01831
-1.9	.02872	.02807	.02743	.02680	.02619	.02559	.02500	.02442	.02385	.02330
-1.8	.03593	.03515	.03438	.03362	.03288	.03216	.03144	.03074	.03005	.02938
-1.7	.04457	.04363	.04272	.04182	.04093	.04006	.03920	.03836	.03754	.03673
-1.6	.05480	.05370	.05262	.05155	.05050	.04947	.04846	.04746	.04648	.04551
-1.5	.06681	.06552	.06426	.06301	.06178	.06057	.05938	.05821	.05705	.05592
-1.4	.08076	.07927	.07780	.07636	.07493	.07353	.07215	.07078	.06944	.06811
-1.3	.09680	.09510	.09342	.09176	.09012	.08851	.08691	.08534	.08379	.08226
-1.2	.11507	.11314	.11123	.10935	.10749	.10565	.10383	.10204	.10027	.09853
-1.1	.13567	.13350	.13136	.12924	.12714	.12507	.12302	.12100	.11900	.11702
-1.0	.15866	.15625	.15386	.15151	.14917	.14686	.14457	.14231	.14007	.13786
-0.9	.18406	.18141	.17879	.17619	.17361	.17106	.16853	.16602	.16354	.16109
-0.8	.21186	.20897	.20611	.20327	.20045	.19766	.19489	.19215	.18943	.18673
-0.7	.24196	.23885	.23576	.23270	.22965	.22663	.22363	.22065	.21770	.21476
-0.6	.27425	.27093	.26763	.26435	.26109	.25785	.25463	.25143	.24825	.24510
-0.5	.30854	.30503	.30153	.29806	.29460	.29116	.28774	.28434	.28096	.27760
-0.4	.34458	.34090	.33724	.33360	.32997	.32636	.32276	.31918	.31561	.31207
-0.3	.38209	.37828	.37448	.37070	.36693	.36317	.35942	.35569	.35197	.34827
-0.2	.42074	.41683	.41294	.40905	.40517	.40129	.39743	.39358	.38974	.38591
-0.1	.46017	.45620	.45224	.44828	.44433	.44038	.43644	.43251	.42858	.42
-0.0	.50000	.49601	.49202	.48803	.48405	.48006	.47608	.47210	.46812	.46414

Table A.1. Standard Normal Distribution *z* Scores

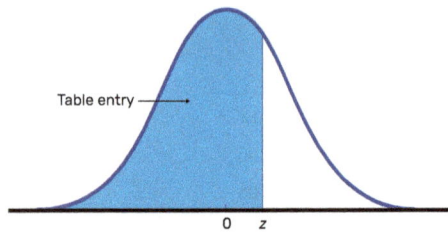

z Score	Area
1.645	0.9500
2.575	0.9950

Common Critical Values	
Confidence Level	Critical Value
0.90	± 1.645
0.95	± 1.960
0.99	± 2.575

z	.00	.01	.02	.03	.04	.05	.06	.07	.08	.09
0.0	.50000	.50399	.50798	.51197	.51595	.51994	.52392	.52790	.53188	.53586
0.1	.53983	.54380	.54776	.55172	.55567	.55962	.56356	.56749	.57142	.57535
0.2	.57926	.58317	.58706	.59095	.59483	.59871	.60257	.60642	.61026	.61409
0.3	.61791	.62172	.62552	.62930	.63307	.63683	.64058	.64431	.64803	.65173
0.4	.65542	.65910	.66276	.66640	.67003	.67364	.67724	.68082	.68439	.68793
0.5	.69146	.69497	.69847	.70194	.70540	.70884	.71226	.71566	.71904	.72240
0.6	.72575	.72907	.73237	.73565	.73891	.74215	.74537	.74857	.75175	.75490
0.7	.75804	.76115	.76424	.76730	.77035	.77337	.77637	.77935	.78230	.78524
0.8	.78814	.79103	.79389	.79673	.79955	.80234	.80511	.80785	.81057	.81327
0.9	.81594	.81859	.82121	.82381	.82639	.82894	.83147	.83398	.83646	.83891
1.0	.84134	.84375	.84614	.84849	.85083	.85314	.85543	.85769	.85993	.86214
1.1	.86433	.86650	.86864	.87076	.87286	.87493	.87698	.87900	.88100	.88298
1.2	.88493	.88686	.88877	.89065	.89251	.89435	.89617	.89796	.89973	.90147
1.3	.90320	.90490	.90658	.90824	.90988	.91149	.91309	.91466	.91621	.91774
1.4	.91924	.92073	.92220	.92364	.92507	.92647	.92785	.92922	.93056	.93189
1.5	.93319	.93448	.93574	.93699	.93822	.93943	.94062	.94179	.94295	.94408
1.6	.94520	.94630	.94738	.94845	.94950	.95053	.95154	.95254	.95352	.95449
1.7	.95543	.95637	.95728	.95818	.95907	.95994	.96080	.96164	.96246	.96327
1.8	.96407	.96485	.96562	.96638	.96712	.96784	.96856	.96926	.96995	.97062
1.9	.97128	.97193	.97257	.97320	.97381	.97441	.97500	.97558	.97615	.97670
2.0	.97725	.97778	.97831	.97882	.97932	.97982	.98030	.98077	.98124	.98169
2.1	.98214	.98257	.98300	.98341	.98382	.98422	.98461	.98500	.98537	.98574
2.2	.98610	.98645	.98679	.98713	.98745	.98778	.98809	.98840	.98870	.98899
2.3	.98928	.98956	.98983	.99010	.99036	.99061	.99086	.99111	.99134	.99158
2.4	.99180	.99202	.99224	.99245	.99266	.99286	.99305	.99324	.99343	.99361
2.5	.99379	.99396	.99413	.99430	.99446	.99461	.99477	.99492	.99506	.99520
2.6	.99534	.99547	.99560	.99573	.99585	.99598	.99609	.99621	.99632	.99643
2.7	.99653	.99664	.99674	.99683	.99693	.99702	.99711	.99720	.99728	.99736
2.8	.99744	.99752	.99760	.99767	.99774	.99781	.99788	.99795	.99801	.99807
2.9	.99813	.99819	.99825	.99831	.99836	.99841	.99846	.99851	.99856	.99861
3.0	.99865	.99869	.99874	.99878	.99882	.99886	.99889	.99893	.99896	.99900
3.1	.99903	.99906	.99910	.99913	.99916	.99918	.99921	.99924	.99926	.99929
3.2	.99931	.99934	.99936	.99938	.99940	.99942	.99944	.99946	.99948	.99950
3.3	.99952	.99953	.99955	.99957	.99958	.99960	.99961	.99962	.99964	.99965
3.4	.99966	.99968	.99969	.99970	.99971	.99972	.99973	.99974	.99975	.99976
3.5	.99977	.99978	.99978	.99979	.99980	.99981	.99981	.99982	.99983	.99983
3.6	.99984	.99985	.99985	.99986	.99986	.99987	.99987	.99988	.99988	.99989
3.7	.99989	.99990	.99990	.99990	.99991	.99991	.99992	.99992	.99992	.99992
3.8	.99993	.99993	.99993	.99994	.99994	.99994	.99994	.99995	.99995	.99995
≥ 3.9	.99995	.99995	.99996	.99996	.99996	.99996	.99996	.99996	.99997	.99997

Table A.2. *t*-Distribution: Critical *t* Values

Degrees of Freedom	Area in One Tail				
	.005	.01	.025	.05	.10
	Area in Two Tails				
	.01	.02	.05	.10	.20
1	63.657	31.821	12.706	6.314	3.078
2	9.925	6.965	4.303	2.920	1.886
3	5.841	4.541	3.182	2.353	1.638
4	4.604	3.747	2.776	2.132	1.533
5	4.032	3.365	2.571	2.015	1.476
6	3.707	3.143	2.447	1.943	1.440
7	3.499	2.998	2.365	1.895	1.415
8	3.355	2.896	2.306	1.860	1.397
9	3.250	2.821	2.262	1.833	1.383
10	3.169	2.764	2.228	1.812	1.372
11	3.106	2.718	2.201	1.796	1.363
12	3.055	2.681	2.179	1.782	1.356
13	3.012	2.650	2.160	1.771	1.350
14	2.977	2.624	2.145	1.761	1.345
15	2.947	2.602	2.131	1.753	1.341
16	2.921	2.583	2.120	1.746	1.337
17	2.898	2.567	2.110	1.740	1.333
18	2.878	2.552	2.101	1.734	1.330
19	2.861	2.539	2.093	1.729	1.328
20	2.845	2.528	2.086	1.725	1.325
21	2.831	2.518	2.080	1.721	1.323
22	2.819	2.508	2.074	1.717	1.321
23	2.807	2.500	2.069	1.714	1.319
24	2.797	2.492	2.064	1.711	1.318
25	2.787	2.485	2.060	1.708	1.316
26	2.779	2.479	2.056	1.706	1.315
27	2.771	2.473	2.052	1.703	1.314
28	2.763	2.467	2.048	1.701	1.313
29	2.756	2.462	2.045	1.699	1.311
30	2.750	2.457	2.042	1.697	1.310
31	2.744	2.453	2.040	1.696	1.309
32	2.738	2.449	2.037	1.694	1.309
34	2.728	2.441	2.032	1.691	1.307
36	2.719	2.434	2.028	1.688	1.306
38	2.712	2.429	2.024	1.686	1.304
40	2.704	2.423	2.021	1.684	1.303
45	2.690	2.412	2.014	1.679	1.301
50	2.678	2.403	2.009	1.676	1.299
55	2.668	2.396	2.004	1.673	1.297
60	2.660	2.390	2.000	1.671	1.296
65	2.654	2.385	1.997	1.669	1.295
70	2.648	2.381	1.994	1.667	1.294
75	2.643	2.377	1.992	1.665	1.293
80	2.639	2.374	1.990	1.664	1.292
90	2.632	2.368	1.987	1.662	1.291
100	2.626	2.364	1.984	1.660	1.290
200	2.601	2.345	1.972	1.653	1.286
300	2.592	2.339	1.968	1.650	1.284
400	2.588	2.336	1.966	1.649	1.284
500	2.586	2.334	1.965	1.648	1.283
750	2.582	2.331	1.963	1.647	1.283
1000	2.581	2.330	1.962	1.646	1.282
2000	2.578	2.328	1.961	1.646	1.282
> 2000	2.576	2.326	1.960	1.645	1.282

Table A.3. Chi-Square (χ^2) Distribution

Degrees of freedom (df)	$\alpha = 0.995$	.99	.975	.95	.9	.1	.05	.025	.01	0.005
					Area to the Right of the Critical Value					
1	—	—	0.001	0.004	0.016	2.706	3.841	5.024	6.635	7.879
2	0.010	0.020	0.051	0.103	0.211	4.605	5.991	7.378	9.210	10.597
3	0.072	0.115	0.216	0.352	0.584	6.251	7.815	9.348	11.345	12.838
4	0.207	0.297	0.484	0.711	1.064	7.779	9.488	11.143	13.277	14.860
5	0.412	0.554	0.831	1.145	1.610	9.236	11.071	12.833	15.086	16.750
6	0.676	0.872	1.237	1.635	2.204	10.645	12.592	14.449	16.812	18.548
7	0.989	1.239	1.690	2.167	2.833	12.017	14.067	16.013	18.475	20.278
8	1.344	1.646	2.180	2.733	3.490	13.362	15.507	17.535	20.090	21.955
9	1.735	2.088	2.700	3.325	4.168	14.684	16.919	19.023	21.666	23.589
10	2.156	2.558	3.247	3.940	4.865	15.987	18.307	20.483	23.209	25.188
11	2.603	3.053	3.816	4.575	5.578	17.275	19.675	21.920	24.725	26.757
12	3.074	3.571	4.404	5.226	6.304	18.549	21.026	23.337	26.217	28.299
13	3.565	4.107	5.009	5.892	7.042	19.812	22.362	24.736	27.688	29.819
14	4.075	4.660	5.629	6.571	7.790	21.064	23.685	26.119	29.141	31.319
15	4.601	5.229	6.262	7.261	8.547	22.307	24.996	27.488	30.578	32.801
16	5.142	5.812	6.908	7.962	9.312	23.542	26.296	28.845	32.000	34.267
17	5.697	6.408	7.564	8.672	10.085	24.769	27.587	30.191	33.409	35.718
18	6.265	7.015	8.231	9.390	10.865	25.989	28.869	31.526	34.805	37.156
19	6.844	7.633	8.907	10.117	11.651	27.204	30.144	32.852	36.191	38.582
20	7.434	8.260	9.591	10.851	12.443	28.412	31.410	34.170	37.566	39.997
21	8.034	8.897	10.283	11.591	13.240	29.615	32.671	35.479	38.932	41.401
22	8.643	9.542	10.982	12.338	14.042	30.813	33.924	36.781	40.289	42.796
23	9.260	10.196	11.689	13.091	14.848	32.007	35.172	38.076	41.638	44.181
24	9.886	10.856	12.401	13.848	15.659	33.196	36.415	39.364	42.980	45.559
25	10.520	11.524	13.120	14.611	16.473	34.382	37.652	40.646	44.314	46.928
26	11.160	12.198	13.844	15.379	17.292	35.563	38.885	41.923	45.642	48.290
.27	11.808	12.879	14.573	16.151	18.114	36.741	40.113	43.194	46.963	49.645
28	12.461	13.565	15.308	16.928	18.939	37.916	41.337	44.461	48.278	50.993
29	13.121	14.257	16.047	17.708	19.768	39.087	42.557	45.722	49.588	52.336
30	13.787	14.954	16.791	18.493	20.599	40.256	43.773	46.979	50.892	53.672
40	20.707	22.164	24.433	26.509	29.051	51.805	55.758	59.342	63.691	66.766
50	27.991	29.707	32.357	34.764	37.689	63.167	67.505	71.420	76.154	79.490
60	35.534	37.485	40.482	43.188	46.459	74.397	79.082	83.298	88.379	91.952
70	43.275	45.442	48.758	51.739	55.329	85.527	90.531	95.023	100.425	104.215
80	51.172	53.540	57.153	60.391	64.278	96.578	101.879	106.629	112.329	116.321
90	59.196	61.754	65.647	69.126	73.291	107.565	113.145	118.136	124.116	128.299
100	67.328	70.065	74.222	77.929	82.358	118.136	124.342	129.561	135.807	140.169

Table A.4. *F*-Distribution (α = 0.05)

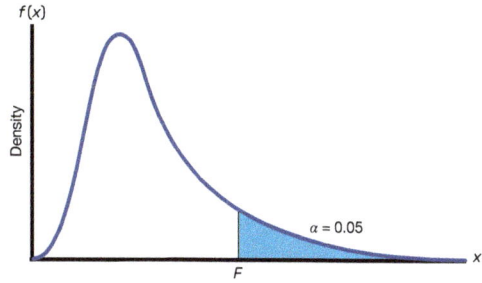

	Numerator degrees of freedom (df_1)																		
	1	2	3	4	5	6	7	8	9	10	12	15	20	24	30	40	60	120	inf
1	161.45	199.5	215.71	224.58	233.99	230.16	236.77	238.88	240.54	241.88	243.91	245.95	248.01	249.05	250.1	251.14	252.2	253.25	254.31
2	18.513	19.000	19.164	19.247	19.296	19.33	19.353	19.371	19.385	19.396	19.413	19.429	19.446	19.454	19.462	19.471	19.479	19.487	19.496
3	10.128	9.5521	9.2766	9.1172	9.0135	8.9406	8.8867	8.8452	8.8123	8.7855	8.7446	8.7029	8.6602	8.6385	8.6166	8.5944	8.572	8.5494	8.5264
4	7.7086	6.9443	6.5914	6.3882	6.2561	6.1631	6.0942	6.041	5.9988	5.9644	5.9117	5.8578	5.8025	5.7744	5.7459	5.717	5.6877	5.6581	5.6281
5	6.6079	5.7861	5.4095	5.1922	5.0503	4.9503	4.8759	4.8183	4.7725	4.7351	4.6777	4.6188	4.5581	4.5272	4.4957	4.4638	4.4314	4.3985	4.365
6	5.9874	5.1433	4.7571	4.5337	4.3874	4.2839	4.2067	4.1468	4.099	4.06	3.9999	3.9381	3.8742	3.8415	3.8082	3.7743	3.7398	3.7047	3.6689
7	5.5914	4.7374	4.3468	4.1203	3.9715	3.866	3.787	3.7257	3.6767	3.6365	3.5747	3.5107	3.4445	3.4105	3.3758	3.3404	3.3043	3.2674	3.2298
8	5.3177	4.459	4.0662	3.8379	3.6875	3.5806	3.5005	3.4381	3.3881	3.3472	3.2839	3.2184	3.1503	3.1152	3.0794	3.0428	3.0053	2.9669	2.9276
9	5.1174	4.2565	3.8625	3.6331	3.4817	3.3738	3.2927	3.2296	3.1789	3.1373	3.0729	3.0061	2.9365	2.9005	2.8637	2.8259	2.7872	2.7475	2.7067
10	4.9646	4.1028	3.7083	3.478	3.3258	3.2172	3.1355	3.0717	3.0204	2.9782	2.913	2.845	2.774	2.7372	2.6996	2.6609	2.6211	2.5801	2.5379
11	4.8443	3.9823	3.5874	3.3567	3.2039	3.0946	3.0123	2.948	2.8962	2.8536	2.7876	2.7186	2.6464	2.609	2.5705	2.5309	2.4901	2.448	2.4045
12	4.7472	3.8853	3.4903	3.2592	3.1059	2.9961	2.9134	2.8486	2.7964	2.7534	2.6866	2.6169	2.5436	2.5055	2.4663	2.4259	2.3842	2.341	2.2962
13	4.6672	3.8056	3.4105	3.1791	3.0254	2.9153	2.8321	2.7669	2.7144	2.671	2.6037	2.5331	2.4589	2.4202	2.3803	2.3392	2.2966	2.2524	2.2064
14	4.6001	3.7389	3.3439	3.1122	2.9582	2.8477	2.7642	2.6987	2.6458	2.6022	2.5342	2.463	2.3879	2.3487	2.3082	2.2664	2.2229	2.1778	2.1307
15	4.5431	3.6823	3.2874	3.0556	2.9013	2.7905	2.7066	2.6408	2.5876	2.5437	2.4753	2.4034	2.3275	2.2878	2.2468	2.2043	2.1601	2.1141	2.0658
16	4.494	3.6337	3.2389	3.0069	2.8524	2.7413	2.6572	2.5911	2.5377	2.4935	2.4247	2.3522	2.2756	2.2354	2.1938	2.1507	2.1058	2.0589	2.0096
17	4.4513	3.5915	3.1968	2.9647	2.81	2.6987	2.6143	2.548	2.4943	2.4499	2.3807	2.3077	2.2304	2.1898	2.1477	2.104	2.0584	2.0107	1.9604
18	4.4139	3.5546	3.1599	2.9277	2.7729	2.6613	2.5767	2.5102	2.4563	2.4117	2.3421	2.2686	2.1906	2.1497	2.1071	2.0629	2.0166	1.9681	1.9168
19	4.3807	3.5219	3.1274	2.8951	2.7401	2.6283	2.5435	2.4768	2.4227	2.3779	2.308	2.2341	2.1555	2.1141	2.0712	2.0264	1.9795	1.9302	1.878
20	4.3512	3.4928	3.0984	2.8661	2.7109	2.599	2.514	2.4471	2.3928	2.3479	2.2776	2.2033	2.1242	2.0825	2.0391	1.9938	1.9464	1.8963	1.8432
21	4.3248	3.4668	3.0725	2.8401	2.6848	2.5727	2.4876	2.4205	2.366	2.321	2.2504	2.1757	2.096	2.054	2.0102	1.9645	1.9165	1.8657	1.8117
22	4.3009	3.4434	3.0491	2.8167	2.6613	2.5491	2.4638	2.3965	2.3419	2.2967	2.2258	2.1508	2.0707	2.0283	1.9842	1.938	1.8894	1.838	1.7831
23	4.2793	3.4221	3.028	2.7955	2.64	2.5277	2.4422	2.3748	2.3201	2.2747	2.2036	2.1281	2.0476	2.005	1.9605	1.9139	1.8648	1.8128	1.757
24	4.2597	3.4028	3.0088	2.7763	2.6207	2.5082	2.4226	2.3551	2.3002	2.2547	2.1834	2.1077	2.0267	1.9838	1.939	1.892	1.8648	1.8128	1.733
25	4.2417	3.3852	2.9912	2.7587	2.603	2.4904	2.4047	2.3371	2/2821	2.2365	2.1649	2.0889	2.0075	1.9643	1.9192	1.8718	1.8217	1.7684	1.711
26	4.2252	3.369	2.9752	2.7426	2.5868	2.4741	2.3883	2.3205	2.2655	2.2197	2.1479	2.0716	1.9898	1.9464	1.901	1.8533	1.8027	1.7488	1.6906
27	4.21	3.3541	2.9604	2.7278	2.5719	2.4591	2.3732	2.3053	2.2501	2.2043	2.1323	2.0558	1.9736	1.9299	1.8842	1.8361	1.7851	1.7306	1.6717
28	4.196	3.3404	2.9467	2.7141	2.5581	2.4453	2.3593	2.2913	2.236	2.19	2.1179	2.0411	1.9586	1.9147	1.8687	1.8203	1.7689	1.7138	1.6541
29	4.183	3.3277	2.934	2.7014	2.5454	2.4324	2.3463	2.2783	2.2229	2.1768	2.1045	2.0275	1.9446	1.9005	1.8543	1.8055	1.7537	1.6981	1.6376
30	4.1709	3.3158	2.9223	2.6896	2.5336	2.4205	2.3343	2.2662	2.2107	2.1646	2.0921	2.0148	1.9317	1.8874	1.8409	1.7918	1.7396	1.6835	1.6223
40	4.0847	3.2317	2.8387	2.606	2.4495	2.3359	2.249	2.1802	2.124	2.0772	2.0035	1.9245	1.8389	1.7929	1.7444	1.6928	1.6373	1.5766	1.5089
60	4.0012	3.1504	2.7581	2.5252	2.3683	2.2541	2.1665	2.097	2.0401	1.9926	1.9174	1.8364	1.748	1.7001	1.6491	1.5943	1.5343	1.4673	1.3893
120	3.9201	3.0718	2.6802	2.4472	2.2899	2.175	2.0868	2.0164	1.9588	1.9105	1.8337	1.7505	1.6587	1.6084	1.5543	1.4952	1.429	1.3519	1.2539
>120	3.8415	2.9957	2.6049	2.3719	2.2141	2.0986	2.0096	1.9384	1.8799	1.8307	1.7522	1.6664	1.5705	1.5173	1.4591	1.394	1.318	1.2214	1

Denominator degrees of freedom (df_2)

Table A.4. F-Distribution ($\alpha = 0.025$)

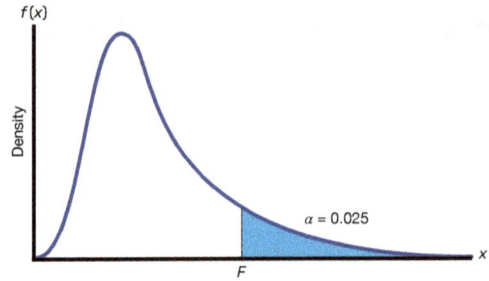

	Numerator degrees of freedom (df_1)																		
	1	2	3	4	5	6	7	8	9	10	12	15	20	24	30	40	60	120	inf
1	647.79	799.5	864.16	899.58	921.85	937.11	948.22	956.66	963.28	968.63	976.71	984.87	993.1	997.25	1001.4	1005.6	1009.8	1014	1018.3
2	38.506	39.000	39.166	39.248	39.298	39.332	39.355	39.373	39.387	39.398	39.415	39.431	39.448	39.456	39.465	39.473	39.481	39.49	39.498
3	17.443	16.044	15.439	15.101	14.885	14.735	14.624	14.54	14.473	14.419	14.337	14.253	14.167	14.124	14.081	14.037	13.992	13.947	13.902
4	12.218	10.649	9.9792	9.6045	9.3645	9.1973	9.0741	8.9796	8.9047	8.8439	8.7512	8.6565	8.5599	8.5109	8.461	8.411	8.36	8.309	8.257
5	10.007	8.4336	7.7636	7.3879	7.1464	6.9777	6.8531	6.7572	6.6811	6.6192	6.5245	6.4277	6.3286	6.278	6.227	6.175	6.123	6.069	6.015
6	8.8131	7.2599	6.5988	6.2272	5.9876	5.8198	5.6955	5.5996	5.5234	5.4613	5.3662	5.2687	5.1684	5.1172	5.065	5.012	4.959	4.904	4.849
7	8.0727	6.5415	5.8898	5.5226	5.2852	5.1186	4.9949	4.8993	4.8232	4.7611	4.6658	4.5678	4.4667	4.415	4.362	4.309	4.254	4.199	4.142
8	7.5709	6.0595	5.416	5.0526	4.8173	4.6517	4.5286	4.4333	4.3572	4.2951	4.1997	4.1012	3.9995	3.9472	3.894	3.84	3.784	3.728	3.67
9	7.2093	5.7147	5.0781	4.7181	4.4844	4.3197	4.197	4.102	4.026	3.9639	3.8682	3.7694	3.6669	3.6142	3.56	3.505	3.449	3.392	3.333
10	6.9367	5.4564	4.8256	4.4683	4.2361	4.0721	3.9498	3.8549	3.779	3.7168	3.6209	3.5217	3.4185	3.3654	3.311	3.255	3.198	3.14	3.08
11	6.7241	5.2559	4.63	4.2751	4.044	3.8807	3.7586	3.6638	3.5879	3.5257	3.4296	3.3299	3.2261	3.1725	3.118	3.061	3.004	2.944	2.883
12	6.5538	5.0959	4.4742	4.1212	3.8911	3.7283	3.6065	3.5118	3.4358	3.3736	3.2773	3.1772	3.0728	3.0187	2.963	2.906	2.848	2.787	2.725
13	6.4143	4.9653	4.3472	3.9959	3.7667	3.6043	3.4827	3.388	3.312	3.2497	3.1532	3.0527	2.9477	2.8932	2.837	2.78	2.72	2.659	2.595
14	6.2979	4.8567	4.2417	3.8919	3.6634	3.5014	3.3799	3.2853	3.2093	3.1469	3.0502	2.9493	2.8437	2.7888	2.732	2.674	2.614	2.552	2.487
15	6.1995	4.765	4.1528	3.8043	3.5764	3.4147	3.2934	3.1987	3.1227	3.0602	2.9633	2.8621	2.7559	2.7006	2.644	2.585	2.524	2.461	2.395
16	6.1151	4.6867	4.0768	3.7294	3.5021	3.3406	3.2194	3.1248	3.0488	2.9862	2.889	2.7875	2.6808	2.6252	2.568	2.509	2.447	2.383	2.316
17	6.042	4.6189	4.0112	3.6648	3.4379	3.2767	3.1556	3.061	2.9849	2.9222	2.8249	2.723	2.6158	2.5598	2.502	2.442	2.38	2.315	2.247
18	5.9781	4.5597	3.9539	3.6083	3.382	3.2209	3.0999	3.0053	2.9291	2.8664	2.7689	2.6667	2.559	2.5027	2.445	2.384	2.321	2.256	2.187
19	5.9216	4.5075	3.9034	3.5587	3.3327	3.1718	3.0509	2.9563	2.8801	2.8172	2.7196	2.6171	2.5089	2.4523	2.394	2.333	2.27	2.203	2.133
20	5.8715	4.4613	3.8587	3.5147	3.2891	3.1283	3.0074	2.9128	2.8365	2.7737	2.6758	2.5731	2.4645	2.4076	2.349	2.287	2.223	2.156	2.085
21	5.8266	4.4199	3.8188	3.4754	3.2501	3.0895	2.9686	2.874	2.7977	2.7348	2.6368	2.5338	2.4247	2.3675	2.308	2.246	2.182	2.114	2.042
22	5.7863	4.3828	3.7829	3.4401	3.2151	3.0546	2.9338	2.8392	2.7628	2.6998	2.6017	2.4984	2.389	2.3315	2.272	2.21	2.145	2.076	2.003
23	5.7498	4.3492	3.7505	3.4083	3.1835	3.0232	2.9023	2.8077	2.7313	2.6682	2.5699	2.4665	2.3567	2.2989	2.239	2.176	2.111	2.041	1.968
24	5.7166	4.3187	3.7211	3.3794	3.1548	2.9946	2.8738	2.7791	2.7027	2.6396	2.5411	2.4374	2.3273	2.2693	2.209	2.146	2.08	2.01	1.935
25	5.6864	4.2909	3.6943	3.353	3.1287	2.9685	2.8478	2.7531	2.6766	2.6135	2.5149	2.411	2.3005	2.2422	2.182	2.118	2.052	1.981	1.906
26	5.6586	4.2655	3.6697	3.3289	3.1048	2.9447	2.824	2.7293	2.6528	2.5896	2.4908	2.3867	2.2759	2.2174	2.157	2.093	2.026	1.954	1.878
27	5.6331	4.2421	3.6472	3.3067	3.0828	2.9228	2.8021	2.7074	2.6309	2.5676	2.4688	2.3644	2.2533	2.1946	2.133	2.069	2.002	1.93	1.853
28	5.6096	4.2205	3.6264	3.2863	3.0626	2.9027	2.782	2.6872	2.6106	2.5473	2.4484	2.3438	2.2324	2.1735	2.112	2.048	1.98	1.907	1.829
29	5.5878	4.2006	3.6072	3.2674	3.0438	2.884	2.7633	2.6686	2.5919	2.5286	2.4295	2.3248	2.2131	2.154	2.092	2.028	1.959	1.886	1.807
30	5.5675	4.1821	3.5894	3.2499	3.0265	2.8667	2.746	2.6513	2.5746	2.5112	2.412	2.3072	2.1952	2.1359	2.074	2.009	1.94	1.866	1.787
40	5.4239	4.051	3.4633	3.1261	2.9037	2.7444	2.6238	2.5289	2.4519	2.3882	2.2882	2.1819	2.0677	2.0069	1.943	1.875	1.803	1.724	1.637
60	5.2856	3.9253	3.3425	3.0077	2.7863	2.6274	2.5068	2.4117	2.3344	2.2702	2.1692	2.0613	1.9445	1.8817	1.815	1.744	1.667	1.581	1.482
120	5.1523	3.8046	3.2269	2.8943	2.674	2.5154	2.3948	2.2994	2.2217	2.157	2.0548	1.945	1.8249	1.7597	1.69	1.614	1.53	1.433	1.31
>120	5.0239	3.6889	3.1161	2.7858	2.5665	2.4082	2.2875	2.1918	2.1136	2.0483	1.9447	1.8326	1.7085	1.6402	1.566	1.484	1.388	1.268	1

Denominator degrees of freedom (df_2)

Table A.5. Critical Values of the Pearson Correlation Coefficient r

n	$\alpha = .05$	$\alpha = .01$
4	.950	.999
5	.878	.959
6	.811	.917
7	.754	.875
8	.707	.834
9	.666	.798
10	.632	.765
11	.602	.735
12	.576	.708
13	.553	.684
14	.532	.661
15	.514	.641
16	.497	.623
17	.482	.606
18	.468	.590
19	.456	.575
20	.444	.561
25	.396	.505
30	.361	.463
35	.335	.430
40	.312	.402
45	.294	.378
50	.279	.361
60	.254	.330
70	.236	.305
80	.220	.286
90	.207	.269
100	.196	.256

Note:
To test $H_0: \rho = 0$ against $H_a: \rho \neq 0$, reject H_0 if the absolute value of r is greater than or equal to the critical value in the table.

Table A.6. Critical Values of the Sign Test

	One Tailed (α)			
	0.005	0.01	0.025	0.05
n				
	Two Tailed (α)			
	0.01	0.02	0.05	0.10
8	0	0	0	1
9	0	0	1	1
10	0	0	1	1
11	0	1	1	2
12	1	1	2	2
13	1	1	2	3
14	1	2	3	3
15	2	2	3	3
16	2	2	3	4
17	2	3	4	4
18	3	3	4	5
19	3	4	4	5
20	3	4	5	5
21	4	4	5	6
22	4	5	5	6
23	4	5	6	7
24	5	5	6	7
25	5	6	6	7

Table A.7. Critical Values of T for the Wilcoxon Signed-Rank Test

	One Tailed (α)			
	0.05	0.025	0.01	0.005
n	Two Tailed (α)			
	0.10	0.05	0.02	0.01
5	1	--	--	--
6	2	1	--	--
7	4	2	0	--
8	6	4	2	1
9	8	6	3	2
10	11	8	5	3
11	14	11	7	5
12	17	14	10	7
13	21	17	13	10
14	26	21	16	13
15	30	25	20	16
16	36	30	24	19
17	41	35	28	23
18	47	40	33	28
19	54	46	38	32
20	60	52	43	37
21	68	59	49	43
22	75	66	56	49
23	83	73	62	55
24	92	81	69	61
25	101	90	77	68
26	110	98	85	76
27	120	107	93	84
28	130	117	102	92
29	141	127	111	100
30	152	137	120	109

Note:
Reject the null hypothesis if the test statistic T is less than or equal to the critical value in the table. Fail to reject the null hypothesis if the test statistic T is greater than the critical value in the table.

Table A.8. Critical Values of Spearman's Rank Correlation Coefficient r_s

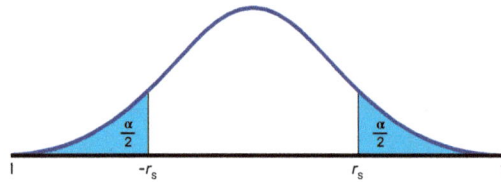

n	α			
	0.10	0.05	0.025	0.01
4	1.000	1.000	-	-
5	0.800	0.900	1.000	1.000
6	0.657	0.829	0.886	0.943
7	0.571	0.714	0.786	0.893
8	0.524	0.643	0.738	0.833
9	0.483	0.600	0.700	0.783
10	0.455	0.564	0.648	0.745
11	0.427	0.536	0.618	0.709
12	0.406	0.503	0.587	0.678
13	0.385	0.484	0.560	0.648
14	0.367	0.464	0.538	0.626
15	0.354	0.446	0.521	0.604
16	0.341	0.429	0.503	0.582
17	0.328	0.414	0.488	0.566
18	0.317	0.401	0.472	0.550
19	0.309	0.391	0.460	0.535
20	0.299	0.380	0.447	0.522
21	0.292	0.370	0.436	0.509
22	0.284	0.361	0.425	0.497
23	0.278	0.353	0.416	0.486
24	0.271	0.344	0.407	0.476
25	0.265	0.337	0.398	0.466
26	0.259	0.331	0.390	0.457
27	0.255	0.324	0.383	0.449
28	0.250	0.318	0.375	0.441
29	0.245	0.312	0.368	0.433

Table A.8. Critical Values of Spearman's Rank Correlation Coefficient r_s

n	0.10	0.05	0.025	0.01
		α		
30	0.240	0.306	0.362	0.425
31	0.236	0.301	0.356	0.419
32	0.232	0.296	0.350	0.412
33	0.229	0.291	0.345	0.405
34	0.225	0.287	0.340	0.400
35	0.222	0.283	0.335	0.394
36	0.219	0.279	0.330	0.388
37	0.215	0.275	0.325	0.383
38	0.212	0.271	0.321	0.378
39	0.210	0.267	0.317	0.373
40	0.207	0.264	0.313	0.368
41	0.204	0.261	0.309	0.364
42	0.202	0.257	0.305	0.359
43	0.199	0.254	0.301	0.355
44	0.197	0.251	0.298	0.351
45	0.194	0.248	0.294	0.347
46	0.192	0.246	0.291	0.343
47	0.190	0.243	0.288	0.340
48	0.188	0.240	0.285	0.336
49	0.186	0.238	0.282	0.333
50	0.184	0.235	0.279	0.329
51	0.182	0.233	0.276	0.326
52	0.180	0.231	0.274	0.323
53	0.179	0.228	0.271	0.320
54	0.177	0.226	0.268	0.317
55	0.175	0.224	0.266	0.314
56	0.174	0.222	0.264	0.311
57	0.172	0.220	0.261	0.308
58	0.171	0.218	0.259	0.306
59	0.169	0.216	0.257	0.303
60	0.168	0.214	0.255	0.301

Note:
To test $H_0: \rho_s = 0$ against $H_a: \rho_s \neq 0$, reject H_0 if the absolute value of r_s is greater than or equal to the critical value in the table.

Table A.9. Runs Test for Randomness: Critical Values for Number of Runs R (Two-Tailed Test, $\alpha = 0.05$)

n_1 \ n_2	2	3	4	5	6	7	8	9	10	11	12	13	14	15	16	17	18	19	20
2	1,6	1,6	1,6	1,6	1,6	1,6	1,6	1,6	1,6	1,6	2,6	2,6	2,6	2,6	2,6	2,6	2,6	2,6	2,6
3	1,6	1,8	1,8	1,8	2,8	2,8	2,8	2,8	2,8	2,8	2,8	2,8	2,8	3,8	3,8	3,8	3,8	3,8	3,8
4	1,6	1,8	1,9	2,9	2,9	2,10	3,10	3,10	3,10	3,10	3,10	3,10	3,10	3,10	4,10	4,10	4,10	4,10	4,10
5	1,6	1,8	2,9	2,10	3,10	3,11	3,11	3,12	3,12	4,12	4,12	4,12	4,12	4,12	4,12	4,12	5,12	5,12	5,12
6	1,6	2,8	2,9	3,10	3,11	3,12	3,12	4,13	4,13	4,13	4,13	5,14	5,14	5,14	5,14	5,14	5,14	6,14	6,14
7	1,6	2,8	2,10	3,11	3,12	3,13	4,13	4,14	5,14	5,14	5,14	5,15	5,15	6,15	6,16	6,16	6,16	6,16	6,16
8	1,6	2,8	3,10	3,11	3,12	4,13	4,14	5,14	5,15	5,15	6,16	6,16	6,16	6,16	6,17	7,17	7,17	7,17	7,17
9	1,6	2,8	3,10	3,12	4,13	4,14	5,14	5,15	5,16	6,16	6,16	6,17	7,17	7,18	7,18	7,18	8,18	8,18	8,18
10	1,6	2,8	3,10	3,12	4,13	5,14	5,15	5,16	6,16	6,17	7,17	7,18	7,18	7,18	8,19	8,19	8,19	8,20	9,20
11	1,6	2,8	3,10	4,12	4,13	5,14	5,15	6,16	6,17	7,17	7,18	7,19	8,19	8,19	8,20	9,20	9,20	9,21	9,21
12	2,6	2,8	3,10	4,12	4,13	5,14	6,16	6,16	7,17	7,18	7,19	8,19	8,20	8,20	9,21	9,21	9,21	10,22	10,22
13	2,6	2,8	3,10	4,12	5,14	5,15	6,16	6,17	7,18	7,19	8,19	8,20	9,20	9,21	9,21	10,22	10,22	10,23	10,23
14	2,6	2,8	3,10	4,12	5,14	5,15	6,16	7,17	7,18	8,19	8,20	9,20	9,21	9,22	10,22	10,23	10,23	11,23	11,24
15	2,6	3,8	3,10	4,12	5,14	6,15	6,16	7,18	7,18	8,19	8,20	9,21	9,22	10,22	10,23	11,23	11,24	11,24	12,25
16	2,6	3,8	4,10	4,12	5,14	6,16	6,17	7,18	8,19	8,20	9,21	9,21	10,22	10,23	11,23	11,24	11,25	12,15	12,25
17	2,6	3,8	4,10	4,12	5,14	6,16	7,17	7,18	8,19	9,20	9,21	10,22	10,23	11,23	11,24	11,25	12,25	12,26	13,26
18	2,6	3,8	4,10	5,12	5,14	6,16	7,17	8,18	8,19	9,20	9,21	10,22	10,23	11,24	11,25	12,25	12,26	13,26	13,27
19	2,6	3,8	4,10	5,12	6,14	6,16	7,17	8,18	8,20	9,21	10,22	10,23	11,23	11,24	12,25	12,26	13,26	13,27	13,27
20	2,6	3,8	4,10	5,12	6,14	6,16	7,17	8,18	9,20	9,21	10,22	10,23	11,24	12,25	12,25	13,26	13,27	13,27	14,28

Note:
To test H_0: Random against H_a: Not random, reject H_0 if either R is less than or equal to the smaller entry in the table, or R is greater than or equal to the larger entry in the table.

Chapter 1 Solutions

Solutions 1.1

1. a. population: all clients of this fitness center
   b. sample: a subset of these clients
   c. parameter: the population mean amount of time of clients who exercise in the center each week
   d. statistic: the sample mean amount of time of clients who exercise in the center each week
   e. variable: X = the amount of exercise time for one client
   f. data: values of X, such as 4 hours, 8.5 hours, and so on

3. a. population: all heart attack patients of the cardiologist
   b. sample: a group of these patients
   c. parameter: the population mean recovery period of all heart attack patients of the cardiologist
   d. statistic: the sample mean recovery period of all heart attack patients of the cardiologist
   e. variable: X = the recovery period of one patient
   f. data: values for X, such as 2 weeks, 3 months, and so on

5. a. population: all the clients of this counselor
   b. sample: a group of clients of this marriage counselor
   c. parameter: the proportion of all clients who stay married
   d. statistic: the proportion of the sample of the counselor's clients who stay married
   e. variable: X = the number of couples who stay married
   f. data: yes, no

7. a. population: all people (maybe in a certain geographic area, such as the United States)
   b. sample: a subset (group) of all people
   c. parameter: the proportion of all people who will buy the product
   d. statistic: the proportion of the sample who will buy the product
   e. variable: X = the number of people who will buy it
   f. data: buy, not buy

9.   a.   variable

11. a.   quantitative discrete, 150 tickets
     b.   quantitative continuous, 37.5%
     c.   qualitative, LA Dodgers
     d.   quantitative continuous, 3.5 minutes
     e.   quantitative discrete, 11,234 students
     f.   qualitative, Cheers
     g.   qualitative, Tom's of Maine toothpaste
     h.   quantitative continuous, 3.56 miles
     i.   quantitative continuous, 52.8 years
     j.   quantitative discrete, 5 packages

13. c.   quantitative continuos

Solutions 1.2

1.   a.   i) The survey was conducted using six similar flights. ii) The survey would not be a true rep-
            resentation of the entire population of air travelers. iii) Conducting the survey on a holiday
            weekend will not produce representative results.
     b.   i) Conduct the survey during different times of the year. ii) Conduct the survey using flights
            to and from various locations. iii) Conduct the survey on different days of the week.

3.   Answers will vary.
     Sample Answer: You could use a systematic sampling method. Stop every tenth person as they
     leave one of the buildings on campus at 9:50 a.m. Then, stop every tenth person as they leave a
     different building on campus at 1:50 p.m.

5.   Answers will vary.
     Sample Answer: Many people will not respond to mail surveys. If they do respond to the surveys,
     you can't be sure who is responding. In addition, mailing lists can be incomplete.

7.   a.   systematic sampling

9. a.   qualitative

   b.   quantitative discrete

   c.   quantitative discrete

   d.   qualitative

Solutions 1.3

1.   Answers will vary.

3.   a.   Explanatory variable: amount of sleep

      b.   Response variable: performance measured in assigned tasks

      c.   Treatments: normal sleep and 27 hours of total sleep deprivation

      d.   Experimental units: 19 professional drivers

      e.   Lurking variables: none; all drivers participated in both treatments

      f.   Random assignment: treatments were assigned in random order; this eliminated the effect of any "learning" that may take place during the first experimental session

      g.   Control/placebo: completing the experimental session under normal sleep conditions

      h.   Blinding: researchers evaluating subjects' performance must not know which treatment is being applied at the time

5.   You cannot assume that the numbers of complaints reflect the quality of the airlines. The airlines shown with the greatest number of complaints are also the ones with the most passengers. You must consider the appropriateness of methods for presenting data; in this case, displaying totals is misleading.

7.   c.   control group

9.   b.   A well-designed experiment

11.   a.   This is an observational study because the data are obtained from the pre-existing medical records of the patients who go to the local clinic.

Chapter 2 Solutions

Solutions 2.1

1.

Number of Times	Frequency
7.5 – 22.5	7
22.5 – 37.5	4
37.5 – 52.5	5
52.5 – 67.5	8
67.5 – 82.5	5
82.5 – 97.5	3

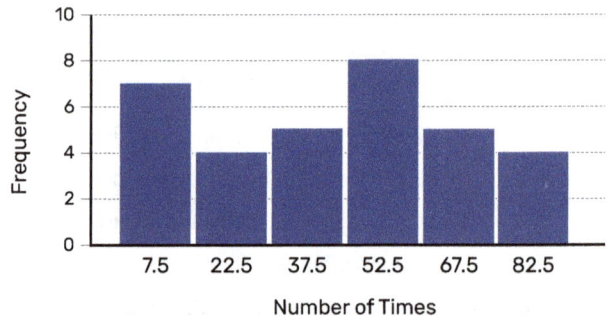

3. a.

Blood Type	Frequency	Relative Frequency
O	216	0.432
A	210	0.42
B	62	0.124
AB	12	0.024

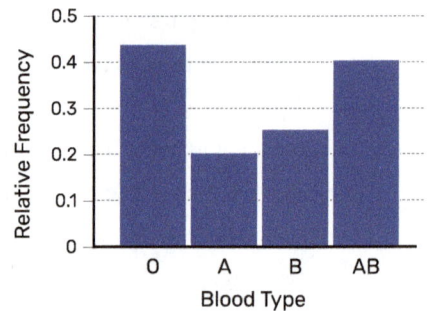

b. Blood type O has the highest relative frequency among these 500 people.

Solutions 2.2

1. a. 19
    b. 127
    c. 7
    d. 49

3. mean = 148.583; median = 149

Solutions 2.3

1. *range* = 9

   $s_2 = 9.0998$

   $s = 3.0166$

3. *range* = 13

   $s_2 = 19.51$

   $s = 4.417$

Solutions 2.4

1. a. Percentile rank of 80 $= \frac{8}{20} = 40\%$

   b. Percentile rank of 92 $= \frac{16}{20} = 80\%$

3. *z*-score: -1.198, -0.599, 0.599, 2.395

5. $x = 7.9$

7. True

Solutions 2.5

1. 1. 68%

3. $k = \frac{\text{the number within}}{\text{standard deviation}} = \frac{2.2}{1.1} = 2$

   The proportion of data that lie between 3 and 7.4 is $(1 - \frac{1}{2^2}) \times 100\% = 75\%$

   The minimum number of data set that must lie between 3 and 7.4 is 0.75*400 = 300

5. a. About 0.68

   b. About 0.16

   c. The length of the longest fish is about 3 standard deviations from the mean, 42 inches

7. 0.023

Chapter 3 Solutions

Solutions 3.1

1.   a.   Find $P(H) = \frac{1}{12}$

      b.   Find $P(N) = \frac{1}{15}$

      c.   Find $P(F) = \frac{1}{5}$

      d.   Find $P(C) = \frac{1}{5}$

3.   a. $P(Y) = \frac{1}{8}$

      b. $P(R) = \frac{1}{2}$

      c.   Odds landing on red $= P(R)/(1 - P(R)) = (\frac{1}{2})(\frac{2}{1}) = 1{:}1$

5.   a. $P(E\backslash M)$ = probability of landing on an even number GIVEN the event of lending on a multiple of three has already occurred.

      b. $P(E \cup M)$ = the probability of landing on an even number OR on a multiple of three

3.2 Solutions

1.   a.   The sample space S={ G1, G2, G3, G4, G5, Y1, Y2, Y3}

      b. $P(G) = \frac{5}{8}$

      c. $P(G\,|\,E) = \dfrac{P(G \cap E)}{P(E)} = \dfrac{2/8}{3/8} = \dfrac{2}{3}$

      d. $P(G \cap E) = \frac{1}{4}$

      e. $P(G \cup E) = \frac{5}{8} + \frac{3}{8} - \frac{1}{4} = \frac{3}{4}$

      f.   G and E *are not mutually exclusive because*

3.   a. $P(C \cap D) = P(C\,|\,D) \times P(D) = 0.3$

      b.   C and D are not mutually exclusive because $P(C \cap D) = 0.3 \neq 0$

      c.   C and D are not independent events because $P(C \cap D) = 0.3 \neq P(C) \times P(D) = 0.2$

      d. $P(C \cup D) = P(C) + P(D) - P(C \cap D) = 0.4 + 0.5 - 0.3 = 0.6$

      e. $P(D\,|\,C) = \dfrac{P(D\,|\,C)}{P(C)} = \dfrac{0.3}{0.4} = \dfrac{3}{4}$

5. a. From P(type O OR Rh-) = P(type O) + P(Rh-) - P(type O AND Rh-) we have

0.52 = 0.43 + 0.15 − P(type O AND Rh-);

Solve above equation, we have

P(type O AND Rh-) = 0.06

b. P(NOT type O AND Rh-) = 1 - P(type O AND Rh-) = 1 − 0.06 = 0.94

7. *Let C = the event that the cookie contains chocolate. Let N = the event that the cookie contains nuts.*

a. P(C OR N) = P(C) + P(N) - P(C AND N) = 0.36 + 0.12 − 0.08 = 0.40

b. P(NEITHER chocolate NOR nuts) = 1 - P(C OR N) = 1 − 0.40 = 0.60

3.3 Solutions

1. A = {2, 3, 5} and B = {1, 3, 5}. The sample space for rolling a fair die is S = {1, 2, 3, 4, 5, 6}.

a. $A \cap B$ = {5}

b. $A \cup B$ = {1, 2, 3, 5}

3. a. *$P(C \cap PT)$ = 0.05*

b. P(C \ PT) = P(C ∩ PT) /P(PT)=0.05/0.5= 0.1

5. a. P(athlete stretches before exercising) = $\frac{35}{80}$

b. P(athlete stretches before exercising | no injury in the last year) =

c. $\dfrac{P(\text{athlete stretches before exercising AND no injury in the last year})}{P(\text{no injury in the last year}}$ $= \dfrac{295/800}{514/800} = \dfrac{295}{514}$

3.4 Solutions

1. $4 \cdot 3 = 12$

3. a. $10 \cdot 9 \cdot 8 \cdot 7 = 5040$

b. $10^4 = 10000$

5. $52 \cdot 51 \cdot 50 \cdot 49 = 6{,}497{,}400$

7. $\dfrac{4!}{2! \cdot 2!} = 6$

9. $\dfrac{10!}{3! \cdot 3! \cdot 2! \cdot 1! \cdot 1!} = 50400$

11. $\dfrac{6!}{2! \cdot 3! \cdot 1!} = 60$

Chapter 4 Solutions

Solutions 4.1

1. a. $P(x = 4) = 0.1$
   b. $P(x \geq 5) = 0.15$
   c. On average, a new hire stays 2.43 years with the company.
   d. The column of $P(x)$ sums to 1.

3. a. $X =$ Number of days Ellen goes to music practice.
   b.

x	P(x)
0	0.03
1	0.04
2	0.08
3	0.85

   c. The other characteristic is that each individual probabilities must be between 0 and 1.

5. $\sigma = \sqrt{\sum (x - \mu)^2 \, P(x)} = \sqrt{3.24} = 1.8$

7. Let $X =$ Event of drawing a face card

   $$P(X) = \tfrac{12}{52} \quad \text{and} \quad P(X') = 1 - P(X) = \tfrac{40}{52}$$
   $$E = \$30 \cdot \tfrac{12}{52} - \$2 \cdot \tfrac{40}{52} \approx \$5.385$$

   The expected value of $5.385 is greater than zero so you should play the game.

Solutions 4.2

1. Let $X =$ Number of babies born with significant hearing loss
   $$P(X = 2) = 0.271$$

3. a. $X \sim B\,(150, 0.09)$

b. $\mu = 13.5$ and $\sigma = 3.505$

c. $P(X = 15) = 0.0988$

d. $P(X \leq 10) = 0.1987$

e. $P(X > 25) = 1 - P(X \leq 25) = 1 - 0.9991 = 0.0009$

5. Let X = number of questions answered correctly and $X \sim B(11, 0.3)$.

We are interested in the probability that the student gets *more than* 75% of 32 questions correct. 75% of 32 is 24. We want to find $P(x > 24)$. The event "more than 24" is the complement of "less than or equal to 24."

$P(x > 24) = 0$

The probability of getting more than 75% of the 32 questions correct when randomly guessing is very small and practically zero.

7. a. X = the number of Oregon residents that have adequate earthquake supplies.

b. $X = 0, 1, 2, ..., 11$

c. The distribution of X is binomial, $X \sim B(11, 0.3)$.

d. $P(X \geq 8) = 1 - P(X < 7) = 1 - 0.9957 = 0.00429$

e. $P(X = 0) = \frac{11!}{(0!)(11!)} (0.3^0)(0.7^{11}) = 0.0198$

$P(X = 11) = \frac{11!}{(11!)(0!)} (0.3^{11})(0.7^0) = 0.00000177$

It is more likely that none of the residents surveyed will have adequate earthquake supplies due to its smaller probability.

9. a. X = Number of kids who prefer broccoli to strawberries.

b. $X = 0, 1, 2, ..., 600$

c. $X \sim B(600, 0.005)$

d. 3

e. $P(X = 0) = 0.0494$

f. $P(X > 4) = 1 - P(X \leq 3) = 1 - 0.6472 = 0.3528$

Solutions 4.3

1. a. X = number of customers a food truck serves in one day.

b. $x = 0, 1, 2, 3, ...$

c. $X \sim P(120)$

$$P(X = 150) = 0.001$$

d. The average number of customers served every 4 hours is $\frac{120}{3} = 40$.

$$P(X = 35) \frac{\mu^x e^{-\mu}}{x!} = \frac{40^{35} e^{-40}}{35!} \approx 0.0485$$

f. The average number of customers served each hour is 10.

$$P(X > 12) = 1 - P(X \le 11) = 1 - 0.697 = 0.303$$

h. The average number of customers served every two hours is 20.

$$P(X < 12) = P(X \le 11) = 0.0214$$

3. a. $X \sim P(5.5)$; $\mu = 5.5$; $\sigma = \sqrt{5.5} \approx 2.3452$

 b. $P(x \le 6) \approx 0.6860$

 c. There is a 15.7% probability that the law staff will receive more calls than they can handle.

 d. $P(x > 8) = 1 - P(x \le 8) \approx 1 - 0.8944 = 0.1056$

5. Let X = the number of defective bulbs in a string.

 Using the Poisson distribution:

 $\mu = np = 100(0.03) = 3$

 $X \sim P(3)$

 $P(x \le 4) \approx 0.8153$

 Using the binomial distribution:

 $X \sim B(100, 0.03)$

 $P(x \le 4) = 0.8179$

 The Poisson approximation is very good—the difference between the probabilities is only 0.0026.

7. a. $P(X = 2) = \frac{7^2 e^{-7}}{2!} = 0.02234$

 b. $P(X \le 2) = P(X = 0) + P(X = 1) + P(X = 2) = \frac{7^0 e^{-7}}{0!} - \frac{6^1 e^{-7}}{1!} - \frac{7^2 e^{-7}}{2!} = 0.0296$

 c. $\sigma = \sqrt{\mu} = \sqrt{7} \approx 2.646$

Chapter 5 Solutions

5.1 Solutions

1. a. $P(2 < X < 5)$

   b. $P(6 < X < 7)$

3. $P(x < 0) = 0$

5.2 Solutions

1. a. $f(x) = \frac{1}{12}$

   b. $\mu = 6$

   c. $\sigma = 3.464$

   d.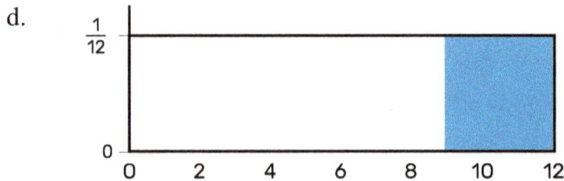

   e. $P(x > 9) = 0.25$

   f. $X = 4.8$.

3. a. Ages of cars

   b. $X =$ The age (in years) of cars in the staff parking lot

   c. Continuous

   d. 0.5 to 9.5

   e. Uniform

   f. $f(x) = \frac{1}{9}$ where x is between 0.5 and 9.5, inclusive

   g.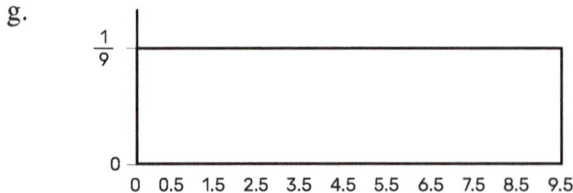

5.3 Solutions

1. a. $z = -1.786$, $z = 1.786$

   b. If your friend's blood pressure was 2 standard deviations below the mean, then his blood pressure would be below 65. This is incorrect based his belief that his blood pressure was between 100 and 150 millimeters. If his blood pressure was between 100 and 150 millimeters, then it would fall within 1.786 standard deviation of the mean.

3. a. One minute is 2 standard deviations below the mean. Values that are more than 2 standard deviations away from the mean μ are unusual values. The probability that it took less than 1 minute to find a parking space $P(X < 1) = 0.0165$.

   b. $P(X \geq 8) = P(X > 8) = 0.0668$

   c. We need to find the x value so that $P(X > x) = 0.7$ or find the x value so that $P(X < x) = 0.3$. Using a TI calculator, we can find $x = $ invNorm $(0.3, 5, 2) = 3.951$. This means that seventy percent of the time, it takes more than 3.95 minutes to find a parking space.

5. $\mu = 28\%$ and $\sigma = 5\%$

   a. $P(X \geq 30) = 0.3446$

   b. We need to find the x value so that $P(X < x) = 0.95$. Using a TI calculator, we can find $x = $ invNorm $(0.95, 0.28, 0.05) = 0.3622$, or 36.22%. This means that the probability of up to 36.22% of 18-34 year-olds spent checking Facebook before getting out of bed in the morning is 0.95.

7. $n = 190$; $p = 0.2$; $q = 0.8$

   $\mu = np = (190)(0.2) = 38$

   $\sigma = \sqrt{npq} = \sqrt{(190)(0.2)(0.8)} = 5.5136$

   a. For this problem: $P(34 < x < 54) = $ normalcdf$(34, 54, 38, 5.5136) = 0.7641$

   b. For this problem: $P(54 < x < 64) = $ normalcdf$(54, 64, 38, 5.5136) = 0.0018$

   c. For this problem: $P(x > 64) = $ normalcdf$(64, 190, 38, 5.5136) = 0.0000012$ (approximately 0)

5.4 Solutions

1.   a.   True by the Central Limit Theorem

      b.   True by the Central Limit Theorem.

      c.   False by the Central Limit Theorem. It be changed to "The standard deviation of the sampling distribution of the means will decrease making it approximately the same as the standard deviation of X as the sample size increases."

3.   a. $\bar{X} \sim N(145, 2)$

      b. $P(142 < X < 146) = 0.1133$

      c.   We want to find the value of x such that
$$P(< x) = 0.8 \Rightarrow \text{invNorm}(0.8, 145, 2) = 146.68$$

      d.   Median of the average running times = 146.68.

5.   64

7.   25

9.   0.927

11.   0.997

Chapter 6 Solutions

6.1 Solutions

1. a. X = the weight of an individual bag.

    b. $\overline{X}$ = the mean weight of bags.

    c.   Use the normal distribution since population distribution of bag weights is normally distributed.

    d.   The 90% confidence interval for true mean weight is between 1.9507 oz and 2.0493 oz.

3. c

5. $n = \dfrac{z_{\alpha/2}^{2}\sigma^{2}}{\text{EBM}^{2}} = \dfrac{z_{0.035}^{2}\sigma^{2}}{\text{EBM}^{2}} = \dfrac{(1.812^{2})(2.5^{2})}{1^{2}} = 20.5 \approx 21$

6.2 Solutions

1. a. X = enrollment numbers. $\overline{X}$ = the mean enrollment numbers of the thirty-five community colleges with $\overline{X} = 8926$

    b. $s_{x} = 6944$

    c.   You should use student's t-distribution as we don't know the population standard deviation and the population is normal.

    d.   i.   CI: (6244, 11,014)

         ii.

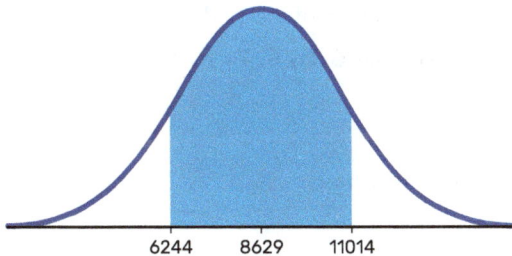

6244   8629   11014

         iii. EBM = 2385

    e.   If 500 community colleges are surveyed, the sample size increases, which causes the error bound to decrease and confidence interval becomes narrower.

3.   92% CI: (11.12, 12.08)

5.   a.   The 90% confidence interval for the population mean grams of fat per serving of chocolate chip cookies sold in supermarkets is (7.6372, 9.3628).
     b.   If you wanted a smaller error bound while keeping the same level of confidence, you would need to increase the sample size, n.

6.3 Solutions

1. $n = \dfrac{(1.645^2)(0.5)(0.5)}{0.05^2} = 270.6 \approx 271$

3.   a.   0.76.
     b. $CL = 0.95$ and $\alpha = 0.05$.
     c. $EBP = 1.96 \sqrt{\dfrac{(0.76)(0.24)}{1005}} = 0.0264$
     d. $CI = (0.7336, 0.7864)$
     e.   We can be 95% confident that about 73.36% to 78.64% of US workers believe that they will continue working past retirement age.

5.   You need to interview at least 385 students to estimate the proportion to within 5% at 95% confidence.

6.4 Solutions

1. $\chi_L^2 = 16.047$ and $\chi_R^2 = 45.722$

3.   Use R function "sd" to find the standard deviation of the 12 data values:
```
> Time = c(4.2, 4.7, 4.8, 5.5, 5.8, 6.5, 6.7, 7.7, 7.9, 8.1, 8.5, 8.8)
> sd(Time)
```
     $s = 1.598$ *and* $s^2 = 2.553$
     90% confidence interval estimate of (1.427, 6.138).

5.   The 99% confidence interval for the population standard deviation of the replacement times is (2.286, 4.688) years.

Chapter 7 Solutions

7.1 Solutions

1. a. $H_0: \mu = 34$; $H_a: \mu \neq 34$
    b. $H_0: p \leq 0.60$; $H_a: p > 0.60$
    c. $H_0: p = 0.29$; $H_a: p \neq 0.29$
    d. $H_0: p = 0.05$; $H_a: p < 0.05$
    e. $H_0: \mu \leq 10$; $H_a: \mu > 10$
    f. $H_0: p = 0.50$; $H_a: p \neq 0.50$
    g. $H_0: \mu = 6$; $H_a: \mu \neq 6$
    h. $H_0: p \geq 0.11$; $H_a: p < 0.11$

3. $H_0: \mu \leq 4.5$; $H_a: \mu > 4.5$

7.2 Solutions

1. a. Type I error: We conclude that the mean is not 34 years, when it really is 34 years.

    Type II error: We conclude that the mean is 34 years, when in fact it really is not 34 years.

    b. Type I error: We conclude that more than 60% of Americans vote in presidential elections, when the actual percentage is at most 60%.

    Type II error: We conclude that at most 60% of Americans vote in presidential elections when, in fact, more than 60% do.

    c. Type I error: We conclude that the proportion of high school seniors who get drunk each month is not 29%, when it really is 29%.

    Type II error: We conclude that the proportion of high school seniors who get drunk each month is 29% when, in fact, it is not 29%.

    d. Type I error: We conclude that fewer than 5% of adults ride the bus to work in Portland, when the percentage that do is really 5% or more.

    Type II error: We conclude that 5% or more adults ride the bus to work in Portland when, in fact, fewer than 5% do.

    e. Type I error: We conclude that the mean number of cars a person owns in his or her lifetime is more than 10, when it is not more than 10.

Type II error: We conclude that the mean number of cars a person owns in his or her lifetime is not more than 10 when, in fact, it is more than 10.

f. Type I error: We conclude that the proportion of Americans who prefer to live away from cities is not about half, though the actual proportion is about half.

Type II error: We conclude that the proportion of Americans who prefer to live away from cities is half when, in fact, it is not half.

g. Type I error: We conclude that the duration of paid vacations each year for Europeans is not six weeks, when in fact it is six weeks.

Type II error: We conclude that the duration of paid vacations each year for Europeans is six weeks when, in fact, it is not.

h. Type I error: We conclude that the proportion is less than 11%, when it is at least 11%.

Type II error: We conclude that the proportion of women who develop breast cancer is at least 11%, when in fact it is less than 11%.

3. a. $H_0: \mu \geq 0.11; H_a: \mu < 0.11$

b. Type I error: We conclude that the mean number of hours of sleep students get per night is less than seven hours when in fact, the mean number of hours is at least seven hours.

Type II error: not to reject that the mean number of hours of sleep students get per night is at least seven hours when, in fact, the mean number of hours is less than seven hours.

7.3 Solutions

1. Normal distribution and t-distribution.

3. Use a Student's t-distribution.

5. A normal distribution for a single population mean.

7. No, for a hypothesis test, the data are assumed to be from a simple random sample.

9. You must increase the number of trials, n.

11. This is a left-tailed test.

13. This is a right-tailed test.

15. This is a two-tailed test.

17. This is a right-tailed test.

7.4 Solutions

1. $H_0: \mu \geq 50{,}000$
   $H_a: \mu < 50{,}000$
   Use normal distribution with the assumption that the population has a normal distribution.
   $z = -2.315$
   p-value $= 0.0103$
   Reject the null hypothesis because the p-value < 0.05.
   Conclusion: At the 0.05 significance level, there is sufficient evidence to conclude that the mean lifespan of the tires is less than 50,000 miles.

3. $H_0: \mu = 10$
   $H_a: \mu \neq 10$
   Use the Student's t-distribution because the sample size is small.
   $t = -1.12$
   p-value $= 0.300 > 0.05$. You fail to reject the null hypothesis.
   Conclusion: At the 0.05 significance level, there is insufficient evidence to conclude that the mean number of sick days is not 10.

5. $H_0: p = 0.517$
   $H_a: p \neq 0.517$
   p-value $= 0.9203$.
   $\alpha = 0.05$.
   Do not reject the null hypothesis.
   At the 0.05 significance level, there is not enough evidence to conclude that the

proportion of homes in Kentucky that are heated by natural gas is not 0.517.

This result is not applicable across the country. First, the sample's population is only of the state of Kentucky. Second, it is reasonable to assume that homes in the extreme north and south will have extreme high usage and low usage, respectively. We would need to expand our sample base to include these possibilities if we wanted to generalize this claim to the entire nation.

7. c

9. H_0: p = 0.54

H_a: p ≠ 0.54

$\hat{p} = \frac{14}{30}$, $p_0 = 0.54$, and $q_0 = 0.46$

$$z = \frac{\hat{p} - p_0}{\sqrt{\frac{p_0 q_0}{n}}} = 0.4667 - \frac{0.54}{\sqrt{\frac{(0.54)(0.46)}{30}}} = -0.806$$

Critical > z = -0.806. Therefore, we fail to reject the null hypothesis.

p-value = 0.42 > 0.05. Therefore, we fail to reject the null hypothesis.

Conclusion: At the 0.05 level of significance, these is not enough evidence that approximately 54% of all fatal auto accidents is not caused by driver error. The AAA proportion seems to be accurate at this level of significance.

11. a. H_0: μ ≥ 150

b. H_a: μ < 150

c.   p-value = 0.0622< 0.01. Therefore, we do not reject the null hypothesis.

d.   At the 0.01 significance level, there is not enough evidence to conclude that first-year students study less than 2.5 hours per day, on average. The student academic group's claim appears to be correct.

Chapter 8 Solutions

8.1 Solutions

1. d.

Suppose that sample information is available on family income and years of schooling of the head of the household. A correlation coefficient = 0 would indicate no linear association at all between these two variables. A correlation of 1 would indicate perfect linear association (where all variation in family income could be associated with schooling and vice versa).

3. a

5. False. The coefficient of determination is r^2 with $0 \leq r^2 \leq 1$, since $-1 \leq r \leq 1$.

7. d

8.2 Solutions

1. A t-test is obtained by dividing a regression coefficient by its standard error, then comparing the result to critical values for Student's t with error df. It tests the claim that $\beta_i = 0$ when all other variables have been included in the relevant regression model.

8.3 Solutions

1. False. Since $H_0: \beta = -1$ would not be rejected at $\alpha = 0.05$, it would not be rejected at $\alpha = 0.01$.

3. d

5. True

7. d

9. a.

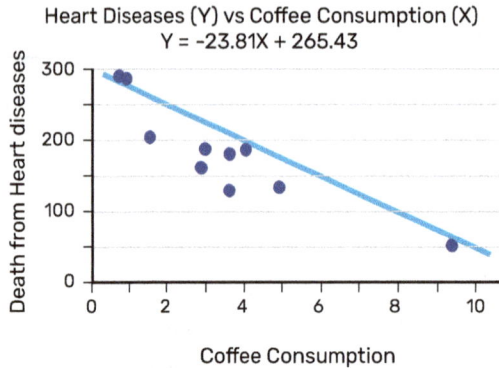

Heart Diseases (Y) vs Coffee Consumption (X)
Y = -23.81X + 265.43

b. $\hat{y} = -23.81x + 265.43$

c. Slope = -23.81 indicates that with each additional liter of coffee consumption, the death from heart disease rate decreases by 23.88 per capita.

y-intercept = 265.43 indicates that the death from heart disease rate is 265.43 per capita without coffee consumption.

d. $r = -0.836$, $r^2 = 0.6987$

The correlation coefficient indicates a linear relationship between yearly coffee consumption and the death rate from heart disease.

e. (0.7, 300) has the largest residual.

$\hat{y} = 249.914$ and $y = 300$

residual $\varepsilon = y - \hat{y} = 300 - 249.914 = 50.086$. The difference between the observed and the predicted value is residual. Here for the 0.7 liters of coffee consumption, the regression equation underpredicts the response value.

f. $H_0 : \rho = 0$ (There is no linear correlation between yearly coffee consumption and the death rate from heart disease)

$H_a : \rho \neq 0$ (There is a linear correlation between yearly coffee consumption and the death rate from heart disease)

p-value $= 0.0026 < 0.05 = \alpha$. Therefore, reject the null hypothesis.

There is convincing evidence that there is a linear correlation between yearly coffee consumption and the death rate from heart disease.

8.4 Solutions

1. Reject H_0 at the significance level $\alpha = .02$ because the p-value $< \alpha$.

3. d

5. Let x = age and y = number of driver deaths per 100,000.

   a.

x	y
17.5	38
22	36
29.5	24
44.5	20
64.5	18
80	28

   b.

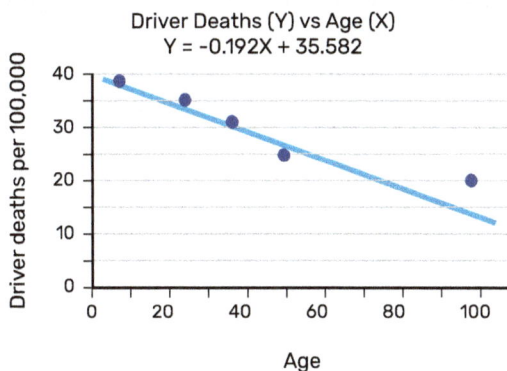

Driver Deaths (Y) vs Age (X)
Y = -0.192X + 35.582

   c. $\hat{y} = -0.192x + 35.582$
   d. $r = -0.57874$
   e. $x_0 = 40, \hat{y} = 27.902$
      $x_0 = 60, \hat{y} = 24.062$
   f. Slope = -0.192, which indicates that for each additional year in age, the number of driver deaths per 100,000 people decreases by 0.192 death per year.

8.5 Solutions

1. The value of R^2 always increases when the number of variables increases. So even if we add a useless variable to the model, the R^2 will still increase. However, adjusted R^2 only increases if the new variable improves the model. This is why we need to use adjusted R^2 so that models with different number of independent variables can be compared based on their adjusted R^2 values.

3. a. $SST = SSR + SSE = 946.181 + 49.773 = 995.954$, we know that $r^2 = \dfrac{SSR}{SST} = \dfrac{946.181}{995.954} = 0.95$

    b. adjusted $R^2 = 1 - \dfrac{(16-1)}{16-(4+1)}\,(1-0.95^2) = 0.867$

    c. $F = \dfrac{0.95(16-4-1)}{4(1-0.95)} = 52.25$

    d. Use R to find the p-value $= P(F > 52.277)$

    > pf(52.277, 4, 11, lower.tail = FALSE)

    [1] 0.0000004338219

    p-value $= 0$

    e. Yes, since p-value$= 0.0000 < \alpha = 0.05$, we conclude that the estimated regression model is significant.

Chapter 9 Solutions

9.1 Solutions

1. Let p_1 = proportion of system failure for OS10
   p_2 = proportion of system failure for OS12
   a. $H_0 : p_1 \leq p_2$
      $H_1 : p_1 > p_2$ (OS12 has fewer crashes)
   b. No, it doesn't appear that OS12 has fewer crashes than OS10 (p-value = 0.207)

3. At the 0.1 level of significance, there isn't enough evidence that the population proportion of drug and alcohol use is higher locally than national. (p-value = 0.2326)

5. At the 0.01 level of significance it seems more women use smartphones than men.

7. At the 0.05 level of significance, there is enough evidence that the percent of student athletes who drive themselves to school is more than the percent of nonathletes (p-value = 0.0178).

9.2 Solutions

1. $\mu_{day} \neq \mu_{night}$

3. μ_1 = the population mean for engine 1, and the population mean for engine 2
   $H_0 : \mu_1 \geq \mu_2$
   $H_a : \mu_1 < \mu_2$
   Since Both populations have normal distributions, and the population standard deviations are known, the normal distribution will be used for the test. The p-value of this one-sided test is always equal 0 indicating that we will reject the null hypothesis. At 0.05 significance level, there is enough evidence that Engine 2 has higher RPM than Engine 1.

5. μ_1 = the mean MPG for hybrid sedans, μ_2 = the mean MPG for non-hybrid sedans
   $H_0 : \mu_1 \leq \mu_2$

$H_a : \mu_1 > \mu_2$

Population standard deviations are known, the normal distribution will be used for the test. The p-value ≈ 0, reject H_0. At the 0.05 significance level, there is convincing evidence that non-hybrid sedan cars have a lower mean miles-per-gallon (mpg) than hybrid ones.

9.3 Solutions

1. Knowing whether samples are independent or dependent before conducting hypothesis tests helps us to determine the right test/distribution to use.

3. The 95% confidence interval for the population mean difference "before" and "after" is -2.764 hours to -0.236 hours. The new medicine appears to be effective in improving sleep at $\alpha = 0.05$ (p-value = 0.0131).

5. At the 0.05 significance level, the population mean difference between the husband's and the wife's is negative, the husband is happier than the wife (p-value = 0.04787).

```
                    Paired t-test

data:   husband and wife
t = -1.8605, df = 9, p-value = 0.04787
alternative hypothesis: true mean difference is less than 0
95 percent confidence interval:
         -Inf -0.007365702

sample estimates:
mean difference
         -0.5
```

9.4 Solutions

1. a. Let σ_1^2 = Population variance of 1st cyclist's speed
   σ_2^2 = Population variance of 2nd cyclist's speed

   b. $H_0 : \sigma_1^2 = \sigma_2^2$
   $H_a : \sigma_1^2 \neq \sigma_2^2$

$$F = \frac{s_2^2}{s_1^2} = \frac{32.1}{23.8} = 1.349$$

c. Using F-table to find the critical value of F:

$F_c = 3.5217 > F = 1.349$ Fail to reject H_0. At the 0.05 significance level, there isn't enough evidence to conclude that the cyclists' variances are not the same.

3. σ_1^2 variances for incomes on the East Coast

σ_2^2 variances for incomes on the West Coast

$H_0 : \sigma_1^2 = \sigma_2^2$

$H_a : \sigma_1^2 \neq \sigma_2^2$

```
              F test to compare two variances

   data:   East_Coast and West_Coast
   F = 0.81173, num df = 7, denom df = 6, p-value = 0.7825
   alternative hypothesis: true ratio of variances is not equal to 1
   95 percent confidence interval:
    0.1425226 4.1549351

   sample estimates:
   ratio of variances
             0.8117333
```

Since p-value $= 0.7825 > 0.05$ Fail to reject H_0. At 0.05 significance level, there is not enough evidence to conclude that the variances for incomes from the two population are not the same.

5. H_0: Engine 2 has a higher population standard deviation than engine 1 $(\sigma_1 \geq \sigma_2)$

H_a: Engine 2 has a higher population standard deviation than engine 1 $(\sigma_1 < \sigma_2)$

p-value $= 0.2104 > 0.05$, fail to reject H_0. At the 0.05 significance level, there is not enough evidence to conclude that Engine 2 has a higher population standard deviation than Engine 1.

Chapter 10 Solutions

10.1 Solutions

1. True

3. True

5. False, the statement should be changed to "In a goodness-of fit test, if the p-value is 0.0113, in general, reject the null hypothesis."

7. H_0: The observed number of businesses that recycle one of the two follows the distribution of the expected values.

    H_a: The observed number of businesses that recycle one of the two doesn't follow the distribution of the expected values.

    Test statistic $\chi^2 = 4.941$

    p-value $= 0.2933 < 0.05$. Fail to reject H_0

    At the 0.05 level of significance, there is not sufficient evidence to conclude that the observed number of businesses that recycle one of the two doesn't follow the distribution of the expected values.

9. a. H_0: The local results follow the distribution of the U.S. overall student population

    H_a: The local results don't follow the distribution of the U.S. overall student population

    Perform the χ^2 GOF-Test in R

```
> observed<-c(113, 94, 136, 10, 604, 43)
> expected<-c(.054, .145, .159, .012, .616, .014)
> chisq.test(x = observed, p = expected)
```

R Display

```
Chi-squared test for given probabilities
data:    observed
X-squared = 146.37, df = 5, p-value < 2.2e-16
```

At the 0.05 significance level, there is sufficient evidence to conclude that the local results don't follow the distribution of the U.S. overall student population based on ethnicity.

b. H_0: The local results follow the distribution of the percentage of students who use mass transit to get to school

   H_a: The local results do not follow the distribution of the percentage of students who use mass transit to get to school

   The test statistic is $\chi^2 = 13.4$

   At the 0.05 significance level, there is sufficient evidence to conclude that the local results don't follow the distribution of those that use mass transit, based on ethnicity.

10.2 Solutions

1. H_0: The location of the best ski area is independent of the level of the skier

   H_a: The location of the best ski area is not independent of the level of the skier

   Performing the Chi-square test in R:

```
                Pearson's Chi-squared test
data:    Skier
X-squared = 10.526, df = 4, p-value = 0.03244
```

   Since p-value $= 0.03244 < 0.05$, reject the null hypothesis

   At the 0.05 significance level, there is sufficient evidence to conclude that the location of the best ski area is not independent of the level of the skier (p-value $= 0.03244$)

3. H_0: student's major in college is independent of starting salaries after graduation

   H_a: Student's majors in college dependent of starting salaries after graduation

   Test statistic $\chi^2 = 33.546 >$ critical value $\chi^2 = 13.362$. Reject H_0

   At the 0.05 significance level, there is sufficient evidence to conclude that student's major in college is dependent of starting salaries after graduation (p-value $= 0.000049$).

10.3 Solutions

1. H_0: The distribution of types of cars for multi-person households and single-person households are the same

    H_a: The distribution of types of cars for multi-person households and single-person households are different

    Test statistic $\chi^2 = 62.912 >$ critical value $\chi^2 = 9.488$

    At the 0.05 significance level, there is insufficient evidence to conclude that the distribution of types of cars for multi-person households and single-person households are different (p-value ≈ 0)

3. H_0: The distribution of breakfast selection for men is the same as the distribution of breakfast selection for women

    H_a: The distribution of breakfast selection for men is different from the distribution of breakfast selection for women

    Test statistic $\chi^2 = 4.013 <$ critical value $\chi^2 = 7.815$. Fail to reject the null hypothesis.

    At 0.1 significance level, there is insufficient evidence to conclude that the distribution of personality types is different for business and social science majors (p-value $= 0.2601$)

5. H_0: The distribution of average energy use in the USA is the same as in Europe between Year 1 and Year 6.

    H_a: The distribution of average energy use in the USA is not the same as in Europe between Year 1 and Year 6.

    Test statistic $\chi^2 = 2.7434 <$ critical value $\chi^2 = 11.071$

    Fail to reject the null hypothesis.

    At the 0.05 significance level, there is insufficient evidence to conclude that the average energy use values in the US and EU are not derived from different distributions for the period from Year 1 to Year 6.

Chapter 11 Solutions

11.1 Solutions

1. One-way ANOVA

3. Answer may vary

5. For a two-side hypothesis test, the critical F value $= 2.1359$

11.2 Solutions

1. H_0: There is no difference in population mean grades among the four sections ($\mu_1 = \mu_2 = \mu_3$)

   H_a: There is a difference in population mean grades among the four sections (Not all μ_k's are equal)

   R Display for One-way ANOVA

	Df	Sum Sq	Mean Sq	F value	Pr(>F)
ind	3	2.877	0.9624	2.23	0.124
Residuals	16	6.904	0.4315		

   Since p-value $= 0.124 > 0.05$ fail to reject the null hypothesis.

   At the 0.05 significance level, there is insufficient evidence to conclude that there is a difference in population mean grades among the four sections.

3. H_0: The populations mean age that teenagers obtain their drivers licenses are the same across the country ($\mu_1 = \mu_2 = \mu_3 = \mu_4 = \mu_5$)

   H_a: Not all the populations mean age at that which teenagers obtain their drivers licenses are the same across the country (Not all μ_k's are same)

   At the 0.05 significance level, there is sufficient evidence to conclude that not all populations mean age at which teenagers obtain their drivers licenses are the same (p-value $= 0.0174$).

5. μ_k = population mean times watching their favorite news station

$k = 1, 2, 3$

$H_0: \mu_1 = \mu_2 = \mu_3$

H_a: Not all the means are equal

Test statistic F = 4.081

p-value = $0.0403 < 0.05$. Reject H_0

At the 0.05 significance level, there is sufficient evidence to conclude that the population mean times (in minutes) that people watch their favorite news station are different (p-value = 0.0403).

7. Define the population means:

μ_1 = population mean for online class

μ_2 = population mean for hybrid class

μ_3 = population mean for face-to-face class

$H_0: \mu_1 = \mu_2 = \mu_3$

H_a: Not all the means are the same (At least one of the means is different from the others)

	Df	Sum Sq	Mean Sq	F value	Pr(>F)
ind	2	20.173	10.087	0.639	0.544
Residuals	13	205.264	15.79		

p-value = $0.544 > 0.05$, fail to reject the null hypothesis.

At the 0.05 significance level, there is insufficient evidence to conclude that at least one of the mean scores of different delivery types is different from the others.

11.3 Solutions

1. With a two-way ANOVA, we can investigate whether the effect of room temperature on sleeping duration depend on the effect of the mattress type. For example, if the mattress type affects the effect of room temperature on sleeping duration, then there is an interaction effect between the room temperature and the mattress type on sleeping duration.

3. H_0 There is no age effect on running time

    H_a: There is age effect on running time

    H_0 There is no gender effect on running time

H_a: There is gender effect on running time

H_0 There is no interaction effect of age and gender on running time

H_a: There is interaction effect of age and gender on running time

Data entry in Excel

	Age		
	20-29	30-39	40 and over
Male	4.11	4.24	4.15
Male	4.42	5.42	5.03
Male	3.97	2.98	6.13
Male	5.62	3.16	4.01
Male	3.62	4.64	3.51
Female	3.02	5.12	5.07
Female	5.84	5.48	5.82
Female	3.82	4.02	6.33
Female	4.24	4.73	4.5
Female	4.55	3.51	3.54

Anova: Two-Factor With Replication

SUMMARY (Male)	20-29	30-39	40 and over	Total
Count	5	5	5	15
Sum	21.74	20.44	22.83	65.01
Average	4.348	4.088	4.566	4.334
Variance	0.58817	1.04772	1.06468	0.812497
SUMMARY (Female)	20-29	30-39	40 and over	Total
Count	5	5	5	15
Sum	21.47	22.86	25.26	69.59
Average	4.294	4.572	5.052	4.639333
Variance	1.07658	0.64557	1.20357	0.94095
SUMMARY (Total)	20-29	30-39	40 and over	
Count	10	10	10	
Sum	43.21	43.3	48.09	
Average	4.321	4.33	4.809	

Variance	0.740699	0.817644	1.073721111	

ANOVA				
Source of Variation	SS	df	MS	F
Sample (gender)	0.699213	1	0.699213333	0.745657
Columns (age)	1.558887	2	0.779443333	0.831216
Interaction (gender*age)	0.484207	2	0.242103333	0.258184
Within	22.50516	24	0.937715	
Total	25.24747	29		

The p-values for three hypotheses are all grater than the significance level. Fail to reject all three null hypotheses.

At the 0.05 significance level, there is insufficient evidence to conclude that age, gender and the interaction effect between age and gender have effect on the outcome, running time.

Chapter 12 Solutions

12.1 Solutions

1. Answer may vary

3. Answer may vary

12.2 Solutions

1. M = population median life of light bulbs

   $H_0: M \geq 730$

   $H_a: M < 730$

   $n = 11$, test statistic $x = 5 > 2$, fail to reject the null hypothesis.

   At the 0.05 significance level, there is no sufficient evidence that the population median life of light bulbs is less than 730 hours.

3. M_1 = population medium gas mileage from cars without fuel additive

   M_2 = population medium gas mileage from cars with fuel additive

   $M = M_1 - M_2$

   $H_0: M \geq 0$

   $H_a: M < 0$ (fuel additive improved gas mileage)

   $n = 8$, test statistic $x = 1 = 1$ reject the null hypothesis.

   At the 0.05 significance level, there is sufficient evidence that the fuel additive improved gas mileage.

12.3 Solutions

1. We will use Wilcoxon Signed-Rank Test for Matched Pairs.

   H_0: The population median stress level before and after music are equal

   H_a: The population median stress level before and after music are not equal

   R Display below

```
         Wilcoxon signed rank test with continuity correction


data:   before and after
V = 35, p-value = 0.1463
alternative hypothesis: true location shift is not equal to 0
```

Since p-value $= 0.1465 > 0.05$, fail to reject the null hypothesis.

At the 0.05 significance level, there is insufficient evidence that the population median stress level before and after music are unequal.

3.　　We will use Wilcoxon Rank-Sum Test for two Independent Populations

H_0: The median number of pain rating of two new pain medications is the same

H_a: The median number of pain rating of two new pain medications is different

```
         Wilcoxon rank sum test with continuity correction

data: drug1 and drug2
W = 166.5, p-value = 0.01253
alternative hypothesis: true location shift is not equal to 0

Warning message:
In wilcox.test.default(drug1, drug2, alternative = "two.sided"):
  cannot compute exact p-value with ties
```

Since p-value $= 0.01253 < 0.05$, reject H_0

At the 0.05 significance level, there is sufficient evidence that the population median number of pain rating of two new pain medications is different.

12.4 Solutions

1.　　The nonparametric test equivalent to one-way ANOVA is the Kruskal-Wallis Test. Discussion may vary.

3.　　The H test is the Kruskal-Wallis Test, it is the nonparametric test equivalent to one-way ANOVA.

One-way ANOVA relies on the assumption of normality of population distribution while the application of H test doesn't require the normality assumption.

5. The Kruskal-Wallis test is appropriate

7. H_0: The median times spent on three news station are the same $(M_1 = M_2 = M_3)$
    H_a: At least one of the median times among three news station is different from the others.
    R code

```
> drug_1<-c(78, 65, 63, 44, 50, 78, 70, 61, 50, 44)
> drug_2<-c(71, 66, 40, 55, 31, 45, 66, 47, 42, 56)
> drug_3<-c(57, 88, 58, 78, 65, 61, 62, 44, 48, 77)
> df<- data.frame(drug_1,drug_2,drug_3)
> df_stacked <- stack(df[1:3])
> kruskal.test(values ~ ind, data = df_stacked)
```

R Display

```
                Kruskal-Wallis rank sum test

data: values by ind
Kruskal-Wallis chi-squared = 6.0183, df = 2, p-value = 0.04933
```

Since p-value $= 0.04933 < 0.05$, reject the null hypothesis.

At the 0.05 significance level, there is sufficient evidence that the population median times spent on watching three news station are different.

12.5 Solutions

1. No, even though the Spearman rank correlation test concludes that there is a correlation between the two variables it is based on ranks. But linear regression results are in terms of actual values and the units. Also, regression has a dependent variable and an independent variable. But Spearman's does not.

3. H_0: $\rho_s = 0$ (There is no correlation between the ranks of temperature and the cricket chips)

H_a: $\rho_s \neq 0$ (There is correlation between the ranks of temperature and the cricket chips)

```
                Spearman's rank correlation rho

    data: Temperature and Cricket_chips
    S = 4.5183, p-value = 3.412e-10
    alternative hypothesis: true rho is not equal to 0
    sample estimates:
          rho
    0.9875871

    Warning message:
    In cor.test.default (Temperature, Crickt_chips,
    method = "spearman"):
       Cannot compute exact p-value with ties
```

Since p-value $= 3.412\text{e-}10 < 0.05$, reject H_0

At the 0.05 significance level, there's correlation between temperature and the cricket chips.

12.6 Solutions

1. H_0: The sequence of male and female passengers entering the train is a random process

H_a: The sequence of male and female passengers entering the train is a non-random process

the test statistic R $= 10$, $n_1 = 15$ (female), $n_2 = 10$ (male)

Check the table for runs test, we the critical values are 7 and 18. Since $7 < 10 < 15$, fail to reject the null hypothesis.

At the 0.05 significance level, there is insufficient evidence to conclude that the male and female passengers entering the train is a non-random process.

Index

www.ingramcontent.com/pod-product-compliance
Lightning Source LLC
Chambersburg PA
CBHW061739210326
41599CB00034B/6726